Geometry

Springer
Berlin
Heidelberg
New York
Barcelona
Hong Kong
London
Milan
Paris
Tokyo

Audun Holme

Geometry

Our Cultural Heritage

With 150 Figures

 Springer

Audun Holme

University of Bergen
Department of Mathematics
Johannes Brunsgate 12
N-5008 Bergen
Norway
e-mail: holme@mi.uib.no

Catalog-in-Publication Data applied for

Die Deutsche Bibliothek - CIP-Einheitsaufnahme

Holme, Audun:
Geometry: our cultural heritage/Audun Holme.-Berlin; Heidelberg; New York;
Barcelona; Hong Kong; London; Milan; Paris; Tokyo: Springer, 2002
ISBN 3-540-41949-7

Mathematics Subject Classification (2000):
Primary: 51-01; Secondary: 01-01, 14-01

ISBN 3-540-41949-7 Springer-Verlag Berlin Heidelberg New York

Springer-Verlag Berlin Heidelberg New York
a member of BertelsmannSpringer Science+Business Media GmbH

http://www.springer.de

© Springer-Verlag Berlin Heidelberg 2002
Printed in Germany

Typeset by the author using a Springer TeX macro package
Cover design: Archytas' famous space geometric construction, as rendered by Maple
Cover production: *design & production* GmbH, Heidelberg

SPIN 10769915   46/3142db - 5 4 3 2 1 0 – Printed on acid-free paper

Preface

This book is based on lectures on geometry at the University of Bergen, Norway. Over the years these lectures have covered many different aspects and facets of this wonderful field. Consequently it has of course never been possible to give a full and final account of geometry as such, at an undergraduate level: A carefully considered selection has always been necessary. The present book constitutes the main central themes of these selections.

One of the groups I am aiming at, is future teachers of mathematics. All too often the geometry which goes into the syllabus for teacher-students present the material as pedantic and formalistic, suppressing the very powerful and dynamic character of this old – and yet so young! – field. A field of mathematical insight, research, history and source of artistic inspiration. And not least important, a foundation for our common cultural heritage.

Another motivation is to provide an invitation to *mathematics in general*. It is an unfortunate fact that today, at a time when mathematics and knowledge of mathematics is more important than ever, phrases like *math avoidance* and *math anxiety* are very much in the public vocabulary. An important task is seriously attempting to heal these ills. Ills perhaps inflicted on students at an early age, through deficient or even harmful teaching practices. Thus the book also aims at an informed public, interested in making a new beginning in math. And in doing so, learning more about this part of our cultural heritage.

The book is divided into two parts. Part 1 is called *A Cultural Heritage*. The section contains material which is normally not included into a mathematical text. For example, we relate some of the stories told by the Greek historian, *Herodotus*, in [17]. We also include some excursions into the history of geometry. These excursions do not represent an attempt at writing the history of geometry. To write an introduction to the history of geometry would be a quite different and very challenging undertaking. Even for the period up to the beginning of the Middle Ages it would vastly surpass what is presently undertaken in [21].

To write the History of Geometry is therefore definitely not my aim with Part I of the present book. Instead, I wish to seek out the roots of the themes to be treated in Part 2, *Introduction to Geometry*. These roots include not only the geometric ideas and their development, but also the historical context. Also relevant are the legends and tales – really *fairy-tales* – told about, for example, *Pythagoras*. Even if some of the more or less fantastic events of *Iamblichus'* writings are unsubstantiated, these stories very much became our *perception* of the geometry of Pythagoras, and thus became part of the *heritage of geometry*, if not of its *history*.

In Chapters 1 and 2 we go back to the beginnings of science. As geometry represents one of the two oldest fields of mathematics, we find it in evidence from the early beginnings. The other field being *Number Theory*, they go back as far as written records exist. Moreover, in the first written accounts from ancient civilizations they present themselves as already well developed and sophisticated disciplines.

Thus we find that problems which ancient mathematicians thought about several thousand years ago, in many cases are the same problems which are stumbling stones for the students of today. As we move on in Chapters 3 and 4, we find that great minds like Archimedes, Pythagoras, Euclid and many, many others should be allowed to speak to the people of today, young and old. They are unsurpassable tutors.

The mathematical insight which Archimedes regarded as his most profound theorem, was a theorem on geometry which was inscribed on Archimedes' tombstone. All of us, from college student to established mathematician, must feel humbled by it. What does it say? Simply that if a sphere is inscribed in a cylinder, then the proportion of the volume of the cylinder to that of the sphere, is equal to the proportion of the corresponding surface areas, counting of course top and bottom of the cylinder. The common proportion is $\frac{3}{2}$. This is a truly remarkable achievement for someone who did not know about integration, not know about limits, not know about... Its beauty and simplicity beckons us. How did Archimedes arrive at this result? Archimedes deserves to be remembered for this, rather for the silly affair of him having run out into the street as God had created him, shouting – *Eureka, Eureka!* But the story may well be true, his absentmindedness under pressure cost him dearly in the end.

Pythagoras and his followers certainly did not discover the so-called *Pythagorean Theorem*. The Babylonians, and before them the Sumerians, not only knew this fact very well, they also had the sophisticated number-theoretical tools for constructing all *Pythagorean triples*, that is to say, all natural numbers a, b and d such that $a^2 + b^2 = d^2$. And the astronomers and engineers – or, if we prefer, the astrologers and priests – of ancient Babylonia, or Mesopotamia, used these insights to construct trigonometric tables. Tables which were simple, accurate and powerful thanks to the *sexagesimal* system they used for representing numbers. What a challenging project for interested college students to understand the math of the Plimpton 322 tablet at Columbia University! And to correct and explain the four mistakes in it. Or finally to construct the successor or the predecessor of this tablet in the series it must have belonged to.

So what did Pythagoras discover himself? We know nothing with certainty of Pythagoras' life before he appeared on the Greek scene in midlife. Some say that he traveled to Egypt, where he was taken prisoner by the legendary, in part infamous, Persian King *Cambyses II*, who also ruled Babylon, which had been captured by his father *Cyrus II*. Pythagoras was subsequently brought to Babylon as a prisoner, but soon befriended the priests, the *Magi*, and was initiated into the priesthood in the temple of Marduk. We tell this story as related by Herodotus and Iamblichus in [17] and [22]. However, the accounts given in these classical books are not always historically correct, the reader should consult the footnotes in [17] to get a flavor of the present state of Herodotus, *The Father of History*, by some of his critics labeled *The Father of Lies!* But Herodotus is a fascinating story-teller, and the place occupied by Pythagoras today has considerably more to do with the legends told about him than with what actually happened. So with this warning, do enjoy the story.

Euclid's Elements represents a truly towering masterpiece in the development of mathematics. Its influence runs strong and clear throughout, leading to non-Euclidian geometry, Hilbert's axioms and a deeper understanding of the foundations of mathematics. The era which Euclid was such an eminent representative of, ended with the murder of another geometer: Hypatia of Alexandria.

In Chapter 5 we describe how the foundation of present day geometry was created. Elementary Geometry is tied to straight lines and *circles*. The theorems are closely tied to constructions with straightedge and compass, reflecting the postulates of Euclid. In *higher geometry* one moves on to the more general class of conic sections, as well as curves of higher degrees.

Descartes introduced – or reintroduced, depending on your point of view – algebra into the geometry. At any rate, he is credited with the invention of the *Cartesian coordinate system*, which is named after him.

In Chapter 6, the last chapter of Part 1, we discuss the relations between geometry and the real world. The qualitative study of *catastrophes* is of a geometric nature. We explain the simplest one among *Thom's Elementary Catastrophes*, the so-called *Cusp catastrophe*. It yields an amazing insight into occurrences of *abrupt events* in the real world.

Also tied to the real world are the fractal structures in nature. Fractals are geometric objects whose *dimensions are not integers,* but which instead have a *real number* as dimension. Strange as this sounds, it is a natural outgrowth of *Felix Hausdorff's* theory of dimensions. Hausdorff was one of the pioneers of the modern transformation of geometry, referred to in his time as the *High Priest of point-set topology*. In the end, this all did not help him. He knew, being a Jew, what to expect when he was ordered to report the next morning for deportation. This was in 1942 in his home town of Bonn, Germany. Instead of doing so, Hausdorff and his wife committed suicide.

The *Geometry of fractals* shows totally new and unexpected geometric phenomena. Amazingly, what was thought of as *pathology*, as useless curiosities, may turn out to give the most precise description of the world we live in.

In Part 2, *Introduction to Geometry*, we take as our starting point the axiomatic treatment of geometry flowing from Euclid.

Euclid's original system of axioms and postulates passed remarkably well the test of modern demands to rigor. But an explanation was nevertheless very much called for, as his original system was set on a somewhat shaky foundation by our current mathematical standards. This clarifying explanation of the foundations was provided by Hilbert. The search for a proof of Euclid's Fifth Postulate at an earlier age, had met with no success. One version of this postulate asserts that there is one and only one line parallel to a given line through a point outside it.

Assuming the converse, one wanted to derive a contradiction. But instead the relentless toil produced alternatives to Euclidian Geometry. This was a highly troubling development for an age in which non-Euclidian Geometry would appear as heretic as "Darwinism" appears in some circles today. We explain non-Euclidian Geometry in Chapter 9. But first we need to do some serious work on foundations. We start with *Logic and Set Theory*. In fact, the Intuitive Set Theory, even as put on a firmer foundation by *Cantor*, turned out to imply grave contradictions. The best known is the so-called *Russell's Paradox*, which we explain in Section 7.3.

Thus arouse the need for *Axiomatic set theory*, to which we give an introduction. The aim is to give a flavor of the field without going into the technical details at all.

We then explain the interplay between *axiomatic theories* and their *models* in Sections 7.3 and 7.4. The troubling result of *Gödel* is explained, in simplified terms, showing that a *mathematical Tower of Babel* as perhaps dreamt of by Hilbert, is not possible: Any axiomatic system without contradictions among its possible consequences, will have to live with some *undecidable* statements. That is to say, statements which are perfectly legal constructions within the system, being inherently undecidable: Their truth or falsehood cannot be ascertained from the system itself.

In Chapter 8 we apply these insights to axiomatic projective geometry. This is an extensive field in itself, and a complete treatment does of course, fall outside the scope of this book. But we give a basic set of axioms, to which other may be added, thus in the end culminating with a set which determines uniquely the *real, projective plane*. This is not on our agenda here. But we do give, in some detail, two important models for the basic system of axioms. The *Seven Point Plane* and the *real projective plane* $\mathbb{P}^2(\mathbb{R})$. That the simple axioms still hold intriguing open problems, is explained in Section 8.2. Use of powerful computers in conjunction with dexterous programming holds great promise for new insights, thus there exist ample possibilities for exiting research.

In Chapter 9 we are ready to explain models for non-Euclidian Geometry. In the *hyperbolic plane* there are infinitely many lines parallel to a given line through a point outside it. In the *elliptic plane* there are no parallel lines: Two lines always intersect. A model for this version is provided by $\mathbb{P}^2(\mathbb{R})$.

Plane non-Euclidian geometries have, of course, their spatial versions. This is best understood by turning to some of the basic facts from *Riemannian Geometry*, which we do in Section 9.5.

Chapter 10 contains some much needed mathematical tools, simple but essential. We need them for constructions to be carried out in following chapters, where we employ these standard techniques. The reader is advised to take the moments needed to ingest this material, which may well appear somewhat dry and barren at the first encounter.

In Chapter 11 we are now able to give coordinates in the projective plane, introduce projective n-space and discuss affine and projective coordinate systems. Again, the material may appear dry, but the reader will be rewarded in Chapter 12. There we use these techniques to give the remarkably simple proof of the theorem of Desargues. We introduce *duality for* $\mathbb{P}^2(\mathbb{R})$ and start the theory of conic sections in $\mathbb{R}^2$ and $\mathbb{P}^2(\mathbb{R})$ discussing tangency, degeneracy and the familiar classification of the conic sections. Pole and Polar belong to this picture, as well as a very simple proof of a famous theorem of Pascal. Using it, we then prove the theorem of Pappus by a classical technique known as *degeneration*, or as the *principle of continuity*. Here we give the first, naive, definition of an algebraic curve.

In Chapter 13 we find the transition to the study of curves of degrees greater than 2. This forms the fundament for Algebraic Geometry, and gives a glimpse into an important and very rich, active and expanding mathematical field. Here we encounter the *cubical parabola*, merely a fancy name for a familiar curve, but also the enigmatic *semi cubical parabola*, so important in modern *Catastrophe Theory*. However, as we shall see in the following chapter, in Chapter 14, from a projective point of view these two kinds of affine curves are the same! This is shown at the end of Section 14.5. We also learn about the *Folium of Descartes*, the *Trisectrix of Maclaurin*, of *Elliptical Curves* – which are by no means ellipses! – and much more. Chapter 14 concludes with Pascal's *Mysterium Hexagrammicum*, which may be obtained as a beautiful application of Pascal's Theorem: Dualizing it the Mystery of the Hexagram is revealed.

In Chapter 15, as the title says, we sharpen the Sword of Algebra. The aim is to show how one finally disposes of the three problems, which have haunted mathematicians and amateurs for two millenia. And unfortunately, still does haunt the latter. The algebra derives in large part from the heritage of Euclid, relying as it does on *Euclid's algorithm*. This mathematics also constitutes the foundation for the important field of *Galois theory* and the theory of equations and their solvability by radicals. That theme is, however, not treated in the present book.

In Chapter 16 we use this algebra for proving that the three classical problems are insoluble: Trisecting an angle with legal use of straightedge and compass, doubling the cube using straightedge and compass, and finally we see how the transcendency of the number π precludes the squaring of the circle using straightedge and compass. Gauss' towering achievement on constructibility of regular polygons conclude the chapter. The solution of this problem by Gauss transformed the final answer to the geometric problem into a number theoretic problem on the existence of certain primes, namely primes of the form $F_r = 2^{2^r} + 1$, the so-called *Fermat primes* . For $r = 0, 1, 2, 3, 4$ the numbers P_r are $3, 5, 17, 257$ and 65537. They are all primes, but then no case of an r yielding a prime is known. Gauss proved that if q is a product of such primes p_r, all of them distinct, then the regular $n = 2^m q$-gon may be constructed with straightedge and compass, and that this is precisely all the constructible cases. Thus for example the regular 3-gon, the regular 5-gon and the regular 15-gon are all constructible with straightedge and compass, as is the regular 30-gon and the regular 60-gon. The first impossible case is the regular 7-con. Now Archimedes constructed the regular 7-gon, but he used means beyond legal use of straightedge and compass. In Section 4.4 we have given Archimedes' construction of the regular 7-gon, the regular *heptagon*, by a so-called *verging construction*. It is not possible by the *legal use* of compass and straightedge, but may be carried out by conic sections or by a curve of degree 3. In fact, such constructions were very much part of the motivation for passing from *elementary geometry* to *higher geometry* in the first place.

In Chapter 17 we take a closer look at the theory of fractals. We explain the computation of fractal dimensions.

Chapter 18 contains a mathematical treatment of introductory Catastrophe Theory. We explain the Cusp Catastrophe as an application of geometry on a cubic surface. For this we also explain some rudiments of Control Theory.

Several variations of courses may be taught from the present text. Two possibilities are outlined below, each with two hours of lecture and preferably one hour of discussion per week, each of one term duration. I have labeled them as follows:

– *Geometry 1: Historical Topics in Geometry*
– *Geometry 2: Introduction to Modern Geometry.*

They may well be merged into one course, then possibly with 3 hours of lecture and two hours of discussion. Geometry 1 would be taught before Geometry 2.

Geometry 1 would comprise all of Part 1, and in addition a somewhat cursory treatment of selected sections from Part 2.

Geometry 2 would essentially consist of Part 2, with an in-depth treatment of the material from Part 2 having been more summarily taught in Geometry 1.

Geometry 1 requires modest background in mathematics, and could be offered to elementary school teachers, possibly in the settings of a Community College.

Sufficient background for Geometry 2 would be high school math with trigonometry and analytic geometry, or a pre calculus course. However, some prerequisites may be dealt with in an extra discussion hour.

With some faculty guidance a freshman seminar combining elements of the two alternatives is also a possibility.

Some of the material in this book has been published in the author's [19] and [21]. The material is included here with the permission of *Fagbokforlaget*, the publisher of [19] and [21]. A large number of the illustrations are created by the marvelous system Cinderella, [28], some of them were made by Ulrich H. Kortenkamp, one of the authors of the system. Others were made by the author, who would like to take this opportunity to thank Professor Kortenkamp for his efforts in making these illustrations, as well as for his valuable advice and assistance during this work. A few illustrations are made with the aid of the Computer Algebra system MAPLE, and finally some were made by Springer's illustrator, based on sketches by the author.

Thanks are also due to Dr. Even Førland of the Norwegian Statoil Corporation, who has red parts of the manuscript and provided valuable comments.

It is a great pleasure to thank Springer Verlag, in particular the Mathematics Editor Dr. Martin Peters as well as Mrs. Ruth Allewelt, for their enthusiasm and support. Mrs. Daniela Brandt of Springer's Production Department has provided great support and assistance in getting the manuscript in shape for the printer.

Bergen,   *Audun Holme*
November 2001

Table of Contents

Part II Introduction to Geometry

Part I

A Cultural Heritage

1 Early Beginnings

1.1 Prehistory

Mankind must have possessed knowledge about geometric phenomena as far back as our historical records take us, and undoubtedly even much further back into the twilight of prehistorical times.

Human conceptions of number and form are documented as far back as the Old Stone Age, the *Paleolithic*. But there is no reason to assume this as being the beginning. Indeed, on the contrary, new and ongoing research would seem to document that monkeys have the ability to think about numbers, as reported in some newspapers.

In some so-called *primitive cultures* there are evidence that not only have such knowledge been present, but it has actually been remarkably advanced and sophisticated.

A famous example of this is the wall paintings found in a huge limestone cave in southern France. The cave contains about 300 remarkable paintings, about 30000 years old. At this time there were made no written records, but these paintings may have served such purposes as well.

So the cave paintings document an artistic level of achievement second to none in later ages. They also indicate an understanding of space and form which certainly would have warranted formation of mathematical and geometrical concepts, had urge or need to do so existed at this time. But that may not have been the case more than 30000 years ago. The people of the caves led difficult lives, collecting their livelihood hunting and gathering. The paintings may very likely have served as a magic vehicle for gaining control over nature, for casting a spell on the game thus ensuring a successful hunt. But a purpose of recording events, enumerating items, describe motion and spacial relationships may also have been present. One could say that such documents contain elements of *protogeometry*. The first stage of representing quantity, space and time, if not yet a truly abstract representation.

1.2 Geometry in the New Stone Age

Mathematics and geometry as we know it today, probably did not develop until the need for it arose. And this happened when the nomadic way of life ended as the sole model for human society, in the New Stone Age, the *Neolithic*. Replacing hunting and gathering by agriculture carried with it a completely altered way of life and a new society. Although taking place over thousands of years, this was a true revolutionary change. Living in one place, with valuable possessions and irreplaceable provisions, the need arouse for protection. Thus early urban centers, villages, were formed. Agriculture depended in many instances on artificial irrigation, through canals and irrigation channels. And protection from flooding necessitated the building of large structures as dykes. Also bulwarks in the form of walls around urban centers became necessary. Thus arose the need for engineering skills and insights.

When early humans lifted their eyes to the nightly sky, they were stirred by the same primordial emotions as their close or remote cousins among other creatures on the Earth. The light from a full moon, the millions of stars scattered across the firmament, the northern or the southern lights. Thunder and lightening in the heavens. But unlike the remote cousins, our forebears wanted to understand. And so they tied the events in the heavens with the life on the earth. The changing of seasons, heralded by certain stars appearing, becoming visible at sunset. In the north life reawakened as the sun gained strength, further south the rain at the onset of fall reawakened nature. All this contributed to the blend of early science, mathematics and astronomy with early religion, astrology and myths.

At the end of the third millennium B.C., northern Europe was still in the Neolithic period, the young Stone Age. From this time dates the Stonehenge at Salisbury Plain, Wiltshire in England. This mysterious structure may have served many purposes, one of them being as an astronomical observatory. Someone who stood at the center of Stonehenge on the morning of the summer solstice 4000 years ago, would see the sun rising directly over a certain stone conspicuously located in the structure.

There is also evidence which documents an impressive insight into astronomy or perhaps some kind of astrology. This has led some writers to speculate that there must have occured visits by extraterrestrials from outer space: No other explanation would seem possible for the presence of such insights in so primitive societies, they argue. This line of reasoning bespeaks the prejudice and cultural arrogance in *our own society*, more than anything else. It is indeed humbling to contemplate how deep insights may have been gained, and then forgotten, time and again throughout our human history. But our brains today are the same as theirs then. The improvement probably lies in the way knowledge was eventually securely recorded and passed on to new generations, even as carefully guarded secrets of a priesthood.

1.3 Early Mathematics and Ethnomathematics

From this period we find numerous patterns of a geometric nature, best preserved on pottery but also from fragments of textiles. Today the study of such geometric activity forms part of what is known as *Ethnomathematics*. Such patterns were also used in artistic works by the Samish population in northern Scandinavia and in Russia. Much of this was destroyed by missionaries, considering it as pagan sorcery.

Several historians of mathematics have speculated on the connection between patterns of art and decoration and the development of numbers and numerology. There may well be close connections, but it is not easy to shed the preconceived ideas our modern mathematical education has provided us with, as we study this material. For example, a pattern of smaller isosceles triangles inside a larger one is esthetically appealing. We may also use the same pattern to define the concept of *triangular numbers*. But does this mean that triangular numbers were known as a mathematical or numerical concept? Or is it fair to assume these patterns being evidence of a developed geometry at this time? Is it really mathematics?

These questions are perhaps more interesting than any proposed answers. They take us into the fascinating realm of the *Philosophy of Science*, perhaps posing insoluble enigmas. For our more practical purpose here, it is better to assume the broadest definition of *what constitutes mathematics*, thereby perhaps offending some purists but not unduly exclude an important facet of our field of study.

In any case there can be no doubt that historians of mathematics have made grave mistakes and misjudgments as to what constitutes *mathematical knowledge*. Thus for instance, even contemporary work on the history of mathematics can be seen as espousing the view that the ancient Mesopotamians "did not know the proof of the Pythagorean Theorem", simply because no clay tablet has been found containing a Greek-style proof of this fact! Is it not true that the Greek conceived of the *novel* idea of a *formal proof* because they considered mathematics as a branch of the *dialectics*, of the art of *debating*? And what if the ancient Mesopotamian Sumerian and Akkadian mathematicians were, not members of some debating-academy, but members of a *Priesthood of the Temple*, who closely guarded their wisdom from falling into the hands of the Unworthy? The mere idea of a *proof*, whose function it should be to convince the incredulous, would strike them at utterly absurd. That they still had secure knowledge in mathematics is firmly documented by the historical record, as we shall see later in this book.

Thus we have to approach the question of what constitutes mathematical knowledge with thoughtful humility.

2 The Great River Civilizations

2.1 Civilizations Long Dead – and Yet Alive

In the first part of the fourth millennium B.C. a group of cultural centers developed in southwestern Asia. Probably emerging through coalescence from a web of small and, it would seem, insignificant neolithic villages, impressive cities formed in the river valleys of the Indus, the Euphrates-Tigris and the Nile. Spreading out to form nets with other, in part more peripheral, urban centers, the classical civilizations bearing their names were born in these river valleys.

Today their once flourishing life is attested to only by the mounds or *tells* covering their ruins. And by no means have we managed to uncover them all. Great finds are awaiting future archaeologists. The fascinating story of how these tells in some cases were persuaded to reveal their secrets, is not the subject of the present narrative. Nor is it our concern here to elaborate on the heroic endeavor and the almost superhuman tenacity, fortitude and resilience shown by those pioneers who managed to decipher the writings left behind by the vanished civilizations, thousands of years dead.

Suffice to recall that the pyramids and ruins in the Nile valley had been the objects of legends for millenia. Tales of treasures and curses.

Beduins were roaming the deserts and fever-infested marshlands which now covered the once flourishing river basins of the Euphrates and the Tigris. They still worshiped with awe the mysterious *tells*, calling them by mystic names, the origins of which were long forgotten.

But in present day Iraq, inside the present city of Baghdad, were the crumbling ruins of the once so marvelous city of *Babylon*, the Gate of The Gods, the *Bab-Ili*. Once a center of science and mathematics, literature and history, astronomy and medicine, astrology and worship. As well as the bearer and center of a complex business structure which extended throughout the known world. With exquisite restaurants and a most sinful and hedonistic nightlife. In the rubble of what once were, there remained a ruin of what evidently had been a gigantic structure. Locally, it still was called reverently by the name of the ancient *Tower of Babel.*

Napoleon's invasion of Egypt led to the discovery of the Rosetta Stone and subsequently the decipherment of its inscriptions by *Jean-Francçois Champollion*. Thus were opened up insights into events extending a thousand years

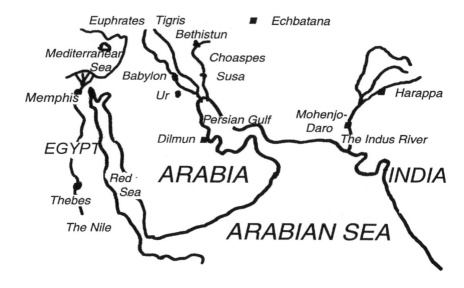

Fig. 2.1. This map shows the areas of the three Great River Civilizations.

The main rivers shown are the Nile, the Euphrates and the Tigris. Euphrates flowed through the city of Babylon at this time, and the two rivers joined before they met the Persian Gulf. Also shown is the river of Choaspes, as well as the Indus River. Trade routes over land and sea connected them, and the presence of trade as well as other kinds of contact and interaction are very much in evidence. The ancient cities of Babylon and Ur of Mesopotamia are shown, Memphis and Thebes of Egypt, and Mohenjo-Daro and Harappa of the Indus Valley Civilization. Susa was the capital city of Persia before Darius founded the magnificent capital of Persepolis (in 518 B.C.). Dilmun was, according to some legends, the location of *the Garden of Eden*, central to three great religions. In ancient times this whole area was much more fertile than today, the climate was better and wildlife was abundant. Lion-hunting was a favorite activity for the Kings of Mesopotamia, as a result of which lions became extinct relatively early there.

beyond those recorded by the Bible and those accounted by the classical Greek historians and travelers. The decipherment of the hieroglyphic script of Egypt provided insights into all aspects of life in this ancient land.

The original script of ancient Egypt was, of course, *the hieroglyphs*. Eventually the need for a more practical and faster way of writing led to the development of the so called *hieratic* (sacred) script, a kind of simplified hieroglyphs. From the eight century B.C. a third type of script, which is even easier, is used, the so called *demotic* script, *the script of the people*. The Rosetta Stone is a fragment dating from 196 B.C, found in 1799 by one of Napoleon's officers. On it a decree had been written in Egyptian with hieroglyphs and with hieratic characters, and in Greek in demotic script. The breakthrough in deciphering the hieroglyphs and hieratic script came when Champollion could locate the names *Ptolemy* and *Cleopatra* in both Egyptian scripts. It was then possible to compare with the Greek text, and piece together the alphabets.

The insights provided by the finds in the pyramids are subject to an important restriction. Almost exclusively all we have is material intended as support for the deceased in after-life. This is not to say that we are lacking texts pertaining to daily life or practical matters. But the criterion for preservation have been that of inclusion into the burial chamber.

Mesopotamia was now part of the once powerful Turk Ottoman Empire which was entering its final decades. The mounds and the areas around them had long been a source of building material of exquisite quality: Bricks of clay, baked at such high temperature that it surpassed by far what could be accomplished with contemporary technology. A number of these bricks were decorated with strange patterns of wedge-formed marks. These ubiquitous *decorated bricks* were excellent as building material, and served useful purposes as landfill as well as for building dykes along the banks of the Euphrates. In fact, as one of the first expeditions arrived at the old cite of Babylon to commence excavations there, workers were busily occupied with constructing huge embankments along the river banks. Using as building material not earth, but the books from the Imperial Library, housed at the Imperial Museum which had been located at the northernmost corner of ancient Babylon.

Indeed, the decorated bricks were books. The key to the decipherment of the *cuneiform* script in – or rather, on – these books, was provided by inscriptions found at Behistun near Persepolis, in the highlands of present day Iran. As in the case of the Rosetta Stone the inscriptions had been made in more than one language, one of which was Old Persian, which was known. An other version of the inscription was in the cuneiform script found on the *"decorated bricks."* Ordered by the Persian King Darius to commemorate his victories, the inscription had been carved on a great limestone cliff near the present village of Behistun, about 300 feet above the ground. Just getting an exact copy of the ancient letters was a strenuous and quite dangerous task. The story about the decipherment of the ancient script of the Sumerians and the Babylonians is suspenseful and fascinating. Building on work by the German philologist George Friedrich Grotefeld (1775-1853), the British diplomat and scholar, later to be called *the Father of Assyriology*, Henry Creswick Rawlinson (1810-1895), finally unraveled the script in 1846.

The assignment undertaken and carried out successfully by the archaeologists and linguists is humbling: A totally unknown script, writing texts in a completely unknown language, several thousand years dead. Nevertheless, the script was deciphered and the language was slowly reconstructed and pieced together. At first the results were viewed with skepticism by the scholarly world, suspecting a conspiracy of swindles. In the end a curious but convincing test was undertaken: A guaranteed new tablet was copied and given to a number of experts. Secluded they were then each required to make their individual versions of translation of the text! They passed. Even though there were discrepancies, there were enough common elements in all the translations to clearly demonstrate that they had indeed red and understood a common text which they had been given, *à priori* unknown to all of them.

The most mysterious of the Great River Civilizations is undoubtedly the *Indus Civilization*, contemporary to the early stages of the Mesopotamian and Egyptian ones. Excavations have uncovered what could be intriguing relations to the *Sumerian* civilization of ancient Mesopotamia.

In the time span between 2700 B.C. and 1500 B.C. the cities in the Indus valley developed into remarkable urban centers, carriers of an advanced civilization second to none from this epoch. Then decline set in, around 1500 it is over. We do not know the cause of the demise of this great human achievement. Over the next 1000 years a different way of life is in evidence, of a totally rural nature.

The three best known cities are *Harappa, Mohenjo-Daro* and *Chanhu-Daro*. The layout of the cities remarkably resemble that of ancient Mesopotamian ones: In Mohenjo-Daro the center is dominated by the *Citadel*, an elevated area surrounded by a wall about 50 feet high. Here we find *the Great Bath*, a watertight pool which may have had a sacred function. Below the Citadel lies the city, with broad avenues and more narrow side streets, arranged in a regular grid. Houses are build with baked brick, are usually two stories high around a central courtyard. All this closely resemble the layout and architecture found in Mesopotamian cities like ancient Ur and Babylon. Running water is supplied, and we find a covered system for drainage and sewage.

An intriguing feature is the lack of imposing structures immediately identifiable as *palaces* of kings or rulers. This has led some scholars to speculate that perhaps no ruling class existed at all, or possibly the ruling class harbored values which made them shun outwardly trappings of their elevated position. Also in evidence are a larger number of female sculptures, leading to hypothesizing of a matriarchal society.

More than 2000 seals and seal impressions have been found. Again we find a close parallel to seals uncovered in Mesopotamia. As in Mesopotamia, they were carved from stone, and probably were used as the signature of the owner on various documents, letters and packets. The script, the *Harappan script*, found on them has not been deciphered as of this writing.

As of this writing no evidence of the mathematics or the geometry of this civilization has been uncovered.

This notwithstanding that evidence of their architecture and technology is everywhere. Also, they had a standardized system for weights and measure.

One may, therefore, speculate that spectacular breakthroughs in revealing their science and mathematics may well lie in the future. This possibility is also borne out by the ubiquity of sophisticated geometric patterns and ornaments in decorations found throughout the Indus Valley area. These finds are of a clear *protogeometric nature*, which constitutes strong evidence of the sophistication required to support geometric ideas. Another intriguing piece of information is the following: In the first Indian mathematical text, presumably of Hindu origin, there are specific geometric rules for constructing *altars*. The tool for doing so is a set of *ropes* or *string*. The title of the work is the *Sulva-Sutra*, which means *"string-rules"*. The methods employed document knowledge of the so-called *Pythagorean Theorem* as well as *similar right triangles*. Now, the altars are supposed to be made of *burned bricks*, a technology the Hindus of that time did not possess, according to knowledgeable sources. But in the cities of the Indus civilization this technique is to be found everywhere. This has led some historians of mathematics to speculate that the Sulva-Sutra may have originated here. It certainly should be admitted that this is a speculative hypothesis, but it should be worth some serious digging at the cites in question.

2.2 Birth of Geometry as we Know it

Some historians have tended to dismiss the early science as *"merely magic and sorcery"*. But others have forcefully espoused the diametrically opposite view: *The ancients employed precisely the same method as modern scientists!* Indeed, the model of explanation they had for events in nature, for disease, for astronomical phenomena, and so on, was tried out. Corrections were attempted for the shortcomings. Eventually, through trial and error, with failures and mistakes, humanity arrived at our state of today. It may have taken a long time. Or did it really? The invention of the wheel, the first written records, may date from around the fourth millennium B.C. That makes 6000 years up to our time. But compare that to the cave paintings of 30000 years ago!

Amazingly, however, the earliest mathematics we encounter is qualitatively of the same nature as the mathematics of today. For no other science can one assert the same.

An important precondition for humans to be able to live in a well organized society, based on agriculture, is the existence of a reliable calendar. Indeed, without secure knowledge on the changes of the seasons, it is not possible to sow the grain or other seeds at the right time. Sowing too early may destroy the crops by nightly frost early on, and sowing too late may not leave enough time for it to ripen.

These needs were of the outmost importance, literally a question of life and death. And knowledge of a calendar is not possible without insights in astronomy, which again requires knowledge of geometry. Geometry and mathematics did also play an important role in measuring land, constructing irrigation channels or dykes along major rivers and in other engineering tasks. Some historians of mathematics speculate that the capricious and often unpredictably violent behavior of the Euphrates and the Tigris accounts for the fact that mathematics seems to have been better developed in ancient Mesopotamia than it was in ancient Egypt, where the more benign Nile behaved with exemplary regularity. But this comparison is not uncontroversial: Other historians argue that we know more about Mesopotamian mathematics than we do of the Egyptian, simply because the former was written on baked clay tablets, a practically imperishable medium, while the Egyptians wrote on papyrus which has a much shorter life under normal circumstances.

2.3 Geometry in the Land of the Pharaoh

The Egyptian civilization erected itself a proliferation of monuments in the form of huge geometric objects: The Great Pyramids. Such is the immenseness of these artifacts that some writers have speculated that they were left behind by extraterrestrials visitors to Earth. How could people without a sophisticated technology make plans for these structures, let alone carry out the actual constructions? And the pyramids themselves have been surrounded by mysticism and speculations by puzzled observers. But the greatest pyramid of them all was not, according to some historians of mathematics, one found in the Egyptian desert. Instead, it is found on an ancient piece of papyrus, named the *Moscow Papyrus*.

The so-called Moscow Papyrus dates from approximately 1850 B.C. The papyrus contains 25 problems or *examples*, already old when the papyrus was written. It was bought in Egypt in 1893 by the Russian collector *Golenischev*, and now resides in the Moscow Museum. The text was translated and published in 1930 by *W. W. Struve*, in [34]. This papyrus may show that the mathematical knowledge of the Egyptians went considerably further than the so-called *Rhind Papyrus* (see below) demonstrates.

In one of the problems treated there, a formula for the volume of a frustum of a square pyramid is given. If a and b are the sides of the base and the top, respectively, and h is the height, then the formula for its volume is

$$V = \frac{h}{3}(a^2 + ab + b^2)$$

This is exactly right, and its beauty and simplicity has led some historians of mathematics to reverently refer to it as *the greatest of all the Egyptian Pyramids*.

It is often asserted that this formula was unknown to the Babylonians, thus documenting a rare instance where Egyptian mathematics surpassed the

Babylonian. But whereas there does exist tablets from Babylonia where the (obviously false) formula

$$V = \frac{h}{2}(a^2 + b^2)$$

is used, there also exists at least one tablet where a formula equivalent to the Egyptian one may have been employed, for a frustum of a cone.

This is according to a controversial interpretation by Neugebauer, see [35], pages 75,76. Much of the interpretation hinges on whether there is an error in the calculation on the tablet. By the way, some of the tablets we find from ancient Babylonia are the "papers" prepared by the students of the *Temple Schools* or the *Business Schools* which could be found in the larger cities, certainly in Babylon itself. So some sources must be treated with caution. On the other hand, there are some 22000 tablets from the Royal Library of the last of the great Assyrian Kings *Ashurbanipal* at Nineveh.

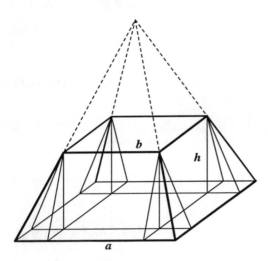

Fig. 2.2. Proof by cutting and reassembling.

But be this as it may, the geometric insights documented by the *Greatest of the Egyptian Pyramids* is surely prodigious. It is instructive to attempt deducing this formula by our present day High School Math. We proceed as illustrated in Figure 2.3.

We let the side of the base be a and the side of the top be b. The height of the big, uncut pyramid is $OC = T$, and the height of the small one, which has been removed, is $OC' = t$. Thus the height of the frustum is $h = T - t$,

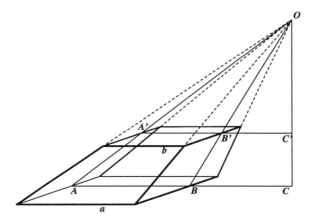

Fig. 2.3. Deducing the formula by our present day High School Math.

in other words the distance between the base and the top. Further $AB = a$ and $A'B' = b$ so that the similar triangles given by O, A', B', and O, A,B yield

$$\frac{T}{a} = \frac{t}{b}$$

We are now ready to compute the volume V of the frustum.

$$V = \frac{1}{3}Ta^2 - \frac{1}{3}tb^2 = \frac{1}{3}(\frac{T}{a}a^3 - \frac{t}{b}b^3) = \frac{1}{3}\frac{T}{a}(a^3 - b^3) =$$

$$\frac{1}{3}\frac{T}{a}(a - b)(a^2 + ab + b^2) = \frac{1}{3}(T - \frac{T}{a}b)(a^2 + ab + b^2) =$$

$$\frac{1}{3}(T - t)(a^2 + ab + b^2) = \frac{1}{3}h(a^2 + ab + b^2)$$

The Moscow-papyrus also contains another problem of great interest. Struve, in [34], claims that in it, Egyptian mathematicians document that they know how to compute the surface area of the sphere (actually, the hemisphere). He interprets the text as computing this area with a value for π implicitly given by the formula

$$\frac{\pi}{4} = (1 - \frac{1}{9})^2,$$

which gives $\pi = 3\frac{13}{81} \approx 3\frac{1}{6}$. Other researchers disagree sharply with Struve's interpretation. Van der Waerden writes as follows in [35]:

The genius of the Egyptians would have been wonderful and indeed incomprehensible, if they had succeeded in obtaining the correct formula for the area of the hemisphere.

The situation is not improved by the presence of an unfortunate hole at a decisive spot in the papyrus. Thus it must be regarded as an open question whether the Egyptians knew the formula for the surface area of the sphere. But the claim is supported by the fact that Papyrus Rhind also does give this value for the number π.

The so-called *Papyrus Rhind* is in fact the most important papyrus for our understanding of Egyptian mathematics. It has been given this name because it was bought in Luxor by the Scottish Egyptologist *A. Henry Rhind* in 1858. Rhind, who was in poor health, had to spend some winters in Egypt. He died on his way home from his last visit there in 1863, and the papyrus was purchased from his executor by British Museum, together with another Egyptian mathematical document known as *the Leather Scroll.*

A more appropriate name for this important papyrus would be *the Ahmes Papyrus*, after the *Egyptian Scribe*[1] who copied it from a considerably older papyrus. This name is now being used more frequently. Ahmes relates on it that the original stems from the Middle Kingdom, which dates to about 2000 to 1800 B.C.

The copy by Ahmes is from around 1650 B.C. Together with the Moscow Papyrus and the Leather Scroll, the Ahmes Papyrus forms our main source for Egyptian mathematics. The Egyptians used the value given above for π, and with this value a computation which appears on the papyrus, uses the correct formula for the area of a circular disc. Altogether the papyrus has the appearance of a practical handbook of math, explaining basic methods by doing a total of 85 examples.

A very beautifully booklet has been published recently with photographs in color of the entire papyrus, transcription of the hieroglyphs and figures on it and explanation of the mathematics in a modern language. Highly recommended reading, [29].

2.4 Babylonian Geometry

When we use the term *Babylonians* we actually mean the civilization residing in the whole of Mesopotamia, not just the citizens of that marvelous city Babylon. This culture was already highly developed at the time from which we find the earliest records, the ancient culture of the *Sumerians.* The main city was not Babylon, until comparatively recent times. The ancient city of *Ur* in southern Mesopotamia was the spiritual and political center for a long

---

[1] *Ahmes* is the earliest individual name associated with mathematics which we know.

time. The Sumerians arrived in this region with their culture already well developed, we do not know from where. The political hegemonies shifted over time, most notably with the arrival of the Akkadians, of which the Babylonians eventually were part. But new rulers carefully preserved the old culture of the Sumerians, and the Kings carefully collected ancient books, baked clay tablets, in Libraries, and made translations into the Akkadian from the Sumerian. In fact we have preserved elaborate dictionaries for the two languages, as well as parallel translations.

The Babylonians had a sophisticated way of representing numbers and computing. They represented numbers to the base 60, in the same way as we represent numbers to the base 10. Thus for instance they would represent the number 61 as (1)(1), while the number 6359 would be represented as (1)(45)(59). Here we have written the *sexagesimal* digits in parenthesis. Those possible digits are of course (0),(1), ...,(59).

Fig. 2.4. Using a stylus usually cut from reed, the Babylonians impressed wedges on clay tablets, which were subsequently baked if the writing was to be preserved. Wedges of different shapes were used, thus making it possible to codify a large set of characters. The digits from 1 to 59 were build up of two types of wedges, in the simplest script in use (others were also present at different epochs). Here we see the rendering of 59. Note the mixture of base, as the individual digits in the base 60-system were represented with symbols for 1's and 10's.

The Babylonians did not directly use the digit (0) in the beginning, but did so indirectly by leaving an open space: Nothing there! But as *scribes*, writers and copiers, copied old tablets to new clay to be baked, mistakes were easily made. So to clarify matters, they started to write a symbol which meant *None* or *Not*. But trailing zeroes were not used. Thus context would have to determine whether (1) meant 3600, 60, 1, $\frac{1}{60}$,... Even though we would find this clumsy, it represented a *numerology*, a representation and understanding of numbers, far superior to that of the Egyptians, Greek or the Romans.

We know a great deal about the mathematics of the Babylonians. This is to a large degree due to the efforts of *Otto Neugebauer* and his collaborators and associates. Like many others Neugebauer had to flee Germany during the Nazi era, and came to the United States. He uncovered and interpreted innumerable tablets from Babylonia, and made the striking discovery of the meaning of the most famous of all tablets which have been found until now.

While realizing that the Babylonians had admirable mathematical insights, historians of mathematics had no clear understanding of the motivation behind Babylonian math. In fact, it was a widespread view that all mathematics prior to the Greek period only consisted of simple practical computations for everyday applications in trade, agriculture and simple engineering tasks. Mathematics as the science we know, they maintained, did not exist until the advent of the Greek. This view would be espoused since it was the Greek who introduced the concept of a mathematical proof.

But it is a fundamental misunderstanding that there can be no mathematics as a science without our modern notion of *proof*. Indeed, the creative process which every research mathematician engages in when mathematics is discovered is almost the complete opposite of a formal proof. Only *à posteriori* do we mathematicians cloak our work in the formal style of *Satz – Beweis*, so beloved by some professors but equally hated by the majority of their students. Of course proofs are necessary so as to ensure correctness of results. And actually *finding* a proof of a conjecture everyone believes to be true is also very much central to mathematics, as in the case of Andrew Wiles' proof of the famous Fermat Conjecture in the last decade of the 20th. century. But it really is not necessary to have produced a formal *Bourbaki-style proof* of a mathematical theorem in order to document complete knowledge of why the theorem is indeed true. [2]

As it happens, a careful analysis of a baked clay tablet from ancient Babylon elucidates this point very well.

The tablet which is perhaps the most famous one, has been given the name *Plimpton 322*. It signifies that it is the tablet numbered 322 in the *A. G. Plimpton* collection at Columbia University in New York. The tablet is written in old Babylonian characters, dating from the period 1900 - 1600 B.C. Unfortunately the tablet is damaged, in that a piece along the entire left edge is missing. Moreover, there is a deep indentation at the middle of the right hand side. Finally it is also somewhat damaged at the upper left corner. Examination has revealed traces of modern glue along the rupture-edge at the left, thus indicating that it was complete at excavation, but broke thereafter in the possession of individuals with access to such amenities as glue, who attempted repairing it.

It would, as we shall see, be extremely interesting if the missing piece could somehow be traced. Quite possibly it resides in one of the many bins of unclassified and unintelligible fragments of Babylonian tablets. As it happens, this was the gravest danger facing the ancient tablets: Destruction at the time of their excavation, which was often – at least in the beginning – done quite crudely.

---

[2] The group of French mathematicians operating under the pseudonym *Nicolas Bourbaki* had a great influence on the way mathematics is presented. The Bourbaki-style is, according to the opinion expressed by some mathematicians, characterized by a rather barren exposition.

The tablet contains a table of numbers, arranged in four columns of 15 numbers each. The rightmost column just consists of the numbers 1, 2, ..., 15. The column to the left is partly destroyed by the missing part.

We shall now follow Neugebauer's reconstruction of what this column contained. The reconstruction has far reaching consequences. It demonstrates first of all that Babylonians knew the so-called Pythagorean Theorem, namely that if we are given a right triangle with hypothenuse equal to d, and a, b denoting the two other sides, then

$$d^2 = a^2 + b^2,$$

the sum of the squares on the two other sides is equal to the square on the hypothenuse. Before we go into this, I shall present the simplest and most beautifully proof I know of this theorem. In all likelihood *this* is the *Babylonian proof*, the proof they knew. Only they would not call it a proof, but regard it as an example of using the *rule by which certain areas may be added.* And, of course, we give it here in modern language and symbolism. But first we give a more conventional proof, which may also well have been known to the Babylonians, in Figure 2.5. See Howard Eves, [5].

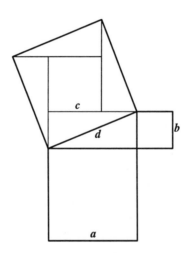

Fig. 2.5. A (hypothetical) Babylonian proof of "Pythagoras' Theorem". The essential part of this figure, namely the subdivision of the largest square, appears in the oldest Chinese mathematical text we know, the *Chóu-pü*, from the second millennium B.C. Thus evidence strongly suggests that this insight formed part of a common wisdom in the ancient world.

In Figure 2.5 the three sides in the right triangle are labeled as above: The hypothenuse as d, the two others as a and b, where $a \geq b$. We then set

$$c = a - b$$

From the figure we now see that the area of the square on the hypothenuse, d^2, is equal to c^2 plus the areas of the four right triangles congruent with the given one. As the area of a triangle is equal to *half the base times the height*, a fact well known to the Babylonians, we get

$$d^2 = c^2 + 4(\frac{1}{2}ab) = c^2 + 2ab$$

But as the Babylonians also knew,

$$(a \pm b)^2 = a^2 \pm 2ab + b^2,$$

which, using the formula in the case of the minus-sign, finally yields

$$d^2 = a^2 + b^2,$$

as desired.

Figure 2.5 and the corresponding proof is one possibility. A variation of the same theme, less familiar to us in our usual thinking concerning "Pythagoras' Theorem", but even more in line with the way the *Babylonians* thought, is a proof derived from Figure 2.6.

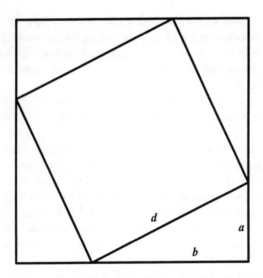

Fig. 2.6. The Putative Babylonian Proof of "Pythagoras' Theorem".

Indeed, the Figure 2.6 yields

$$(a + b)^2 = d^2 + 2ab,$$

from which follows $d^2 = a^2 + b^2$.

Such methods for dealing with sums of squares is well documented from Babylonian tablets. From [33] pages 27 and 28 we reproduce the following example, to be found on a tablet in Strasbourg's *Bibliothèque National et Universitàire*. Phrased in modern language:

An area A, consisting of the sum of two squares, is 1000. The side of one square is $\frac{2}{3}$ of the side of the other square, diminished by 10. What are the sides of the square?

The Babylonians would solve this as follows, again presented in modern language: The sides of the respective squares are denoted by x and y. We then have $x^2 + y^2 = 1000$, as well as the relation $y = \frac{2}{3}x - 10$. Squaring the latter yields

$$y^2 = \frac{4}{9}x^2 - 2 \cdot \frac{2}{3}x \cdot 10 + 10 \cdot 10 = \frac{4}{9}x^2 - \frac{40}{3}x + 100.$$

Substitution into the first equation yields

$$\frac{13}{9}x^2 - \frac{40}{3}x - 900 = 0.$$

Having thus transformed the geometric problem into an algebraic one, the Babylonian scholars and scribes – rather, in the present case presumably students doing their homework – could find the solution utilizing their knowledge about equations and systems of equations. The answer to the present problem is 30, the one positive solution of the equation.

The presentation of the solution starts like this: "*Square 10, this gives (1)(40) (i.e, 100.) Subtract (1)(40) from (16)(40) (i.e, 1000), this gives (15)(0)(i.e, 900)...*"

We return to the Plimpton 322 tablet. The rightmost column only serves to *number* the entries in the other columns. But the two next columns look at first rather haphazard and arbitrary. This led some to assume that the tablet merely constituted a fragment of some business-files, which are actually very much present in quantity among the ancient tablets from Babylon. But the column to the left bears the heading "*diagonal*", while the next has the heading "*breadth*". As with most of the numbers in the first row, to the left, also the heading here is illegible.

Thanks to Neugebauer and Sachs, [26], we now fully understand the contents of this tablet.

The tablet documents that the Babylonians knew how to construct *all Pythagorean triples*: A Pythagorean triple is a triple of natural numbers (a, b, d) such that $a^2 + b^2 = d^2$. Clearly, given any such triple (a, b, d), we get another by multiplying each number by the same natural number r, obtaining (ra, rb, rd). Thus we need only to generate the so-called *primitive* Pythagorean triples, that is to say the triples where the numbers do not have a common factor > 1. Now there is an elegant way of generating all possible primitive Pythagorean triples. Plimpton 322 yields convincing evidence for the claim that the Babylonians knew this method. Usually the method and its proof is attributed to *Diophantus*, as it is explained in his *Arithmetica*. But recent detective work might indicate that *Hypatia of Alexandria* deserves some of the credit for this work, see Chapter 4, Section 4.13, as well as [4]. But as it happens, the Babylonians may have known the proof long before them.

We now give an account of the method for finding the primitive Pythagorean triples, according to Diophantus as reconstructed by *Fermat and Newton* much later, still.

Let (a, b, d) be a Pythagorean triple. We then have

$$\left(\frac{a}{d}\right)^2 + \left(\frac{b}{d}\right)^2 = 1,$$

i.e., the point $(x, y) = \left(\frac{a}{d}, \frac{b}{d}\right)$ lies on the unit circle which has the equation

$$x^2 + y^2 = 1$$

So the problem is equivalent to finding all points with rational coefficients on this circle. We now pull one of today's standard tricks, taught in every class of first year calculus: We wish to find *a rational parameterization of the circle*, that is to say, to find rational expressions in some variable t, $x = \varphi(t)$, $y = \psi(t)$, such that when t varies, then $(\varphi(t), \psi(t))$ runs through all points on the circle. The trick is to let t be the slope of the line through the point $(-1, 0)$, see Figure 2.7.

The equation of this line is

$$y = t(x + 1),$$

which we substitute into the equation for the circle, thus obtaining

$$x^2 + t^2(x + 1)^2 = 1,$$

and hence

$$(1 + t^2)x^2 + 2t^2x + t^2 - 1 = 0,$$

which, as $1 + t^2$ is never zero, may be written as

$$x^2 + \frac{2t^2}{1 + t^2}x + \frac{t^2 - 1}{1 + t^2} = 0.$$

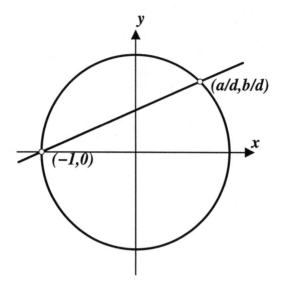

Fig. 2.7. Finding all rational points on the circle.

Now the formula for the roots of the general second degree equation,

$$x^2 + px + q = 0,$$

is

$$x = -\frac{p}{2} \pm \sqrt{\frac{p^2}{4} - q},$$

which when applied to the equation in question here yields

$$x = -\frac{t^2}{1+t^2} \pm \sqrt{(\frac{t^2}{1+t^2})^2 - \frac{t^2-1}{1+t^2}} = -\frac{t^2}{1+t^2} \pm \frac{1}{1+t^2},$$

after a short computation. We thus obtain

$$x = -1 \text{ or } x = \frac{1-t^2}{1+t^2}$$

Substituting the last solution into the equation for the line, we get

$$y = t(\frac{1-t^2}{1+t^2} + 1) = \frac{2t}{1+t^2}$$

Since the points (x, y) have rational coordinates, we may write $t = \frac{u}{v}$ for natural numbers u and v. Here we must have $v > u$ since the slope t of our line lies in the interval $< 0, 1 >$. Substituting this into the expressions for x and y, we obtain the following formulas:

$$x = \frac{a}{d} = \frac{v^2 - u^2}{v^2 + u^2}, \text{ and } y = \frac{b}{d} = \frac{2vu}{v^2 + u^2}.$$

We have essentially completed all ingredients needed to prove the following:

Theorem 1 (Ancient Wisdom of Babylon). *All primitive Pythagorean triples (a, b, d) are given by*

$$a = v^2 - u^2, b = 2uv \text{ and } d = v^2 + u^2,$$

where u and v are positive integers, $v > u$, without a common factor > 1. Moreover, u and v are not both odd numbers.

Proof. First of all, numbers of the form $a = v^2 - u^2, b = 2uv, d = v^2 + u^2$ where u and v are natural numbers do form a Pythagorean triple, as is seen by computing $a^2 + b^2$. If we assume that u and v have no common factor > 1, then the triple is also primitive, *except for the possibility that $a = 2\bar{a}, b = 2\bar{b}$ and $d = 2\bar{d}$*. Indeed, a, b, d can have no other common factor than 2, and it is easily seen that this happens if and only if u and v are both odd numbers. Then the overlined numbers do form a primitive Pythagorean triple. In this case we introduce new versions of u and v by putting

$$\bar{v} = \frac{v + u}{2}, \bar{u} = \frac{v - u}{2},$$

from which we find

$$2\bar{u}\bar{v} = \frac{v^2 - u^2}{2} = \bar{a}, \bar{v}^2 - \bar{u}^2 = uv = \bar{b} \text{ and } \bar{v}^2 + \bar{u}^2 = \frac{v^2 + u^2}{2} = \bar{d}$$

It is not difficult to verify that $\bar{v}, \bar{u}$ have no common factor > 1, and are not both odd numbers.

Now, given a primitive Pythagorean triple (a, b, d). From the considerations preceeding the formulation of the theorem, we can always find natural numbers v and u, such that

$$\frac{a}{d} = \frac{v^2 - u^2}{v^2 + u^2}, \text{ and } \frac{b}{d} = \frac{2vu}{v^2 + u^2}.$$

Unless u and v are both odd numbers, we therefore have that a, b, and d must be as claimed in the theorem. If u and v are both odd, then we proceed as above, obtaining new u and v's, $\bar{v} = \frac{v+u}{2}, \bar{u} = \frac{v-u}{2}$, also without common factors, but now not both odd numbers, such that

$$2\overline{vu} = \frac{v^2 - u^2}{2}, \overline{v}^2 - \overline{u}^2 = uv, \text{ and } \overline{v}^2 + \overline{u}^2 = \frac{v^2 + u^2}{2}.$$

Thus the primitive Pythagorean triple a, b, d is described as in the theorem, but with the roles of a and b interchanged. □

By means of the theorem we may explain the numbers on Plimpton 322. It is generally accepted that it contains four errors. Three of them are easy to explain as a simple mistake with the stylus, whereas the fourth is more mysterious. Several explanations have been offered, but as long as we only have this one table of this type, and in view of the missing part, it is difficult to decide what the correct explanation is.

At any rate, except for these presumed errors the second and third column from the right consists of the numbers b and d described above, for the choices of u and v shown in the table presented as Figure 2.8.

b	a "Breadth"	d "Diagonal"	No.	v	u
120	119	169	1	12	5
3456	3367	4825 (11521)	2	64	27
4800	4601	6649	3	75	32
13500	12709	18541	4	125	54
72	65	97	5	9	4
360	319	481	6	20	9
2700	2291	3541	7	54	25
960	799	1249	8	32	15
600	481 (541)	769	9	25	12
6480	4961	8161	10	81	40
60	45	75	11	2	1
2400	1679	2929	12	48	25
240	161 (25921)	289	13	15	8
2700	1771	3229	14	50	27
90	56	106 (53)	15	9	5

Fig. 2.8. The reconstructed Plimpton 322.

We have written the corrected numbers with the presumed erroneous ones in parenthesis. Also note that these are all *primitive* triples corresponding to the given values of v, u, with the exception of two entries. We start with entry number 11: The values 2 and 1 should give $(b, a, d) = (4, 3, 5)$ which is not shown. Instead this triple is multiplied with 15, to give more palatable digits in the Babylonian number system. Next we note that the last entry, in line number 15, is not a primitive triple: To wit, u and v are in this case both odd numbers. However, they of course give a Pythagorean triple. The corresponding primitive one would be $(45, 28, 53)$, and as the only two numbers which are legible on the tablet, and not are reconstructed, are 56 and 53 this could lead to the question if perhaps the entry 56 represents the mistake: Should it be 28 instead? At any rate, the primitive triple corresponds to $(v, u) = (7, 2)$. But beware: The roles of a and b are now interchanged, in that $2uv = 28$ and $v^2 - u^2 = 45$.

Why did the Babylonians do it this way? They had good reason for it, this is no mistake! Namely, 28 is not a regular sexagesimal number, it does not have a finite sexagesimal fraction as an inverse. But $\frac{1}{90} = (0).(0)(40)$. All the listed values for b are regular sexagesimal numbers.

The values of u and v are not given on the tablet, but included here. Also, we have included the values of $b = 2vu$ in the column to the left, but the damaged and unreadable column to the left on the tablet itself can not have consisted of these numbers. Rather, a careful study of the readable fragments reveals that this column must have listed the values for $\frac{d^2}{b^2}$, that is to say $\sec^2 v = \frac{1}{\cos^2 v}$ for the angle v between the hypothenuse and the side with length b in the right triangle.

The numbers u and v are carefully chosen. First, they are all regular sexagesimal numbers: Their inverses are finite sexagesimal fractions. That such choices are possible at all for the entire table is due to the choice of base 60, which has the prime factors $2, 3, 5$, whereas base 10 only has $2, 5$. Thus for instance, in Babylon they would have $\frac{1}{3} = (0).(20)$ and $\frac{1}{15} = (0).(4)$. Then the tricky long division in the sexagesimal system could be avoided in many cases, and replaced by multiplication, which they easily performed using multiplication tables on baked clay tablets.

With our base of 10, we have a special relationship to the numbers $3, 7$ and 13, as being, respectively, *lucky, sacred and unlucky*. The Babylonians do not seem to have offered 3 much thought, but 7 was sacred and 13 was very unlucky, *The Number of the Raven*.

Thus one finds that this tablet would have been convenient to use in computations, and probably have served as an equivalent to a table over $\cos(v)$ for v between $44°46'$ and $31°53'$. The decrement in the values of v are not constant, however $\sec^2 v$ decreases by almost exactly $\frac{1}{60}$ from one line to the next.

Thus not only did the Babylonians know "Pythagoras' Theorem", but they knew the theory of primitive Pythagorean triples as well, and may have used it to compile trigonometric tables for use in their astronomy and engineering.

The Babylonians did most of their number-work relying on tables. For example, multiplication could be carried out using the tables of squares by the formula

$$xy = \frac{1}{4}((x+y)^2 - (x-y)^2)$$

Moreover, one should note that

$$\frac{1}{4} = (0).(15),$$

and multiplication with this number is especially simple in base 60, much like multiplying by 0.2 or 0.5 in base 10.

3 Greek and Hellenic Geometry

3.1 Early Greek Geometry. Thales of Miletus

The word *geometry* is derived from two Greek words, namely $\gamma\eta$, gē, which means *earth* and $\mu\varepsilon\tau\rho o\nu$, metron, which means *measure*. Our sources on early Greek geometry – and mathematics in general, for that matter – are sparse. Indeed, as far as mathematical contents is concerned we have to rely on the work of the first serious *historian of mathematics*, namely *Eudemus of Rhodes*, 350 – 290 B.C. He was, probably, a student of Aristotle, at any rate a close associate and collaborator of him. But Eudemus of Rhodes should not be confused with *Eudemus of Cyprus*, another philosopher associated with Aristotle. In any case, our Eudemus is known to have written three works on the history of mathematics, namely *The History of Arithmetic, of Geometry* and *of Astronomy*. All three are lost now, but were available to Hellenistic mathematicians and used to the extent that at least some of their contents is known to us today. In particular Eudemus reports, in his History of Astronomy, that *Thales of Miletus* predicted a solar eclipse, which is presumed to be the one which occured on May 28, 585 B.C. But most historians of mathematics tend to be skeptical to this claim. The reason for this is that Thales is generally agreed to have been the first Greek astronomer, and that such abilities would have been unlikely at this early stage of Greek astronomy. However, it appears that the most plausible explanation is offered by *van der Waerden* in [35], where he writes:

> *The conclusion is inescapable that he must have drawn upon the experience of Oriental astronomers.*

By the way, the Greek historian *Herodotus* also makes this assertion concerning the prediction by Thales. The solar eclipse occured during a battle fought between the Lydians, under their King *Alyattes*, and the Medians under their King *Astyages*. The war had been going on for five years, and when the eclipse occured during an ongoing battle, the belligerent parties found it prudent to end the fighting and make peace. The Gods, evidently, did not approve of what they were doing. Thales had ties to the Lydian kingdom, and when Alyattes' son Croesus later went to war against the Persian King Cyrus II, who had meanwhile conquered the Median kingdom, Thales went along as an advisor to the Lydian King. Thales is credited with a clever scheme for splitting the river Halys, so that the Lydian troops could pass over.

Eudemus' historical works are lost. But their contents are, to some extent, known through later summaries. The last Greek philosopher and mathematician was *Proclus Diadochus*. He was head of the Neoplatonic Academy in Athens late in the fifth Century A.D., one of the last holdouts of classic civilization. At that time Eudemus' books were still extant. As all research towards the end of the classic civilization, Proclus' research is not very original. But as part of his work at the Academy he wrote a summary of Eudemus' History of Geometry, as an introduction to his own *Comments on Euclid's Elements*, Book I. This is essentially the only surviving source on early Greek geometry, frequently referred to as *The Eudemian Summary*. There can be no doubt that Proclus amply deserves a honorable place in the history of geometry and mathematics for preserving this knowledge for posterity. Another important contribution by Proclus was the formulation of Euclid's Fifth Postulate as we state it today, usually referred to as *Playfair's Axiom*. See Section 4.1.

Thales is the first Greek mathematician whose name we know. He lived and worked in Miletus, a Greek city in Asia Minor, now in Turkey. He was born about 625 B.C. and died around 545 B.C., in Miletus. We may regard Thales as the *Founding Father of Greek Geometry*. His mother was *Cleobulina*, the first woman philosopher in Greece. Thales referred to her as *The Wise One*.

There are reports that Thales was *of Phoenician descent*, but others refute this by asserting that *"... the majority opinion considered him a true Milesian, and of a distinguished family."* Do we sense a trace of bigotry here? Perhaps the infusion of some Phoenician blood through Thales did the Greeks and their science some real good...

Thales is supposed to have estimated the height of a pyramid in Egypt by measuring its shadow at the time when the shadow cast by himself was equal in length to his own height. Eudemus ascribes to Thales a method for finding the distance between two ships at sea. We do not know exactly what this method was, but van der Waerden in [35] supposes that it might be something like the method described by the Roman surveyor *Marcus Junius Nipsius*, which goes as follows:

> In order to find the distance from A to the inaccessible point B, one erects in the plane a perpendicular AC to AB, of arbitrary length, and determines its mid point D. On C one constructs a line CE perpendicular to CA, in a direction opposite of AB, and one extends it to a point E, collinear with D and B. Then CE has the same length as AB.

The rule is illustrated in Figure 3.1.

Thales also is credited for discovering that the base angles of isosceles triangles are equal, and that vertical angles are equal. He is also said to have

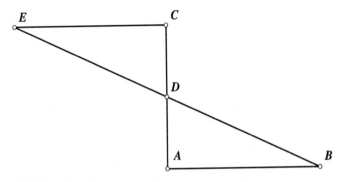

Fig. 3.1. To find the distance to an inaccessible point.

discovered that a diameter of a circle divides it in two equal parts. In what sense Thales "discovered" these geometrical facts is not clear, it does seem reasonable to assume that this knowledge would have predated Thales by perhaps more than a thousand years, in Egypt, Mesopotamia, and elsewhere in the East. He may, however, have studied this material, providing some sort of proofs for the above statements.

According to Aristotle, Thales was ridiculed by some Milesians for directing a lot of energy to activities which had no useful applications, and from which he made no profit. Thales then decided to show them that if he had thought it worthwhile, he could do better than most of them in this regard as well. Thus, noticing signs that a bumper crop of olives was in the comings, he bought up all the presses. When the bumper crop then subsequently *did* materialize, the growers had to buy or rent presses from him, at a substantial price.

3.2 The Story of Pythagoras and the Pythagoreans

Pythagoras of Samos is a rather enigmatic figure. It is frequently asserted in texts on the history of mathematics that we know practically nothing of his life and work prior to the time when he founded the *school of the Pythagoreans* in Croton, at which time Pythagoras may have been in his mid 50's. We do know however, that he appears at this precise point, and that he undoubtedly possessed extensive knowledge of mathematics in general and geometry in particular. Prior to that time this kind of knowledge is only very sparsely documented in Greece, and all of it comes to us from Thales. But in the East, in Egypt, Mesopotamia, in India and even, perhaps, in the early Indus valley civilization, as well as in China, we find evidence of extensive insights into these matters. Add to this the many stories which are told concerning his travels in Egypt and more widely. We have to realize, however, that for now there is no solid evidence on which the legends of Pythagoras' travels can be accepted as historical facts. So until some new papyrus is found in Egypt, or a tablet uncovered from ancient Babylon, relating the tale of the Greek *visiting priest* at the Temple, we might as well sit back and enjoy the stories. Some of them simply are too good not to be true!

Pythagoras was born about 570 B.C. in Samos, one of the most fertile Greek islands, just off the coast of Asia Minor. It seems to be general agreement that he died in the Greek city of Metapontium, in southern Italy, probably some time during the first decades of the fifth Century B.C., one estimate being approximately 480 B.C. At any rate there are reports that he died at the advanced age of 90.

Some historians of mathematics think that Pythagoras was a student of Thales. Others feel that the age-gap between them makes this unlikely. But with the – admittedly hypothetical – dates of birth and death we have put down, Pythagoras would have been 25 at the time of Thales' death. This does not preclude him having been a student of Thales, but it is probable that Pythagoras at least also had other teachers, working in the same mathematical environment as Thales. In fact, Samos and Miletus were geographically close.

Iamblichus relates in [22] that Pythagoras *"...went to Pherecydes and to Anaximander, the natural philosopher, and also he visited Thales at Miletus. All of these teachers admired his natural endowments and imparted to him their doctrines. Thales, after teaching him such disciplines as he possessed, exhorted his pupil to sail to Egypt and associate with the Memphian and Diospolitan priests of Jupiter by whom he himself had been instructed, giving the assurance that he would thus become the wisest and most divine of men."*

So according to this source, Pythagoras followed in Thales' footsteps. Not only did he take up his geometry, he also made extensive travels in the known civilized world. In Samos *Polycrates* assumed dictatorial powers, but he was in many ways an enlightened ruler, and at least in the beginning Pythagoras may have had good relations with him.

Polycrates had allied himself with Amasis, the King of Egypt. Polycrates was very successful in the beginning, and he established Samos as a naval power, he build temples, harbors and aqueducts and he encouraged art and science including mathematics. Herodotus relates how Polycrates became worried when he received a message from his Egyptian ally, warning him that his good fortune would eventually make the Gods envious, thus bringing some kind of disaster down on him. The advice he gave was for Polycrates to throw away his most valued possession. The grief this would cause him, should suffice to placate the envious Gods. After thinking about it, Polycrates decided that a precious ring he owned would be a suitable object to loose, and he went out to sea on a boat, where threw his ring into the water. Some days later, however, a local fisherman caught a big fish. The fish was so extraordinary that the fisherman brought it to Polycrates, expecting to be rewarded lavishly. Polycrates was very pleased, and showed it by inviting the fisherman to his supper, where the fish was to be served. The cook started the preparations and cut the fish open, and in its stomach he found the ring. He brought the ring to Polycrates, who was not exactly overjoyed. When Amasis learned about this, he realized that Polycrates could bring him nothing but bad luck, and canceled his alliance with him. And in fact, towards the end of his reign Polycrates engaged in some ill-conceived schemes, trying to ally himself with the Persians against the Egyptians. This failed because of mutiny among the men he sent, who with good reason suspected that Polycrates really wanted to get rid of them. He himself was later lured into an ambush by the Persians and suffered a shameful death.

Returning to Pythagoras, he went to Egypt, some say around 535 B.C. Polycrates had supplied him with letters of recommendation, so he could gain access to the Temples there.[1] He visited many temples where he had discussions with the priests. He tried to gain admittance to the Order of the Temples, and finally succeeded when he was admitted into the Temple and Priesthood at Diospolis, near Memphis. Here he stayed for some time, and absorbed their customs and their geometry, as well as their magic and astrology.

But this quiet life was interrupted when there appeared on the scene a Persian King and warlord by the name *Cambyses*. He invaded Egypt in 525 B.C., and capturing Memphis came across Pythagoras in the Temple, as Iamblichus relates in [22]. Pythagoras was then taken prisoner by Cambyses, and if this story is true, he must have had some very exiting and interesting years, under Cambyses' rather heavy hand. In the beginning it would not have been too bad, Cambyses himself respected the Egyptians and showed great interest in their traditions and customs. He even had himself designated a *Pharo* under the name *Ramesut*. He also had himself initiated into the priesthood, and if Pythagoras were around this, he might have had something to do with it. In fact, Cambyses' father was *King Cyrus* II or *Cyrus the Great*. He could possibly have met Thales, Pythagoras' mentor, under the following circumstances: According to Herodotus, Thales accompanied King Croesus when he went to war against the Persians under King Cyrus. Croesus lost, and after several dramatic events he was saved from being burned alive on a pyre erected by the victorious Persians. These same events also led him to become a trusted friend and advisor of King Cyrus. This happened in 547 B.C., admittedly late in Thales' life, if not after his death.

Cyrus was one of Persia's great Kings, who went on to capture the marvelous ancient City of Babylon, in 539 B.C. He is the *Cyrus the King* referred to in the Old Testament, who restored the Jews to Palestine and ordered the Temple of Jerusalem to be rebuild. Unfortunately for him, however, he did not rest on his laurels. Instead, he marched with his troops across *the Araxes*, the river now named *Araks* which flows east to the Caspian Sea. He

---

[1] Some say this, others claim that Pythagoras feared Polycrates, and fled because of him.

went against the *Massagetic queen Tomyris*, she ruled over a kingdom in that area. His advisor Croesus was with him, and the crossing of the Araxes was undertaken on his advice. This was a disastrous move. Tomyris defeated Cyrus, who was slain in a battle 529 B.C. Then his son, Cambyses II, succeeded him on the Persian throne. On his fathers advise, he retained Croesus as an aide and advisor, in spite of the sad outcome of his last service to his father. And Croesus accompanied Cambyses to Egypt. Thus Pythagoras and Cambyses' aide would have some points of contact.

At any rate, the good state of affairs for Pythagoras in Egypt did not last. Cambyses continued his milliary expansion, and now he met with some very serious, humiliating setbacks and defeats. Without going into details, let us just relate that he turned into a paranoid man, suspicious of everything. When he arrived back from one of his ill-fated expeditions, his troops decimated and starved, having been reduced to cannibalism, he unfortunately came just in time for a big celebration in Memphis. Feeling that the people rejoiced because of his own misfortune, he ordered the leading citizen rounded up and executed. The most repulsive incident occured when it was explained to him that the celebration was on occasion of the appearance of a very special calf, the latest incarnation of the God Apis. On his orders the calf was brought into his presence. Cambyses, in a fit of senseless rage, grabbed his sword and dealt the Holy Calf a powerful blow, wounding it in the thigh, in front of all the terrified Egyptians. The Holy Calf fell to the ground, and it died some time afterwards from the infected wound.

He also committed various other acts of sacrilege, like several instances of outrageous profanation of temples, killings of priests, he broke up ancient tombs and examined the bodies, burned them in some cases, and so on.

Matters worsened. Cambyses appears to have gone completely mad. According to Herodotus one of the misdeeds he committed was to have his own brother, *Smerdis*,[2] murdered. Smerdis had been a member of his Egyptian expedition, but Cambyses had sent him back to Persia because of jealousness caused by his brother's physical strength. Some time after Smerdis' return, Cambyses had a dream which caused him great worry: He dreamt that a messenger arrived from Persia, telling him that Smerdis was sitting on the royal throne and that his head was touching the sky. Interpreting this to mean that his brother would kill him and seize the throne of Persia, Cambyses sent his most trusted Persian friend *Prexaspes* back to Persia to do away with Smerdis. Prexaspes dutifully did what he had been ordered. And then he informed the people that His Royal Highness the Prince spent all his time in seclusion at the palace, praying for the success of his brother the King during his campaign abroad. Cambyses later rewarded him for his services by murdering his son in front of his very eyes, in order to prove his marksmanship with bow and arrow and ability to hold his liquor.

---

[2] According to the Persian sources Cambyses murdered a brother by the name *Bardiya*.

Now Herodotus relates that Cambyses had left the control of his household with a man who belonged to the caste of *the Magis*, his name was *Patizeites*. Patizeites had a brother, named Smerdis, like the prince. This brother also looked a lot like the murdered prince, and as Patizeites knew of Cambyses' foul deed regarding his brother, he hatched a rather obvious plan: He had his own brother usurp the throne, claiming to be Cambyses' brother![3]

The Magis constituted the hereditary caste of priests among the ancient Persians. They interpreted dreams and performed sacred rituals, being devoted to the Gods. In the New Testament the astrologers who divine the birth of the King of the Jews by the appearance of a star in the east are called *Magis*. The priests of Babylonia are also frequently called *Magis*, and of course the term is preserved today in our word *magic*.

Heralds were sent out proclaiming the change of regent, and one of them happened to encounter Cambyses and his men in Ecbatana in Syria. When brought before the rightful, if incompetent, King, the herald was questioned about the situation. Cambyses suspected that Patizeites had double-crossed him, but the latter had the explanation ready: *"I think, my lord, that I know what happened. The rebels are the two Magi brothers you left in charge of your household. One of the brothers is named Smerdis, as you may recall."* Cambyses now realized the true meaning of his dream. The Smerdis on the throne was really Smerdis the Magi! The murder of his brother had served no purpose, in fact it had made the prophesy of the dream come true, rather than preventing it from happening. As sanity started to return, he understood the depths to which he had fallen, and he bitterly lamented the abysmal situation in which he found himself. Finally he resolved to march back to Persia at once, to attack the Magi. But as he leaped into the saddle, the cap fell off the sheath of his sword. The exposed blade cut his tight, at the very spot where he had struck the sacred Egyptian Bull of Apis. Cambyses now felt that he was mortally wounded, and asked his men for the name of the town they were in. Being told that the name was Ecbatana, he realized the true meaning of a prophecy from the oracle at Buto: Namely, that he should die at Ecbatana. He had thought this to be *Median Ecbatabna*, his capital city, and that he should therefore die at home of old age. Now he realized that the oracle meant Ecbatana in Syria. At this point sanity fully returned to Cambyses, and he said no more. After twenty days he called the leading Persians together, and explained the situation to them. In tears he bitterly lamented his cruel fate, and the Persians tore their cloths, crying and

---

[3] Persian sources give the name of the Magian usurper, or pretender, as *Gaumata*. Thus there is no homonymy in the Persian version of this story, as well as other discrepancies with the account as given by Herodotus. It is generally accepted among historians that Herodotus' version of the story is far from accurate. A political intervention by priests of the temples in the face of a ruler who was obviously incompetent and mentally disturbed, as well as a political rivalry between Medes and Persians with economic and social ramifications, has undoubtedly taken place. But the details are lost today.

groaning. Shortly after, gangrene and mortification of the thigh set in, and Cambyses died.

However, his men really did not believe him. They suspected another malicious lie, to set the country against his brother Smerdis.

Thus no obvious course of action seemed to present itself, and about one year of political strife followed in Persia, with the Magi on the throne. Prexaspes originally decided to side with the Magis, out of fear for punishment and also his bad feelings towards the house of Cyrus and Cambyses. Thus he changed his story about having murdering Smerdis the Prince. The Magi rule ended when a young and ambitious nobleman by the name *Darius*, himself of royal descent, headed a successful *coup d'etat*. Prexaspes, repenting his treason to the Persian cause (the Magi were originally a Median caste), confessed his crime to an assembled crowd from the main tower, and then leaped to his death.[4] Darius then assumed power, to become the famous Darius I, *Darius the Great.*

The story of the false Smerdis, the usurpation of power by the Magis and finally the accession of Darius plays an important role in the history of mathematics, at least indirectly. In fact, the Persian version of it, as told to us by Darius himself, forms part of the inscription at Behistun, described in section 2.1, and thus provided the basis for Rawlinson's decipherment of the cuneiform script. This again led to our present insights into the mathematics in Mesopotamia, of the Sumerians, Assyrians and the Babylonians. As already noted, the inscription by Darius himself differs considerably from the tale as told by Herodotus. For more details, see note 25 on page 571 in [17].

Pythagoras, however, had been brought to Babylon by Cambyses' troops. At least so the story goes. The political situation in the Persian Empire being somewhat murky, he sought refuge in the Temple, where he was once more initiated into the Priesthood. *Iamblichus* writes as follows in [22], in the fourth Century A.D.:

"Here the Magi instructed him in their venerable knowledge and he arrived at the summit of arithmetic, music and other disciplines. After twelve years he returned to Samos, being then about fifty six years of age."

There are some ancient busts claiming to show what Pythagoras may have looked like. One is a bronze copy of an original believed to be from the fourth century B.C., which is displayed at Villa dei Papiri in Herculaneum, Museo Nazionale, Neapels. Here Pythagoras is shown wearing turban and oriental dress, absolutely compatible with our story. A photo of the bust is shown in [21] and in [35].

Iamblichus has Pythagoras' stay in Egypt to last for 22 years, plus 12 years in Babylon, altogether 34 years abroad. At any rate he spent many years in Egypt and in Babylon, working and learning in the temples.

---

[4] Still according to Herodotus, the Persian story runs differently. There is no character by the name *Prexaspes* in that version.

Cambyses had died in 522 B.C., and Polycrates, the tyrant of Samos, was killed by the Persians about the same time. King Darius I took over in 521 B.C., and after Polycrates death Samos came under his rule. Exactly when Pythagoras returned to Samos is uncertain. Some say that he returned at a time when Polycrates was still alive and in power, others assert that he returned at a time when Samos had fallen under Persian rule. In any case, after the fall of the Magi from power, it would seem to make sense for Pythagoras to leave Babylon, since he presumably had close ties with that group.

Iamblichus reports that Pythagoras formed a school in the city of Samos, called *the semicircle*. He also reports that Pythagoras made a cave outside the city, where he did his teaching, and spent both nights and days doing research in mathematics. But then Iamblichus goes on to tell how Pythagoras attempted to employ the same didactical principles he had learned in the temples of Egypt and Babylon, to teaching the Samians. This did not work too well, they found his teachings too abstract and symbolic. Pythagoras did not like such attitudes any better than some present day college professors do, and decided to leave. At least this is the reason Pythagoras himself is supposed to have given for leaving Samos.

Actually, the Samians were by no means ignorant of geometry. Herodotus relates how they constructed, at the order of Polycrates, an aqueduct for bringing drinking water to the capital city by the same name as the island. They had to dig a tunnel through a mountain, and started to dig at both ends simultaneously. And in fact, they met in the middle of the mountain with remarkable accuracy! The direction of the tunnel had to be found by reasoning with similar triangles. Also a fairly sophisticated use of a diopter had to be employed. *Heron of Alexandria* explains the method in his work *Dioptra*, about 600 years later, around 60 A.D. For details, see [21] or [35]. The engineers, some of them quite possibly being *slaves*, who worked on the tunnel at Samos certainly knew quite sophisticated geometry. But this knowledge was part of their practical work in the field, not necessarily as an object of the *"refined contemplation"* considered worthy of *free men*.

So Pythagoras left for Croton, a Greek city on the coast of southern Italy. Here he formed his school or brotherhood, *The Pythagoreans*. The society consisted of an *inner circle*, whose members were called *mathematikoi*, and an outer circle whose members were known as *the akousmatics*.

The mathematikoi lived permanently with the Society, they had no personal belongings, were vegetarians and practiced celibacy, did not eat *beans*, and did not wear cloths made of animal skin. Presumably this was the way of life Pythagoras had picked up at the temples in Egypt, although *Herodotus* does report on ample supplies of meat and wine for the Egyptian priests.

It should be noted that there are marked similarities between the practices of the Pythagoreans and those associated with *the Orphic Cult*. Orpheus of Thrace was the founder of this cult. He played so divinely on the lyre that all nature stopped to listen. When his wife *Eurydice* died, he went to the nether world, to *Hades*, to bring her back. By the music from his lyre he succeeded in obtaining her release, but on the condition that he would not look at her until they were clear of the world of death. However, he could not bear to refrain from looking, and she had to return to Hades for good.

The akousmatics, however, were allowed to live normal lives. Both men and women were allowed to be Pythagoreans, and there are some reports of women Pythagoreans who became well known mathematicians and philosophers.

There are accounts to the effect that *Pythagoras had a wife*. Her existence would seem to contradict the claimed practice of celibacy, but this particular kind of contradiction should not disturb historians too much. Her name was *Theano*, and she had three daughters with Pythagoras. Together with them she is said to have continued Pythagoras' school after his death. Her most important mathematical work is supposed to have been a treatise on *the Golden Section*. We refer to [39], [23] and to [36]. As far as this author's information goes, this is the first known, or claimed, individual name of a woman mathematician. Pythagoras' three daughters also were Pythagoreans. *Damo* is said to have been entrusted the responsibility for her fathers works, which she refused to sell and therefore had to live in poverty. The two other daughters *Arignote* and *Miyia* were also Pythagoreans, and are credited with several works on a variety of subjects. Other women Pythagoreans were *Themistoclea*, priestess of Apollo at Delphi and said to be Pythagoras' sister, and *Melissa*, thought to have been one of the very first Pythagoreans.

The Pythagoreans were in opposition to the *democratic* movement in Greece. The followers of the philosophical school of the *Sophists* were democrats, while the Pythagoreans believed in *oligarchy*, the rule by a small political elite. Some of the Greek geometers did in fact belong to the democrats. They did not get along too well with the main stream Pythagoreans, who were very influential. Thus for example *Hippasus of Metapontium*, who was a Pythagorean, and nevertheless democrat, made known the findings that not all line segments have a common measure, that there are *incommensurable* line segments. We say more about this below, but the Pythagoreans did not take lightly to this breach of secrecy! In fact, he was severely denounced for having described the *Sphere of the Twelve Pentagons*, in other words the *dodecahedron* and for having revealed *the nature of the non-mensurable* to *the Unworthy*.

To the Pythagoreans the regular pentagon with the inscribed pentagram, the 5-pointed star formed by all the diagonals, was a sacred symbol. There is a story about a Pythagorean who became seriously ill while traveling, far from home. The keeper of the inn where he stayed was a compassionate man, and had his servants nurse him as best they could. The money of our traveling Pythagorean expended, he was reduced to nothing: Seriously ill, at the mercy of foreigners, far from home. Nevertheless the inn-keeper stood by him, providing for him at his own expense. As the unfortunate Pythagorean realized that his *Earthly Goal* for the present incarnation was approaching, he called for his benefactor. Not being able to leave behind any significant earthly values, he told him to paint the symbol of the pentagon with the inscribed pentagram on his door, *but to paint it right, not upside down.* If ever a Pythagorean came this way again, he would generously return the favor. And so the man did, after his foreign guest had passed away. Not that he had much belief in the benefits to be reaped from this undertaking. But years later a rich Pythagorean traveled through the area, saw the pentagon with the inscribed pentagram, and did indeed repay the local good Samaritan generously.

Returning to Hippasus, his treasonous publication may have happened towards the end of Pythagoras' life, maybe after his death. Hippasus was expelled from the Brotherhood, and one version of what happened afterwards is this: The Pythagoreans made a grave monument for him, as he was to be considered dead. Soon afterwards he perished at sea, and this was seen as punishment from the Gods: *He died as a godless person at sea.* Another version of the story is that he was murdered by Pythagoreans, who threw him overboard from a ship at sea. Be this as it may, during this time the opposition to the Pythagoreans grew, Pythagoras himself had to move from Croton to Metapontium. A prominent citizen of Croton by the name *Cylon* is said to have been refused entry into the Pythagorean Brotherhood by Pythagoras, presumably because he was lacking in the spiritual qualities required, and as a result the same Cylon mobilized his followers against Pythagoras and the Pythagoreans. Others report the events differently, but at any rate Pythagoras had to move to Metapontium, not too far from Croton, as the situation became difficult. He died in Metapontium soon afterwards. According to some accounts he was murdered, killed by arson at the house of his daughter Damo.

In Croton the Pythagoreans continued to exist as an organization, but increasingly surrounded by controversy. Finally mobs emanating from the democratic party killed a large number of Pythagoreans when they set fire to the house in which they were assembled, the house of an athlete named *Milo*, a famous wrestler. As many as 50 or 60 Pythagoreans are said to have been killed at that time. The surviving Pythagoreans fled from Croton, and thus, ironically, the ideas of Pythagoras were spread more widely in the Greek domain. Later still the Pythagoreans reappeared in the area, the last important of them being *Archytas of Tarentum*, 438 - 365 B.C. His best known work is probably an ingenious 3-dimensional construction which accomplishes the doubling of the cube. We shall explain this in Section 3.11.

Again we should reiterate the warning that the story of Pythagoras' life which we have told here is regarded by some as being highly unreliable. Contradicting ones are in circulation as well. The indisputable fact, however, is that these stories and legends about him do exist, and have been told for 2500 years.

3.3 The Geometry of the Pythagoreans

No work by Pythagoras is extant, and in fact the practice of the early Pythagoreans was to ascribe all their findings to the master himself, to Pythagoras. But it is well documented from later sources that the Pythagoreans viewed mathematics as basic to the very fabric of reality, and that certain fundamental doctrines were important to their thinking and teaching. One such doctrine was that *numbers*, that is to say, the natural numbers, formed the basic organizing principle for everything. The motion of the planets could be expressed by ratios of numbers. Musical harmonies could be expressed so as well. The right angle was fixed by ratios like 3:4:5, as a triangle with sides in these proportions is a right triangle.

This takes us to the *geometry* of the Pythagoreans. Several discoveries have traditionally been attributed to the Pythagoreans, but at least some of them are without question of a much earlier origin. We reproduce a list of such discoveries in geometry, together with some comments. See [15] and [37].

> 1. The Pythagoreans knew that the sum of the angles of a triangle is equal to two right angles. They also knew the generalization to any polygon, namely, that in any n-gon the sum of all the interior angles is equal to $2n - 4$ right angles, while the sum of all exterior angles is equal to four right angles.

The last assertion may be viewed as completely obvious, as far as the mathematical realities are concerned. As for the first, that the sum of the angles in a triangle equals two right angles, Egyptian, Babylonian, Chinese and Indian geometers knew well the properties of similar triangles. It is, therefore, hard to believe that the realities behind such properties of triangles were not known before the Pythagoreans. However, the precise formulation as a mathematical proposition, as well as a formal proof may well have been first supplied by them.

> 2. The Pythagoreans knew that in a right triangle the square on the hypothenuse is equal to the sum of the squares on the two sides containing the right angle.

This theorem, the so-called *Pythagorean Theorem*, was certainly known to the Babylonians at least 1000 years before Pythagoras. As we have seen, not only did the Babylonians know this, they also knew how to generate *all the so called Pythagorean triples*, namely triples (a, b, c) of integers such that $a^2 + b^2 = c^2$. Whether the Babylonians also knew proofs of the Pythagorean

Theorem is more hypothetical. But proofs based on a simple figure combined with some algebraic manipulation could well have been known to the Babylonians, who were superb algebraists.

3. The Pythagoreans knew several types of constructions by straightedge and compass of figures of a given area. They also solved what we would call algebraic problems by *geometric means*.

Again, much of this would be known long before the Pythagoreans. Thus for instance *the Sulva Sutra*, the oldest source of Indian mathematics, contains rules for constructing altars of a given area. Typical assignments would include the following:

1. Construct a square altar table, the area of which is twice that of a given square alter table. Solution: Use the diagonal of the given one as the length of the sides of the new one. We will return to this assignment in Section 3.8.

2. Given a rectangular altar table. Construct a square one of the same area. Solution: Let the sides in the rectangular table be a and b, the unknown side of the square be x. Then $x^2 = ab$, thus $a : x = x : b$, in other words, x is the *mean proportional* of a and b. We then draw a half circle of diameter $a + b$, erect a line normal to this diagonal where a is joined to b, and find x as the half-cord. See Figure 3.2.

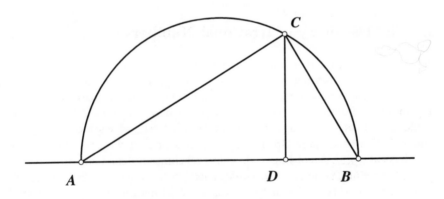

Fig. 3.2. Construction of the mean proportional.

By the way, we may also use 2. to solve 1., of course. But the first method is simpler.

Finally we come to a discovery which is universally credited to the Pythagoreans, if not to Pythagoras himself. There are some who think that the discovery was made by a woman mathematician, *Theano*, who was

Pythagoras' wife. It is arguably one of the most profound piece of mathematics discovered by the Greek classical school, and brought the Greeks almost to the point of discovering the system, or *the field,* of real numbers, as we would say in modern language.

But somehow the decisive last step was never taken, and the discovery of the field of real numbers as a powerful extension of the rationals would have to wait for about 2000 years. Perhaps one of the reasons for this was that the Greeks did not possess any good algebraic notation. Only towards the end of the Hellenistic epoch do we see a movement in this direction, in the work of *Apollonius.* Also, the Greeks were really *true geometers,* and not algebraists. They considered geometry to be a more complete science than algebra, in fact they did their "algebra" in terms of geometry, we would call it *Geometric Algebra.* Perhaps it was this philosophical prejudice which prevented them from taking the last definitive step and discovering the system of real numbers as an extension of the rationals. But even to say that the Greeks worked with rational numbers, is somewhat misleading. To them, what we would understand as the number $\frac{3}{2} = 1.5$ would be the proportion $3 : 2$.

However, when this is said it has to be added that some historians of mathematics seem to have underestimated the sophistication and power of Greek computing abilities. Especially towards the end of the Hellenistic Epoch such abilities to an impressive degree are documented in the work of *Claudius Ptolemy* and others. See Section 4.11.

3.4 The Discovery of Irrational Numbers

Presumably the Pythagoreans would early on work from the assumption that given any two line segments a and b, then their *proportion* $a : b$ would always be equal[5] to the proportion between two *numbers,* i.e., in our present language be equal to a fraction $\frac{r}{s}$ where r and s are positive integers. Arguably, this would be the position taken by Pythagoras himself, at least originally. Of course at this time many Greek philosophers espoused the *atomistic* view of the physical world. According to this idea, all things are made up of *incredibly many,* but a finite number, of *incredibly small,* but of a definite size, *indivisible atoms.* In fact, this *model* for the physical world became generally accepted all the way up to our own times. Some of the early Pythagoreans applied this idea to geometry and mathematics as well. For *numbers* they had the atom in the number 1, from which all other numbers were built.

In accordance with this general way of thinking, *lines* would consist of small chained *line elements.* In particular two line segments a and b would have *a common measure*: There would exist some line segment c such that c would fit exactly an integral number of times, say r, in a, and exactly an

---

[5] The Greek concept of equality for proportions will be explained below.

integral number of times, say s, in b: Of course this would be true, at the very worst one would have to take *one of the miniscule line elements*, which would work since the two line segments were made up of whole numbers of such line elements. The line element would *always* constitute a common measure, for any two line segments. Now, for convenience one would let c be the largest such *common measure*. This situation is illustrated in Figure 3.3.

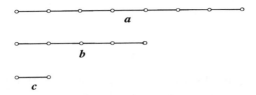

Fig. 3.3. c is the largest common measure of a and b.

How would we go about finding the biggest common measure of two given line segments a and b? The procedure is an ancient method, which the Greeks called *antanairesis*, meaning successive subtractions. Literally, given the two line segments a and b, the smallest is subtracted from the biggest. Of the remaining, the smallest is again subtracted from the biggest. This subtraction-procedure is repeated again and again, until the two segments are *equal in length*. Note that if you believe in the *atomistic nature of lines*, then this will occur sooner or later, at the very worst when you are left with two line-atoms, two line elements discussed above. Then a moment of contemplation will convince you that these two equal line segments are indeed the greatest common measure of the original line segments a and b.

This method of *successive subtractions* was very useful in ancient times. It allowed amazingly exact mensurations of an unknown distance, using only a measuring rod without subdivisions, and a good sized compass. It is no accident that the Master Builder so frequently is depicted with the measuring rod and the compass! He would proceed as follows. Let's say that the measuring rod would be, anachronistically, one meter long. First, as carefully as possible he would count the number of times the whole measuring rod could be subtracted from the unknown distance, i.e., find the number of whole meters. Let's say he gets 50. Then he would take the residue, the left over piece, in his compass, and count the number of times *it* could be subtracted from the length of the *measuring rod* itself. Let's say he gets 2, and a new left over piece, a new residue. He now successively repeats the procedure, counting the number of times the new residue can be subtracted from the previous one, and writing down the numbers. Let us say he repeats this 4 more times, getting 1, 1, 4 and 2, at which point there is nothing left, at least as far as he can see: Then, of course, he has to stop. Denoting the length to be measured by L, the measuring rod (here of 1 meter) by m, the first residue by r_1, the second by r_2, then r_3 and finally, r_4, we obtain

$$L = 50m + r_1$$
$$m = 2r_1 + r_2$$
$$r_1 = r_2 + r_3$$
$$r_2 = r_3 + r_4$$
$$r_3 = 4r_4 + r_5$$
$$r_4 = 2r_5$$

To find L in terms of m, we substitute $r_5 = \frac{1}{2}r_4$ from the sixth relation into the fifth relation, obtaining $r_4 = (\frac{1}{4+\frac{1}{2}})r_3$, which substituted into the fourth yields

$$r_3 = (\frac{1}{1 + \frac{1}{4+\frac{1}{2}}})r_2$$

and so on, until we finally get

$$L = (50 + \frac{1}{2 + \frac{1}{1+\frac{1}{1+\frac{1}{4+\frac{1}{2}}}}})m$$

We show the process in Figure 3.4.

As m is supposed to be one meter, we find after some computing of fractions that the length L is $50\frac{20}{51}$ meters, or 50.39 meters, in present days decimal notation. Of course the number is deceptive, as counting the 50 meters to begin with could introduce an error of around 5 cm. But using a longer rod, or a longer string of a known length, this measuring error would be reduced.

Now, it is generally thought that the first "irrational number", discovered by the Pythagoreans, was $\sqrt{2}$. But first of all, the Pythagoreans, as indeed all Greek mathematicians of this time, did not think of this as a *number*. Rather, it was a question about the *proportion* between the lengths of two line segments *not being equal to the proportion of two numbers*, we would say not being a rational number, the fraction of two integers. It is presumed, by some, that the first such pair of line segments found was the diagonal and the side of a square. It is also asserted, frequently, that the so called Pythagorean Theorem should have been essential in realizing this. Others find this questionable. First of all, at the time of Pythagoras proving that two line segments are *incommensurable* would consist in showing that the *process of repeated subtraction applied to these two particular line segments never stops.* Later more sophisticated methods were developed by geometers like Theodorus of Cyrene, (465 – 398 B.C), pupil of Pythagoras and teacher of

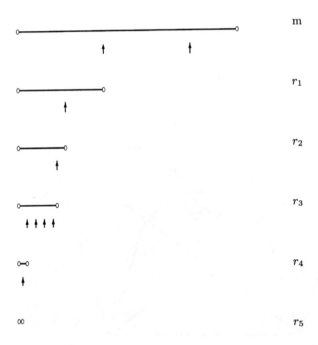

Fig. 3.4. Measuring by repeated subtractions.

Plato, and by Theaetetus. They are the principal characters in two of Plato's famous dialogues, one of them dealing with square roots.

At Pythagoras' time the simplest case to consider would be *the diagonal and the side of the regular pentagon*. This certainly appears surprising, since we would view the regular pentagon as considerably more complicated than a square. But from the point of view of repeated subtraction of the side and the diagonal it is the absolutely simplest figure in existence. A look at Figure 3.5 will explain this.

Indeed, the diagonal is AC and the side is AB. Now AB = AD, as elementary considerations yield the equality of the angles ∠ABD = ∠ADB. Thus subtracting AB from AC we are left with DC, and the subtraction can only be performed once. In the next step CD is to be subtracted from AB. Now CD = AD' and AB = AD, thus in this next step we may also only subtract once, and the remainder is D'D. But as CD = CE = ED', the third step will be to subtract the side of *the inner pentagon from the diagonal of the inner pentagon*! Thus, magnifying the inner pentagon and turning it upside down, we are back to the starting point. Hence the process evidently repeats itself

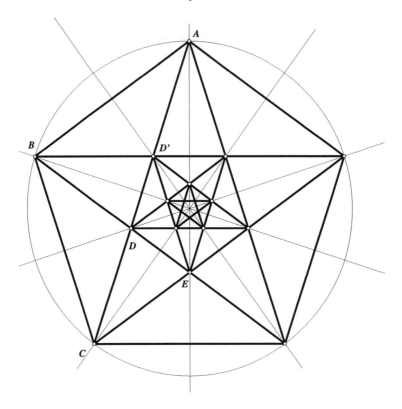

Fig. 3.5. The pentagon and the pentagram.

without ever stopping. Thus the incommensurability of diagonal and side of the regular pentagon is proven.

A similar procedure may be carried out for the diagonal and the side of a square, but it is considerably more complicated. And in view of the special relationship the Pythagoreans had to the regular pentagon, it is a very plausible guess that this is how they arrived at the conclusion that *not all line segments are commensurable*.

A final point to be made is this: If we put $x = $ AC:AB, then we obtain

$$x = 1 + \cfrac{1}{1 + \cfrac{1}{1+\dots}} = 1 + \frac{1}{x},$$

which yields the equation

$$x^2 - x - 1 = 0,$$

Indeed, this follows in the same manner as the computation carried out on the basis of Figure 3.4. Hence $x = \frac{1}{2}(1 + \sqrt{5}) \approx 1.6180$. This number is often referred to as the *Golden Section*.[6]

3.5 Origin of the Classical Problems

There are three problems occupying a special position in Greek geometry, namely the so-called *classical problems*. They are all insoluble in their strictest interpretation. However, they may be solved by various creative procedures and they have generated an enormous amount of mathematics. Their attraction on mathematical amateurs is perhaps paralleled only by the famous *Fermat Conjecture*, which was finally proven not too many years ago by *Andrew Wiles*. The first of these problems we encounter in the history of mathematics is the

> **Squaring the Circle.** Given any circle. Then construct a square with the same area as the one enclosed by the circle.

The first time we find this problem mentioned, is in connection with the Greek philosopher *Anaxagoras*. Anaxagoras lived at a time when Athens stood at the summit of its power, politically and intellectually.

After Athens and Sparta had won the protracted war against Persian invaders, there followed half a century of peace and prosperity. This was a time of flourishing cultural life in Athens. Many of the expelled Pythagoreans found their way to Athens, and *Socrates* played an important role in the intellectual life of the city state.

Athens had an enlightened leader in *Pericles* for a great part of this time, from about 460 B.C until he died in the great plague in the year 429 B.C., two years after the peace had been broken and the devastating Peloponnesian war with Sparta had broken out. Unfortunately Pericles must bear a large part of the responsibility for this fratricidal struggle. In fact, he transformed the alliance of the Greek cities against the Persians, *the Delian Alliance*, into an instrument for Athenian dominance. This worked fine for Athens, but the Spartans and others were considerably less pleased. Athens now had more than 300000 inhabitants, one third were slaves and about 40000 were male citizens enjoying full rights. The city wall also enclosed the port city of *Piraeus*, and their fleet was the dominating power at sea.

Pericles erected the magnificent buildings at Acropolis, and showed great interest in mathematics and philosophy. He belonged to *the democrats*, from the aristocratic wing of the party. He was succeeded by *Cleon* when he died, also a democrat but from the less aristocratic wing.

---

[6] Other names include the Golden Mean, the Golden Number and the Golden Ratio.

Pericles' teacher and close friend was *Anaxagoras*. Anaxagoras was born about 500 B.C., in Clazomenae (now Izmir), in Ionia, presently Turkey. He died 428 B.C. in Lampsacus in the Troad, where he had sought refuge for persecution by his enemies in Athens, who continued to press charges for *impiety* against him.

He was more a natural philosopher than a mathematician. Nevertheless he played an important role in Greek geometry, and indeed in the development of mathematics, since he was, apparently, the first to be tied to one of the great problems of antiquity, *the Squaring of the Circle.*

In his teachings, he had denied that the heavenly bodies were divinities. Instead, he explained them as stones torn from the earth, the sun being red hot from its motion. The sun was as big as all of the Peloponnese, he asserted, and the moon reflected the light from the sun. The moon was an inhabited world, like the earth, according to Anaxagoras.

These ideas were hard to swallow, any right-thinking Athenian would be disgusted at such impiety. Consequently Anaxagoras was incarcerated. According to *Plutarchus* Anaxagoras spent the time in prison by attempting to square the circle.

Pericles had to be cautious, since he had many powerful enemies in Athens. But he also stood by his friends, and he finally managed to get Anaxagoras out of prison. But Athens certainly was not a safe place for him any more, and he therefore moved to Lampsacus where he founded his own Academy. Aristotle speaks highly of the reputation he enjoyed there.

The Peloponnesian War broke out in 431 B.C., and two and a half years later Pericles died in the great plague which had started to ravage Athens. One year later Anaxagoras also died.

The plague had broken out for full in 427 B.C., the presumed year of Plato's birth. The plague weakened Athens considerably, one fourth of its population is said to have perished. According to the legend, the citizens of Athens sent a delegation to the oracle of Apollo at Delos, to ask for advice on how to emerge form the dire circumstances in which they found themselves: A war with Sparta which would be very difficult to win, now that they also had this debilitating pestilence to cope with. The answer delivered by the priestess of Apollo was enigmatic: *The cubic Altar of Apollo should be doubled.* They may also have received other instructions as well, since Athens carried out extensive *purifications* of the island in the year 426 B.C.: Among other things all graves on the island were opened, and the remains which were buried there removed and reburied on the neighboring island of Rheneia. Doubling the cubic altar proved more difficult. The Greek geometers realized of course that the purest, and most pleasing way to Apollo, would be using compass and straightedge. In other words to perform a geometric construction which for a given cube would render another with volume twice the given:

Doubling the Cube, or the Delian Problem. Given any cube, construct with straightedge and compass the side of another cube, the volume of which is twice that of the given one.

It must have been quite intriguing to the geometers in Athens that this problem proved so hard, since the corresponding assignment for *a square* was so easy. More on that below, in Section 3.8.

There are accounts to the effect that *Plato*, when consulted about the problem later, voiced the opinion that Apollo had not offered the oracle because he wanted his altar doubled, but that he had intended to censure the Greeks for having turned their back on mathematics and geometry: By paying more attention to science and philosophy instead of making war, things would start to go better for them.

Greek geometers were fully aware that all circles are similar, as are all squares and cubes. Thus the problems stated above for *any* circle and for *any* cube is equivalent to the same problem stated for *one* circle or for *one* cube: If you can square one circle you can square them all, if you can double one cube, then you can double them all. Not so with the *third problem*, which also circulated in Athens about this time:

Trisecting the Angle. Given any angle, divide it in three equal parts using straightedge and compass.

In this last case the situation is different: There is an infinite number of angles which may be *trisected* using ruler and compass. We show the construction for *a right angle*, that is to say an angle of $\frac{\pi}{2}$ radians or 90°, in Figure 3.6.

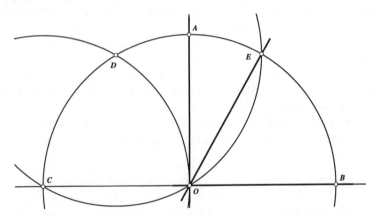

Fig. 3.6. Trisecting a very special angle by straightedge and compass.

We start with the right $\angle AOB$, and draw a circle with O as center passing through B. Producing BO we find the point C. With C as center draw the circle passing through O. The latter circle intersects the former in D. With D as center draw the circle passing through O, this circle intersects the one about O in E. Then $3 \times \angle AOE = \angle AOB$. Thus there are angles which may be trisected by compass and straightedge, and there are infinitely many such angles: Namely, we may by continued bisection divide $\angle AOB$ in 2^n equal parts for any n, and the resulting small angle may then be trisected by similarly bisecting $\angle AON$ in 2^n equal parts. Of course these are not all, there are several other kinds of angles which may be trisected by compass and straightedge as well.

3.6 Constructions by Compass and Straightedge

Another remark to be made concerning the little construction in Figure 3.6 is this: The construction illustrates the *legal use of compass and straightedge*. The legal use of compass and straightedge is tied to what later was codified as *Euclid's axioms*. Many complex constructions may be performed under these rules, but the three classical problems are not soluble in this way. This led Greek geometers to introduce other methods, like the use of *conic sections*, also *curves of higher degrees*, even *transcendental curves*, as we would say in modern language: The transition from elementary to *higher geometry* was initiated as a consequence of the struggle with the *classical problems*. The transition is not as unnatural as one might think, since employing conic sections or higher curves is equivalent to solving the problem *by an infinite number of steps* using ruler and straightedge, at each stage in a completely legal manner, according to the rules. We now state these rules.

> **Legal Use of Compass and Straightedge.** A finite set of points is given. A point is constructed if it is a point of intersection between two lines, two circles or a line and a circle as produced according to 1. and 2. below:
> 1. The straightedge may be used to draw a line passing through two given or previously constructed points, and to produce it arbitrarily in both directions.
> 2. The compass may be used to draw a circle with a given or already constructed point as center, passing through a given or already constructed point.

We note that according to 2. above, the compass may *not* be used to move a distance. A compass which may only be used in this restricted way, is frequently referred to as a *Euclidian compass*. We may imagine that the compass *collapses immediately when either end is lifted from the paper*.

Using these two procedures is also referred to as *constructing by the Euclidian tools*. By Euclidian tools we may easily perform tasks like *dividing any angle in two equal parts*, drop the normal to a given line from a given

point or erect the normal at a given point on a given line. This is shown in
the three top constructions in Figure 3.7.

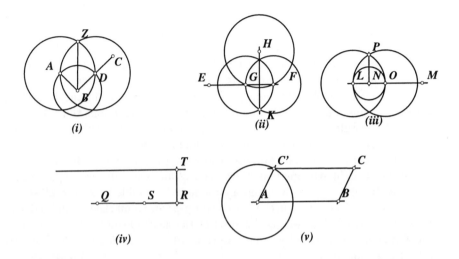

Fig. 3.7. Simple but essential constructions which may be carried out using
straightedge and the Euclidian compass.

An angle is given by the points A, B and C. We wish to bisect ∠ABC.
Draw the circle with B as center through A, D is the point of intersection
between this circle and the line (possibly produced) BC. Then circles are
drawn with A and D as centers, passing through, respectively, D and A.
These circles intersect in a point Z such that a line AZ bisects the angle in
two equal parts. Next, the line EF is given, as well as the point H outside it.
To drop the perpendicular from H to EF, a circle through F is drawn with
H as center, intersecting EF in another point G. With F and G as centers,
circles are drawn through G and F, respectively, intersecting in K. Then HK is
perpendicular to EF, its *foot* is the point of intersection with EF. Finally, we
erect the perpendicular to a line LM in the point N. We leave the explanation
of this construction to the reader.

In the lower part of the figure, we show how to construct a parallel to a
given line QS through a given point T, by first dropping the perpendicular
from T to QS (produced), its foot being R, then erecting the perpendicular
to RT at T.

We now find a pattern, similar to *proving complex theorems from simpler
propositions or axioms*: The construction in (iv) is obtained by appealing to
the two previous ones in (ii) and (iii), without having to start from scratch.
This becomes even more striking by including construction (v).

Namely, if we allow the compass to be used to draw a circle about a given or constructed point with radius equal to the distance between two other points in the construction, then this is strictly speaking is not allowed according to the rules above. But actually, we may nevertheless do this, since we have the construction (v). Here the points A, B and C are given, and we wish to draw a circle with A as center and radius BC. Proceed as follows: Draw the line BC. Through C construct the parallel to AB, and through A the parallel to BC. They intersect in C'. Now the length of BC equals the length of AC', so draw the circle with center A passing through C'.

The parallel to AB through C is unique since we are in the Euclidian world. The possibility and the uniqueness of the constructions thus hinge on the Fifth Postulate of Euclid. It might be interesting to contemplate what constructions would be like in a non-Euclidian plane.

But to mark off a distance on *the straightedge* is prohibited. By such illegal use of the straightedge one may indeed trisect any angle in three equal parts, as we shall see in Section 3.9, and a cube may be doubled, as we shall see in Section 4.6. In fact, constructions with compass and a marked straightedge is equivalent to including among the Start Data one single higher curve, namely the *Conchoid of Nicomedes*, which we treat in detail in Section 4.6. We also refer to Section 16.8.

We now turn to some specifics on the three problems. Even though ideally they should be solved with ruler and straightedge, Greek geometers of course soon realized that this would be very difficult. So they came up with a variety of solutions, ranging from rather simple but effective mechanical schemes, in some cases constructing various kinds of instruments, to very sophisticated geometric constructions like *Archytas'* famous three-dimensional construction for the doubling of the cube, using a cylinder, a cone and a torus. Also employed were a variety of higher algebraic, as well as transcendental, curves in the plane. We shall give some glimpses of these prodigious efforts in the following three sections.

3.7 Squaring the Circle

We have already mentioned that if you can square one circle, then you can square them all. In fact, suppose that a circle of the fixed radius r may be squared, that is to say that we may construct a square of side s such that its area equals that of the circle. The situation is shown in Figure 3.8.

Here we have a fixed circle, together with a fixed square with side KQ, known to have the same area as the area enclosed by the circle. These two being given, we may square *any* circle as follows: We construct a right triangle VWX, where the side VW is equal to the diameter of the given circle, while WX is equal to the side KQ of the given square. VW and WX are the sides containing the right angle. Now consider an arbitrary, new circle, shown in the lower left corner. Mark off VY on VW equal to its diameter, and let YZ

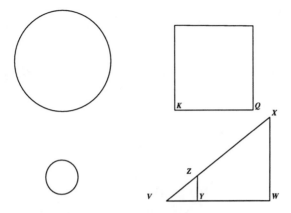

Fig. 3.8. If you can square one circle, you can square them all.

be parallel to WX, Z falling on XV. Then YZ is the side of the square of area equal to the that of the new circle.

3.8 Doubling the Cube

We first look at the much simpler problem of *doubling the square* by straightedge and compass. This construction is shown in Figure 3.9.

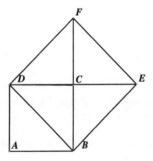

Fig. 3.9. Doubling the square by straightedge and compass.

Here we have the square ABCD. We now perform the *doubling of the square* in a way very much in the spirit of Greek geometry as follows: Produce

the line DC, and mark the point E such that DC = CE. Similarly produce BC
and mark F such that BC = CF. Then the square BEFD will have twice the
area of ABCD. Indeed, the former consists of four congruent right triangles
while the latter only requires two.

But this observation is just the beginning of what led *Hippocrates of Chios*
to a most remarkable discovery: Namely, we notice that the triangles ABD
and BDF are similar, thus

$$AB : BD = BD : BF$$

Thus

$$AB : BD = BD : 2AB$$

so the side of the double square is the mean proportional between the side
and the double side of the given square. Thus putting AB = 1 and using
modern notation, we find the side x of the double square by

$$1 : x = x : 2 \text{ or } \frac{1}{x} = \frac{x}{2}$$

so $x = \sqrt{2}$. A construction for doubling the cube which has much of the
same flavor, while of course not being possible by straightedge and compass,
is attributed to Plato and will be explained in the next section.

It was Hippocrates who realized that the enigma of doubling the cube was
but one very special case of a much more general and much more interesting
problem: Namely that of *constructing a continued proportionality:*

Construction of a continued proportionality. Let a and b be
two line segments. For a given integer n, construct n line segments
$x, y, z, \ldots, u, v, w$ such that

$$a : x = x : y = y : z = \cdots = u : v = v : w = w : b$$

x, y z etc. are referred to as the *mean proportionals* of the continued
proportionality. A double mean proportionality is one with two mean
proportionals, a triple has three, etc.

He saw that doubling a cube of side a is equivalent to constructing *a*
double continued proportionality between a and $2a$: To construct x and y
such that

$$a : x = x : y = y : 2a$$

We check this with modern notation. We have

$$\frac{a}{x} = \frac{x}{y} = \frac{y}{2a}$$

This gives

$$ay = x^2 \text{ and } 2ax = y^2$$

Squaring the former and substituting y^2 from the latter yields $2a^3x = x^4$, i.e. $x = a\sqrt[3]{2}$.

Recall the following construction of the mean proportional between two line segments $a \geq b$. We refer to Figure 3.2: First draw a semicircle with diameter $AB = a$, then mark the point D such that $AD = b$.[7] We then have similar triangles ABC and ACD, thus

$$AB{:}AC = AC{:}AD$$

and so $AC = x$ is the mean proportional.

There is a continuation of this construction to a double continued proportionality, and indeed to any continued proportionality. In fact, from D in Figure 3.2 we construct a line perpendicular to AC, see Figure 3.10.

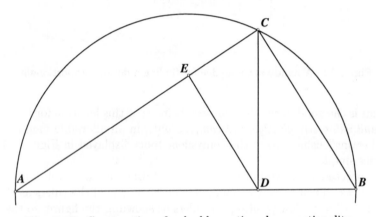

Fig. 3.10. Construction of a double continued proportionality.

Letting $\sim$ denote the relation of being similar triangles, we have

$$\varDelta ADE \sim \varDelta ACD \sim \varDelta ABC$$

from which it follows that

$$AB : AC = AC : AD = AD : AE$$

Thus if we wish to construct the double continued proportionality between the line segments $a \geq b$,

---

[7] Note that this is a slightly different construction from the one explained when we first encountered Figure 3.2.

$$a : x = x : y = y : b$$

then first draw the semicircle with diameter AB = a, and then observe what happens as the point C on the semicircle moves from B to A: In the right $\triangle$ABC draw the perpendicular to AB through C, meeting AB in D. Then through D draw the perpendicular to AC, meeting it at E. As now C moves, starting with the degenerate case of C = B where AE = a, AE will decrease to 0 when the other degenerate case of C=A is reached. Therefore at some unique location for C on the semicircle, AE = b. There we take AC = x and AD = y, which solves our problem.

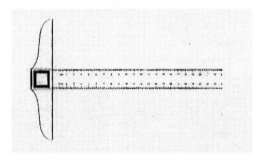

Fig. 3.11. A handy straightedge for finding a double proportionality.

This is the good news, the *bad news* being that this location for C *can not be found using straightedge and compass only*, in an allowable manner. But by "cheating" using two of the convenient tools displayed in Figure 3.11, it becomes simple.

We proceed as shown in Figure 3.12: First draw the semicircle with diameter AB = a. Then mark the points A' and E' on one of the rulers as shown, so that A'E' = b. Now position the rulers as shown in the figure, so that the vertical, unmarked, straightedge meets the marked one in a point on the line AB, where A' coincides with A, and C is found as the point where the marked straightedge crosses the semicircle. E' on the marked straightedge gives us the point E in our figure. We then have the construction from Figure 3.10.

This construction is of course completely illegal as a construction with straightedge and compass. In fact, it is even illegal as a version of the already illegal *insertion principle*, which we will explain in the next section. However, in its pure form the insertion principle was much used by Greek geometers, this is also known as a *verging construction*.

3.9 Trisecting any Angle

The construction of *bisecting any angle* was, as we have seen, very simple. And subdividing a line segment in any number of equal pieces is also a very

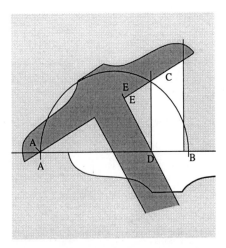

Fig. 3.12. Two rulers, one of them marked, used for finding a double proportionality.

simple construction. To Greek geometers it must therefore have been a source of frustration and bewilderment that the problem of *dividing any angle into three equal pieces* turned out to be so difficult.

This problem began to attract attention at about the same time as the problem of Doubling the Cube. Some special angles could easily be trisected, as the construction we display in Figure 3.6.

Greek geometers found solutions to the trisection-problem by solving what they referred to as a *Verging Problem*. We shall not attempt to give a general definition of this concept, but in Figure 3.13 we present the solution to the trisection problem as being reduced to one variety of such a Verging Problem. Another kind is represented by the famous construction of the regular *7-gon* found by Archimedes, treated in Section 4.4. Of course, neither the trisection problem nor the construction of the regular 7-gon are possible by legal use of compass and straightedge.

Now for Figure 3.13. To the left we have the angle $v = \angle ABC$, we draw the circle about B through C, and then we find the point E on that circle such that the line EC produced meets AB produced in a point D such that the segment DE is equal in length to the radius BC. This is the verging-part of the construction, it is possible by marking off the length BC on the straightedge. Denote the angle at D by u. Then $\angle CEB = 2u = \angle ECB$, thus $v = 3u$. To the right we have the same construction, essentially, but we do not use the circle, nor a marked straightedge, to find the point E such that $AB = BE = ED$. There are simple mechanical devises which may be used, however, based on the construction we have given here.

There are various algebraic curves of degrees higher than 2, so called *Higher Curves*, by means of which the verging problem may be solved. The

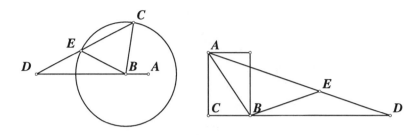

Fig. 3.13. Two Verging Constructions solving the Trisection Problem.

most famous of these are probably the *Conchoid of Nicomedes*, which we treat in detail in Section 4.6.

There is also another famous curve which may be used to trisect any angle, *and* to square the circle as well, in fact it may be used to divide any angle in any number of equal parts and to construct a regular $n-$gon for *any* number n. A truly marvelous curve! It is the *Quadratrix of Hippias*, treated here in Section 4.6 and explained in Figure 4.24. This is not an *algebraic curve*, however. Like the Archimedian Spiral, it is what we call a *transcendental curve*.

3.10 Plato and the Platonic Solids

Plato was born in 427 B.C. in Athens and died there in 347. Although he made no original contribution to geometry himself, he has had an immense influence on the subject. In 387 B.C. he founded the *Academy* in Athens, devoted to philosophy and geometry as well as other sciences. Plato had been engaged in the Peloponnesian war as a young man, and he saw his esteemed teacher and friend Socrates condemned and executed. He felt that one reason why the Greek civilization in general, and the one in Athens in particular, was in decline, had to be sought in the disregard of philosophy and geometry. To Plato the problem of *Doubling the Cube*, for example, was a question of developing insights into geometry. Thus it was not a question of finding some practical means for carrying out the physical labor involved, like devising some mechanical instruments or "cheating" with the straightedge. Instead it was a question of understanding the mathematics involved. Therefore Plato would regard highly the doubling-constructions involving higher curves or space-geometric constructions, even if these were of lesser practical value in the actual work of doubling any given cubical altar!

Of course this is exactly how we enjoy this problems today, as well as the one of trisecting any angle or squaring any circle. We understand them in terms of properties of *algebraic numbers*. We return to this in Chapter 16.

To Plato geometry was part of the ideal world, whereas the physical world would only represent imperfect approximations. He ascribed a special significance to the *regular convex polyhedra*, as symbolizing *the four elements Earth, Fire, Air* and *Water*. The fifth one, namely the dodecahedron, stood for the whole *Universe*.

In our modern language a *polyhedron* is a surface enclosing a solid figure composed of (plane) polygons. These are called the *faces* of the polyhedron. The sides of the polygons are called the *edges*, and the corners where the edges meet, are called the *vertices*. At each vertex there a configuration as shown in Figure 3.14: The vertex P from which the edges a, b, c and e emanate.

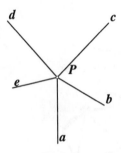

Fig. 3.14. A polyhedral angle.

This configuration is referred to as *the polyhedral angle at* P, so a polyhedral angle is a point in space with a certain number of half lines emanating from it.

A convex polyhedron is one where a plane containing any face does not cut the other ones. See Figure 3.15 for an illustration of the property of *convexity*.

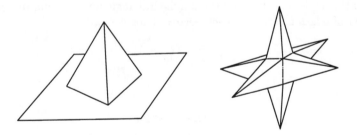

Fig. 3.15. To the left a convex polyhedron. Any plane containing one of the faces, does not cut any other. To the right evidently this property does not hold.

We say that a polyhedron is *regular* if all the edges are of equal length, and all polyhedral angles are *congruent*, that is to say that all the configurations of rays at the vertices are the same. We also require that the faces are *regular polygons of the same kind*, i.e., all are equilateral triangles, all are squares etc. Finally, we require that it be *convex*. Then there are exactly *five such polyhedra*, they are shown in Figure 3.16.

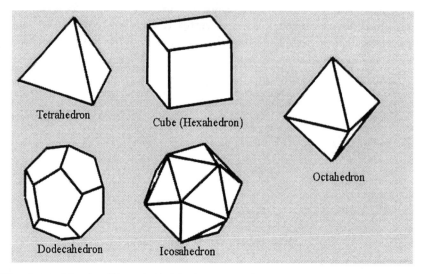

Fig. 3.16. The five Platonic Polyhedra, or as they are also known, the *Platonic Solids*.

We can show that these are the only such polyhedra as follows. Let P be a polyhedron of this type, consisting of regular n-gons. Let v be the angle at each vertex. For any convex n-gon, in particular any regular one, the sum of the angles contained by adjacent sides is $(n-2)\pi$. This is easily seen by subdividing it into $n-2$ triangles. Thus $v = \frac{n-2}{n}\pi$. On the other hand the sum of the angles constituting the polyhedral angle must be mv, m being the number of edges meeting at each vertex. Thus we have

$$m\left(\frac{n-2}{n}\right)\pi < 2\pi$$

and so

$$m(n-2) < 2n$$

For $n=3$ this leaves the possibilities $m=3,4$ or 5, $n=4$ leaves only $m=3$, as does $n=5$. For $n \geq 6$ no value for m is possible. The values for m listed above are indeed realized, and yield the five Platonic Solids.

3.11 Archytas and the Doubling of the Cube

Archytas of Tarentum was born 428 B.C. and he died in 365 in a shipwreck near his home city of Tarentum. Tarentum is located not far from Croton and Metapontium. After the events when the Pythagoreans had been driven out of Italy, things had quieted down to the effect that they had been able to reestablish themselves in the area. He is considered the last great Pythagorean, and in fact Book VIII of Euclid's Elements is generally attributed to him.

He had been a student of another Pythagorean, namely *Philolaus of Tarentum*. Philolaus had studied with some of the expelled Pythagoreans, and he was interested in number magic and mysticism. But he had been allowed to write about the ideas of the Pythagoreans, and the book he wrote is supposed to have been Plato's source of information on the mathematics of Pythagoras and the Pythagoreans.

Archytas made it to the top of Tarentum's politics, he was elected admiral, never lost a battle, and became the ruler of Tarentum with unlimited power. But he is supposed to have been an enlightened ruler, who had a deeply rooted belief in the virtues of philosophy and rationality in politics. He thought that these forces would lead to enlightenment and social justice.

In spite of his political and military work, he also managed to pay attention to mathematics in general and geometry in particular. He lectured extensively, Plato studied under his direction in Tarentum.

Another important Greek geometer who studied under Archytas' direction was *Eudoxus of Cnidus*. Eudoxus had ideas which were precursors to fundamental concepts in our calculus and analysis of today. He probably did the work contained in Euclid's Elements, Book V. See Section 4.1.

Archytas' significant contributions to the didactics of mathematics include its division into four subjects: *Arithmetics* constitute the numbers at rest, *Geometry* is the magnitudes at rest, *Music* is the numbers in movement and *Astronomy* is the magnitudes in movement. Later *the mathematical quadrivium* was seen as constituting the *seven free arts*, jointly with a *trivium* which consisted of the subjects *Grammar, Rhetoric and Dialectic*. These ideas were important in didactical practice up to our times.

It is told that Plato once became a prisoner of the notorious tyrant of Syracuse, *Dionysus I*, who ruled with an iron fist, while at the same time writing poems and tragedies. Archytas, who was concerned about the safety of his student and friend, sent a letter to his colleague in Syracuse. In it, he explained to Dionysus that Plato was one of his students and also a dear friend, and that he, Archytas of Tarentum who had never yet lost a single battle, would not like it if his friend should come to harm.[8] This saved Plato's life. A quite significant contribution to philosophy from the admiral in Tarentum.

---

[8] Others say that Archytas sent a warship to Syracuse.

Archytas solved the problem of *doubling the cube* by a general construction of the *second continued proportionality* between $a > b$, applied to the case $a = 2b$. His marvelous construction uses an analogy to constructions with straightedge and compass, in the form of finding points in space as intersections of *tori, cylinders and cones*. We show the situation in Figure 3.17, with the torus, the cylinder and the cone sketched in the first and the second octant, anachronistically including coordinate axes.

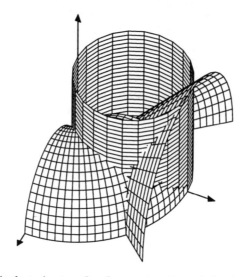

Fig. 3.17. Archytas' setup for the construction of the double continued proportionality, by intersecting a cylinder, a cone and a torus.

We now explain *Archytas' 3-dimensional construction* of the double continued proportionality between $a > b$ in Figure 3.18. The whole point of the construction is to obtain the right triangles in Figure 3.10, without using the extended version of the *insertion principle* we employed with our two rulers in Section 3.8.

One might say that Archytas' construction appears as a clear cut space-geometric generalization of constructions with straightedge and compass, employing higher dimensional versions of the compass.

In order to describe the construction, we introduce, anachronistically, a Cartesian coordinate system with x, y and z axes. We denote the origin by A. The following description is a slightly edited and commented version of the one given by Archytas himself, as related by Proclus in the *Eudemian Summary*. Of course Archytas did not use terms like *"the xy-plane"* and the like. The situation is visualized in Figure 3.17, while Figure 3.18 shows the exact geometry of the construction.

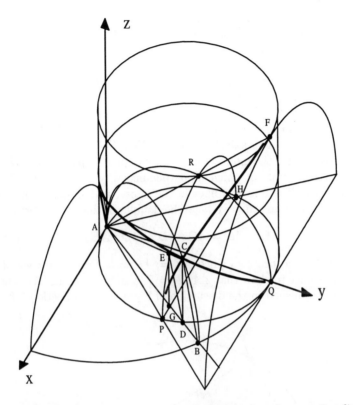

Fig. 3.18. Archytas' construction of the double continued proportionality, by intersecting a cylinder, a cone and a torus. The torus is shown in the third octant only, the cone and the cylinder in the first and the second octants.

Let $a > b$ be the two given line segments, let Q be a the point on the y-axis such that $AQ = a$. Draw a circle with AQ as diameter in the xy-plane and a semicircle with the same diameter in the first quadrant of the yz-plane. Draw a chord AP of length b to the former circle. On this circle also construct a right cylinder above the xy-plane. The semicircle in the yz-plane is now rotated about the z-axis from Q towards P. While being rotated the semicircle meets the cylinder in a moving point which traces out a curve on the cylinder. In Figure 3.18 this curve is indicated from Q to A. (In other words, this is the curve of intersection between the cylinder and the torus produced by rotating the circle.) On the other hand, when the prolongation of the chord AP is rotated about the y-axis, then it also meets the cylinder in a moving point, tracing out a curve, which is indicated in the figure from F through the point C pointing towards P. (This is the curve of intersection between the cone and the cylinder.) Evidently these two curves, one sloping upwards from Q and the other sloping downwards from F, will meet in a unique point. (In other words, the three surfaces, the torus, the cylinder and the cone, have exactly one point in common in the first octant.) In Figure 3.18 this is the point denoted by C. Drop the perpendicular from C to the xy-plane. Denote its foot by D. Now CD of course lies on the cylinder, and thus D lies on the circle in the xy-plane. The moving semicircle through C meets the xy-plane in the point B.

Draw the line PH parallel to the x-axis, it intersects AB in the point G. The line AC meets the circular arc from P to H via R, which P describes as AP is rotated about the y-axis, in a point E. Now EG is perpendicular to the xy-plane: Indeed, it is the intersection of the two planes spanned by ABC and PER, respectively, both of which are perpendicular to the xy-plane. We have now established all points and lines in the figure, and shown their relevant properties. The claim is that we have the double continued proportionality

$$AB : AC = AC : AD = AD : AE$$

which will solve the problem since $AE = AP = b$. From what we already know, it will suffice to show that $\angle AED = \frac{\pi}{4}$, a right angle. In fact, that this suffices was established in the discussion of Figure 3.10 in Section 3.8.

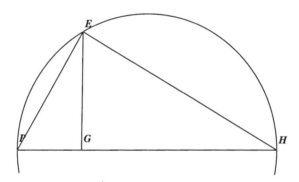

Fig. 3.19. $HG : EG = EG : PG$.

First, from Figure 3.19 we conclude that $HG : EG = EG : PG$. or in other words, $HG \cdot PG = EG^2$.

But from Figure 3.20 we find $HG \cdot GP = AG \cdot GD$ since $\triangle DPG \sim \triangle HAG$. Thus we conclude that $AG : EG = EG : GD$.

We now finally use this information on the detail from Archytas' construction shown in Figure 3.21.

Indeed, we have that $\triangle AGE \sim \triangle EGD$: They have one angle equal, namely the right angle at G, and the sides containing it are pairwise proportional. Hence in particular $\angle EAD = \angle DEG$. But as the corresponding pair of lines AD and EG of these two angles are perpendicular, so must be the case for the other pair. Thus DE is perpendicular to AC, as claimed.

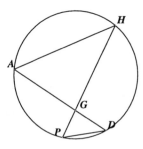

Fig. 3.20. $HG \cdot GP = AG \cdot GD$

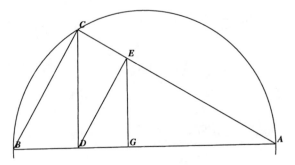

Fig. 3.21. The final argument.

This completes the proof of Archytas' construction of the double continued proportional between $a > b$.

Having completed Archytas' argument, we shall now carry it out by methods which he did not have at his disposal, namely by *algebraic geometry*. Putting the two arguments side by side we are better able to appreciate Archytas' geometric genius, as well as the power and convenience of algebra in geometry. One may even sympathize with those in the beginning of the twentieth century, who resisted the algebraic methods in geometry, feeling that geometry was defaced and destroyed in this way! Plato, incidentally, had similar misgivings about the use of mechanical tools in solving problems like the duplication of the cube. Comparing Archytas' solution to our crude and illegal use of the two rulers, a procedure very probably well known to Archytas, we may safely conclude that these misgivings were shared by Archytas himself.

The equation of the cylinder in Figure 3.18 is

$$x^2 + y^2 = ay,$$

the equation of the torus is obtained by putting $r = \sqrt{x^2 + y^2}$, the equation of this surface is then

$$z^2 + r^2 = ar$$

Thus the equation for the torus is

$$x^2 + y^2 + z^2 = a\sqrt{x^2 + y^2}$$

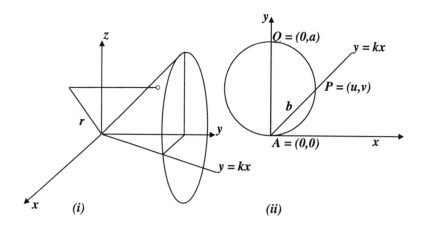

Fig. 3.22. The figure shows how we deduce the equation of the cone in Archytas' construction.

Finally, in (ii) of Figure 3.22 we see how the cone is produced by rotating the line $y = kx$, where $P = (u, v)$, so that $u^2 + v^2 = av$. Thus $b^2 = av$, and hence

$$k = \frac{v}{u} = \frac{b}{\sqrt{a^2 - b^2}}$$

When the line in (ii) is rotated about the y-axis, the cone in (i) is generated. With r given by $r^2 = x^2 + z^2$, the equation of the cone becomes

$$y = kr = k\sqrt{x^2 + z^2}$$

i.e.,

$$y^2 = k^2(x^2 + z^2)$$

and when the expression for k, namely $k = \frac{b}{\sqrt{a^2 - b^2}}$, is substituted into this equation for the cone, we finally obtain that the cone is given by

$$x^2 + y^2 + z^2 = \frac{a^2}{b^2}y^2$$

where the only constants occuring are a and b. We are now ready to state the

Claim: *With $\alpha = AC$ and $\beta = AD$ we have*

$$a : \alpha = \alpha : \beta = \beta : b$$

We put $C = (p, q, r)$, so that

$$\alpha = \sqrt{p^2 + q^2 + r^2} \text{ and } \beta = \sqrt{p^2 + q^2}$$

The equation for the torus yields $\alpha^2 = a\beta$, which gives the first proportionality.

From the equation for the cylinder we have $\beta^2 = aq$, while the equation for the cone yields $\alpha = \frac{a}{b}q$, so that

$$b\alpha = \beta^2,$$

which gives the last proportionality.

4 Geometry in the Hellenistic Era

4.1 Euclid and Euclid's Elements

Alexandria was founded where the Nile meets the Mediterranean by Alexander the Great, in the year 331 B.C. The city became the capital of Egypt, and rapidly developed into one of the richest and most beautiful cities in the world. That is to say, in the world known to the antique.

Alexandria developed into a center for civilization, science, art and culture in general, and remained so for more than three quarters of a millennium. The city possessed many magnificent buildings and awe-inspiring structures. The lighthouse at Faros was counted as one of the worlds seven wonders.

Alexandria was well positioned for trade. It bristled with a lively exchange of valuable goods and commodities between Europe, Asia and Africa. The flourishing city also developed a diverse industrial sector, with products including glass, paper and priceless fabric and cloth. Art and science continued to find fertile soil here, with the most eminent schools of mathematics, astronomy, philology and philosophy.

Euclid – or *Eucleides* which was his real name – was a Greek mathematician who lived around the year 300 B.C. and worked in Alexandria. He should not be confused with another ancient by the same name, a certain *Euclid from Megara*, who was one of the disciples of the philosopher *Socrates,* and appears in Plato's dialogue *Theaetetus.*[1] The latter Euclid has in no way left a comparable legacy to that of the former, disregarding the point of view explained in the footnote.

Euclid collected and systematized the entire body of mathematics known to his time. First and foremost stood *geometry*. Here we must understand that to the Greek mathematical tradition, geometry was in a sense more perfect as a science than computing with numbers. They had no concept of *irrational numbers*, whereas they could "compute" with a quite large class

---

[1] However, when Euclid's Elements were reintroduced to Europe towards the end of the Middle Ages, this confusion of the two Euclids did happen. And some historians espouse the theory that *Euclid* was a pseudonym inspired by this dialogue and used by a group of mathematicians working in Alexandria, much in the same way as the name *Nicolas Bourbaki* has been used by a group of French mathematicians in our days.

of such numbers via geometry and geometric constructions with straightedge and compass. So maybe we can view straightedge and compass as the calculator of the ancient Greeks! And we should realize why the Greeks laid such tremendous importance to the *classical problems*. The problems were all of the same nature: *"What can our calculator do?"*

Euclid based his work on a fundamental idea, which without question was one of the most important ideas in mathematics. It representing a watershed in the understanding of how mathematical insights are gained and secured, and how mathematical activity should be conducted. Essentially taken for granted today by everyone engaged in activities of a mathematical nature, it had emerged through the development of Greek geometry. Philosophers like Plato and Aristotle had also contributed. This fundamental principle is the following:

> **The Hypothetical-Deductive Method** All known geometric facts or theorems should be deduced by agreed upon logical rules of reasoning from a set of initial, self evident truths, called *postulates*.
> These postulates should be such that every informed person would agree on their validity, to the extent that they did not require proof. The set of postulates should be kept as small as possible, thus one should endeavor to construct proofs of assertions which, even though self evident, could be deduced from other even more fundamental self evident ones.

In a similar manner Euclid defined the more complicated figures and concepts using fundamental ones like points and lines. He even gave definitions of these, for example asserting that *a point is that which has no parts*, and defining a line as *that which has only length*. Even though his method has stood up through more than 20 centuries, the postulates themselves have had to be refined and made more precise.

Euclid based geometry on five *axioms* or common notions, and five *postulates*. The former were supposedly more obvious than the latter, nevertheless all were considered as *self evident truths*, not requiring proofs. They could be taken for granted.

Euclid's work was in no way easy to read! When his powerful mentor, King Ptolemy I, asked if there did not exist an easier way to learn geometry than to read all this, Euclid answered:

– *"No, to the geometry there is no separate road for kings, there is no Royal Road to Geometry."*

Euclid's work *The Elements* has had an enormous influence on mathematics in general and geometry in particular. But it was not confined to geometry alone. It also contains a substantial body of algebra.

Almost up to our time it has been used as textbook, and then been replaced by works which did not always represent improvements.

Geometry was, for a long time, synonymous with *Geometry according to Euclid*. The Elements of Euclid was elevated to a level almost comparable to *The Holy Bible* itself throughout the Middle Ages. The *Euclidian Geometry* was the very basis for the understanding of Space as God had created it. Questioning the truth of the assertions upon which Euclidian Geometry rested came, in the end, to be regarded as heretics. Of course, such attitudes are anathema to any kind of scientific inquiry.

4.2 The Books of Euclid's Elements

We follow Heaths edition of Euclid's Elements, [16]. In Book I the foundations for geometry is laid out. Here we find the fundamental definitions and axioms. Their conciseness and precision are remarkable, even today. Of course there has been critical remarks, and as an axiomatic system it has required a considerable amount of work over the more than 2000 years which have elapsed since these statements were written. But let us enjoy Euclid's terse and precise style! Book I opens with

Euclid's Definitions

1. A point is that which has no parts.[2]
2. A line is breathless length.
3. The extremities of a line are points.
4. A straight line is a line which lies evenly with the points on itself.
5. A surface is that which has length and breath only.
6. A plane surface is that which lies evenly with the straight lines on itself.
7. A plane angle is the inclination to one another of two lines in a plane which meet one another and do not lie in a straight line.
8. And when the lines containing the angle are straight, the angle is caller rectilinear.
9. When a straight line set up on a straight line makes the adjacent angles equal to one another, each of the angles is right, and the straight line standing on the other is called the perpendicular to the one on which it stands.
10. An obtuse angle is an angle greater than a right angle.
11. An acute angle is an angle less than a right angle.
12. A boundary is that which is an extremity of anything.
13. A figure is that which is contained by any boundary or boundaries.
14. A circle is a plane figure contained by one line such that all the straight lines falling upon it from one point from those lying within the figure are equal to one another.
15. And the point is called the center of the circle.

---

[2] An alternative translation from the Greek original would be: "A point is that which is indivisible into parts".

16. A diameter of the circle is any straight line drawn through the center and terminated in both directions by the circumference of the circle, and such a straight line also bisects the circle.

17. A semicircle is the figure contained by the diameter and the circumference cut off by it. And the center of the semicircle is the same as that of the circle.

18. Rectilinear figures are those which are contained by straight lines, trilateral figures being those contained by three, quadrilateral those contained by four, and multilateral those contained by more than four straight lines.

19. Of trilateral figures, an equilateral triangle is that which has its three sides equal, an isosceles triangle that which has two of its sides alone equal, and a scalene triangle that which has its three sides unequal.

20. Further, of trilateral figures, a right-angled triangle is that which has a right angle, an obtuse-angled triangle that which has an obtuse angle, and an acute angled triangle that which has its three angles acute.

21. Of quadrilateral figures, a square is that which is both equilateral and right-angled, an oblong that which is right angled but not equilateral, a rhombus that which is equilateral but not right-angled, a rhomboid that which has its opposite sides and angles equal to one another but is neither equilateral nor right-angled. And let quadrilaterals other than these be called trapezia.

22. Parallel straight lines are straight lines which, being in the same plane and being produced indefinitely in both directions, do not meet one another in either direction.

It is interesting to note that the term *line* does not signify only *straight line*, and that the straight lines are of finite length, but may be infinitely produced in either direction. The (curved) lines of Euclid do define angles, and they do have *length*. Today, in an axiomatic treatment of geometry, it is customary to take the terms *point* and *line* as undefined, as well as the relation of *incidence* between points and lines. This will be our approach in Chapter 8. Now Euclid does not specify his *undefined terms*, but he nevertheless makes his definitions using such terms. One is the word *part*, and perhaps *divisible*. Another term would be *length*, and so on. Many eminent mathematicians have worked on the project of understanding Euclid's definitions and postulates. It is fair to say that Euclid's achievement represents the single most fruitful set of ideas in the entire history of mathematics!

David Hilbert has treated Euclid's geometry in his classical work *Grundlagen der Geometrie*, [18]. This work is viewed by many as the final word in mathematically securing Euclid's axioms and postulates.

Having thus formulated the basic definitions, in a language and style which summons our admiration even today, Euclid proceeds:

Euclid's Postulates
Let the following be postulated:

1. To draw a straight line from any point to any point.

2. To produce a finite straight line continuously in a straight line.
3. To describe a circle with any center and distance.
4. That all right angles are equal to one another.
5. That, if a straight line falling on two straight lines make the interior angles on the same line less than two right angles, the two straight lines, if produced indefinitely meet on that side on which are the angles less than the two right angles.

These are the *five postulates of Euclid*. Among them the *fifth postulate* occupies a special position. First of all, as it stands it appears considerably *less obvious* than the others. The remarkable fact is, of course, that Euclid must have concluded that the assertion was *indemonstrable*, and that he could not come up with some much simpler statement, equivalent to it in the presence of the other four postulates. But the Fifth Postulate of Euclid continued to haunt mathematicians for more than two thousand years, until it was realized that the Fifth Postulate is indeed independent of the other four. With the discovery of non-Euclidian geometry, fully valid geometries revealed themselves, in which the Fifth Postulate is no longer true. More on that later.

But in the process of attempting to prove the Fifth Postulate, many statements were discovered which in the presence of the other four is equivalent to the Fifth. We shall again follow Heath in [16], and list the most important among these statements.

Alternative version of Euclid's Fifth Postulate

1. Through a given point only one[3] parallel can be drawn to a given straight line.
   Due to *Proclus*. It is commonly known as *Playfair's Axiom*, but was not a new discovery.
2. There exist straight lines everywhere equidistant from one another.
3. There exists a triangle in which the sum of the three angles is equal to the sum of two right angles.
   Due to *Legendre*.
4. Given any figure, there exists a figure similar to it of any size we please.
   This form is due to Legendre, Wallis and Carnot.
5. Through any point within an angle less than two thirds of a right angle a straight line can always be drawn which meets both sides of the angle.
   Due to Legendre.
6. Given any three points not on a straight line, there exists a circle passing through them.
   Due to Legendre and Bolyai.
7. There exists a triangle, the contents of which is greater than any given area.

---

[3] Meaning one and only one

Due to Gauss in a letter to Bolyai in 1799.[4]

Finally Euclid formulated five statements, called *Common Notions* or *Axioms*. These were statements of a general nature, viewed as being universally valid in all fields of human thought. They were the following:

Euclid's Common Notions or Axioms

1. Things which are equal to the same thing are also equal to one another.
2. If equals be added to equals, the wholes are equal.
3. If equals be subtracted from equals, the remainders are equal.
4. Things which coincide with one another are equal to one another.
5. The whole is greater than the part.

On the basis of the Definitions, the Postulates and the Common Notions, Euclid builds the whole of geometry. Now this material was due to many different Greek geometers of course, and several books in the Elements are thought to have been written in the entirety by others than Euclid. The point is that Euclid collected and systematized essentially the complete body of mathematics known at that time.

The last two propositions in Book I, namely I.47 and I.48, treat the "Pythagorean Theorem". I.47 reads as follows, quoted from [16]:

Pythagoras According to Euclid. *In right angled triangles the square on the side subtending the right angle is equal to the (sum of the) squares on the sides containing the right angle.*

The parenthesis is tacitly assumed in the Elements.

The last proposition of Book I is I.48, which is the converse to the Pythagorean Theorem:

The Converse Pythagoras *If in a triangle the square on one of the sides be equal to the squares on the remaining two sides, then the angle contained by the remaining sides is right.*

We shall now render Euclid's famous and very elegant proof of the Pythagorean Theorem. We consider Figure 4.1.

It is not practical to give the proof in the original form, using Euclid's own words. The reason for this is that Euclid proceeds rigorously from the *First Principles*, in other words from the Definitions, Postulates and Common Notions, using also propositions already proven. Thus for example, the fact that the points G, B and C lie on the same straight line requires a proof. In [16] the whole proof fills almost two printed pages, namely pp. 349 and 350

---

[4] Gauss wrote, according to [16]: "If I could prove that a rectilinear triangle is possible the contents of which is greater than any given area, I am in a position to prove perfectly rigorous the whole of geometry."

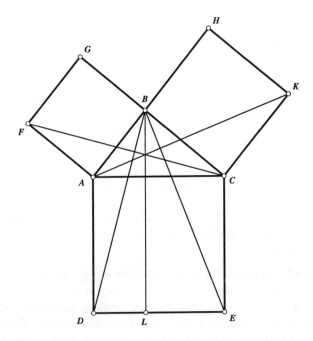

Fig. 4.1. Euclid's illustration to his proof of the Pythagorean Theorem.

of Volume 1. Here we give a modern version of the geometric contents, in the style we find in textbooks of elementary geometry.

The produced normal from B to the hypothenuse AC divides the square on AC in the two rectangles with base DL and LE and height equal to DA. It will suffice to prove that

1) the former rectangle has area equal to the square on AB, and
2) the latter rectangle has area equal to the square on BC.

$\triangle ACF$ is congruent with $\triangle ADB$ since the sides are pairwise equal. $\triangle ACF$ is a triangle with base AF and height AB, its area therefore is half of that of the square on AB. $\triangle ADB$ has base AD and height DL, its area therefore is equal to half of the rectangle with base DL and height DA, and 1) is proven. 2) follows analogously.

The geometry continues in Book II, which deals with *Geometric Algebra*. In Greek mathematics Geometric Algebra played a similar, but less prominent, role to the one played by algebra today. Thus for example, the two first propositions are equivalent to the following formulas:

1) $a(b + c + d + \ldots) = ab + ac + ad + \ldots$
2) $(a + b)a + (a + b)b = (a + b)^2$

The first formula must be interpreted as asserting that the area of the big rectangle to the left is equal to the sum of the many small ones to the right. The other formula expresses the sum of the areas of the two rectangles on the left as the area of the square to the right. But in the Elements the formulas were expressed *verbally*, they were formulated as follows, quoted from [16]:

Formula 1 If there be two straight lines, and one of them be cut into any number of segments whatever, the rectangle contained by the two straight lines is equal to the (sum of the) rectangles contained by the uncut straight line and each of the segments.

Again, the parenthesis is tacitly assumed in the Elements.

The next formula is quoted literally:

Formula 2 If a straight line be cut at random, the rectangle contained by the whole and both of the segments is equal to the square on the whole.

This should give the reader a taste of the *verbal nature* of Greek geometry. In fact, the absence of a good algebraic notation and the verbal form of exposition made it very difficult to resume work in geometry once the line of transition from person to person had been interrupted. When the last great geometer had been dead for fifty or a hundred years, it was not easy to continue the work only with written sources! This was no problem in Euclid's times, when the research community in Alexandria and elsewhere consisted of many individuals, and was robust in that it consisted of people at all levels. But towards the end of the Hellenistic Epoch it probably was a contributing factor to the end of Greek geometry and mathematics. When a new Genius is born, a devoted teacher is essential for it to flourish. A teacher, perhaps, who has had personal contact with the last great master of the subject.

The word *Gnomon* means in some contexts an ancient astronomical instrument. In Greek mathematics it is the *Carpenter's Square*, and using figures of this shape Greek geometers carried out arguments where we would use algebra. In Figure 4.2 we show the equivalent of the deduction of a familiar formula:

$$(a + b)(a - b) = a^2 - b^2$$

In Book II we also find a generalization of the Pythagorean Theorem, in propositions II.12 and II.13.

Book III deals with properties of circles and circle segments, and ends with the following two propositions:

Two Cords If in a circle two straight lines cut one another, the rectangle contained by the segments of the one is equal to the rectangle contained by the segments of the other.

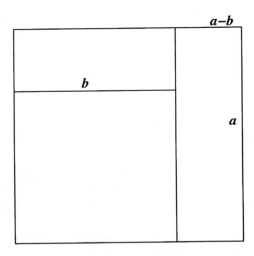

Fig. 4.2. The large square with side a has area a^2, the small one with side b has area b^2. The smaller square is placed at the lower left corner of the bigger one, then the difference gets the shape of the gnomon, or the carpenters square. This gnomon has the breadth $a - b$, and consists of two pieces: One piece has length a, and the other one has length b, a total of $a + b$. The total area is therefore the common breadth, $a - b$, multiplied with the sum of the lengths, $a + b$.

A Line Cutting a Circle If a point be taken outside a circle and from it there fall on the circle two straight lines, and if one of them cut the circle and the other touch it, the rectangle contained by the whole of the straight line which cuts the circle and the straight line intercepted on it outside between the point and the convex circumference will be equal to the square on the tangent.

Book IV contains only problems concerning figures given by straight lines which may be inscribed or circumscribed circles. Book V is probably due to , and contains his theory of proportions between *"magnitudes of the same kind"*. It treats the problem of commensurability in this light.

Two magnitudes are said to be of the *same kind* if they are capable of exceeding one another when suitably multiplied. For two magnitudes of the same kind defines their *ratio*, and the notion of *equality* for ratios is defined in terms of the property of exceeding by multiplication. We shall not go into the details of this here, we have written more about it in [21]. The definition used by Aristotle was to consider two ratios to be equal if the repeated subtraction, the *antanairesis*, had the same pattern in the two cases. Thus to Greek geometers it vas possible to say that the ratio between two *line segments* was equal to the ratio between two certain *areas* or *volumes*. A line segment and an area would not be magnitudes of the same kind, but two line segment would. This latter assertion is some times referred to as the *Axiom of Archimedes*: Two line segments are capable of exceeding one another when suitably multiplied.

Book VI applies the theory from V to geometry. Here it is proven that the areas of two triangles or two parallelograms of the same heights are to each other as their bases.

Books VII, VIII and IX deal with what we would call *elementary number theory*. Book VII opens with *the Euclidian Algorithm*, we treat it in detail in Section 15.3. The point is to find the *greatest common divisor* of two (positive) integers.

Book VIII treats *continued proportionalities*, such as the *double proportionality* which Hippocrates was led to consider from his attempts of solving the problem of *doubling the cube*: $a : b = b : c = c : d$. This book is believed to be due to *Archytas*.

Proposition 14 in Book IX is equivalent to the fundamental result in number theory, that any integer may be factorized, essentially uniquely, into a product of prime numbers. Here we also find Euclid's famous proof that there are infinitely many prime numbers.

Book X treats *incommensurable magnitudes*. This is where the foundations for the important *Principle of Exhaustion* is established. We find it already in Proposition 1, quoted here from [16]:

Euclid X.1 Two unequal magnitudes being set out, if from the greater there be subtracted a magnitude greater than its half, and from that which is left a magnitude greater than its half, and if this process be repeated continually, there will be left some magnitude which will be less than the lesser of the magnitudes.

This proposition is used to prove Proposition XII.2, namely that the areas of two circles are to one another as the squares on their diameters.

Book XI treats basic solid geometry, in Book XII we find applications of the Principle of Exhaustion, for example proving what we would call the formulas for the volumes of a pyramid and a cone. Finally, the aim of Book XIII is to construct the five regular polyhedra and their circumscribed spheres.

All of this material is due earlier Greek geometers, among them Archytas and Eudoxus.

4.3 The Roman Empire

The Roman Empire, *Imperium Romanum*, developed from the city-state of Rome. The tradition has it that Rome was founded 753 B.C., at the outset ruled by kings. The last of them was *Tarquinius Superbus*, Tarquin the Proud, who was driven out around 510 B.C., having overstepped his powers. The Romans then designed a republican constitution, to provide a safeguard against such abuse of power. The republican system lasted from 510 B.C. to 31 B.C., usually subdivided in three epochs.

The Old Republic lasted till about 300 B.C., and is often labeled *The Struggle of the Orders*. The nobility, the *patricians*, stood against the common people, the *plebeians*. Finally a ballance between the two classes was worked out, and shortly thereafter Rome became the master of all of Italy.

The next epoch is often labeled *The Classic Republic*, and lasted from about 300 B.C. to about 130 B.C. During this time Rome had a stable government, dominated by the Senate. Now Rome became the leading Mediterranean power. The conquest of states surrounding the Mediterranean Ocean brought about the wars with Carthage. Carthage was a Phoenician colony originally, the name meant *The New City* in the Phoenician language. This semitic people had an advanced culture, the Greek took their alphabet from them, and according to the Eudemian Summary, it was the Phoenicians who brought the *numbers* to the Greek as well. The rich culture of Carthage was obliterated when the city and its people were annihilated by the Romans at the conclusion of the *Third Punic War*. *Punians* was the Roman term for Phoenicians.

The conquest in the Mediterranean area led to Rome gradually taking control over the Hellenistic world. The city of *Pergamon*, a cultural and scientific center rivaling Alexandria, was taken over by Rome on the pretext of a purported will left by the late King. Macedonia was defeated, as was Antiochus III of Syria.

The success came with a price, however. The Roman army had a core consisting of free Italian peasants, who fought mainly during the summer. As long as the wars were fought close to home, the soldiers could still farm their land in between the expeditions. But when the action moved away far from home, this became difficult. The ensuing economic hardship caused many among them to loose their land, rich landowners bought it up and became even richer. At the same time the conquests brought taxes from the new provinces to Rome, this was mostly in the form of grain and other agricultural produce. The surplus thus created destroyed the profit of the domestic farmers. And, finally, the conquests also brought large numbers of *slaves* to the center of the Empire, and in addition the Eastern slave markets became available. So the disenfranchised peasantry had no way of finding decent employment on the estates of the great landowners, but had to move to Rome, where they now became a growing impoverished proletariat.

Appealing to the people by offering them *bread and circus* became the standard way of campaigning for public office. In this game the ruthless populist gained the upper hand to the decent thinker. A man like *Cicero* would be no match to someone like *Caesar*. And the large sums of money required corrupted the political system in Rome.

The final phase of the Republic is the *Century of Civil War*. It lasts from 130 B.C. to 31 B.C. This phase opens with the passing of laws which limited how much state land a Roman could own. In this way the landless proletariat should be provided with land to farm. The Senate lost much of its power, and its stabilizing effect was greatly reduced. Then, when Germanic tribes attacked, the army proved too weak to hold the borders. This urgent problem was solved by allowing men without land to become soldiers. Thus the army could be strengthened and the Germanic tribes pushed back for now. But the soldiers became more dependent on their generals. The road lay open to a state of affairs where warlords with money could muster loyal troops in their bids for political power. This situation was almost a prerequisite for the civil war which was to commence with full force, as well as for the low quality of some of the emperors who came to more or less short-lived power towards the end of the Roman Empire. The sad story of how the Republic was ended will be told in Section 4.8.

4.4 Archimedes

One of the very greatest mathematician, scientist and engineer of antiquity was *Archimedes*. He lived and worked in Syracuse, as a very close associate of King Hiero. Some have speculated that Archimedes must have been one of the Kings relatives. Be that as it may, the King certainly had ample reason for appreciating Archimedes' friendship.

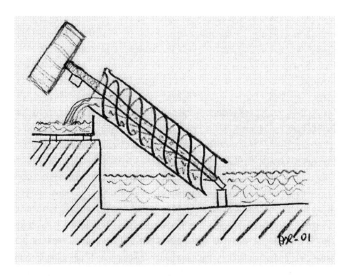

Fig. 4.3. This ingenious pumping device is credited to Archimedes. Pumps of this kind have been in use up to our time i Egypt. At Archimedes' time the pump might be operated by a slave, who would turn the shaft by a treadmill. Sketch by the author.

Indeed, not only did Archimedes provide innumerable inventions, and conduct major engineering enterprises to the advantage of his King and his city, but he also was the sole reason why Syracuse could hold its own for four years against the overwhelming Roman forces. In the so called Second Punic war between Rome and Carthage, Syracuse had sided with Carthage, it had strong historical ties to that Phoenician colony. Under the leadership of the Roman consul *Marcus Claudius Marcellus*, later given the name of honor "Rome's Sword", Syracuse was under siege for two years, from 214 to 212 B.C. And the only reason why Syracuse could hold out for so long, and was finally taken only because of treason from within, was the technological superiority which Archimedes' war machines represented.

Archimedes' war machines included catapults which threw boulders at the enemy forces. He constructed pulleys with ropes and chains, and huge hooks which were used to grab ahold of the Roman vessels from the city walls of Syracuse. The hapless ship would then be hoisted up, half of it reaching considerably out of the water, to be subsequently dropped down again and thus destroyed. Lenses and mirrors were employed in setting approaching warships on fire before they could even get close.

It is not surprising that Marcus Marcellus had ordered his troops to take this man alive, and not to harm him in any way. He should be treated with respect and honors, and brought back to the Emperor in Rome.

One can only speculate what would have happened to Roman science and mathematics if Marcus Marcellus' plan had succeed. *The Romans were pretty short on math,* to say the least, and their science was not much better. They were, however, good soldiers and lawyers, for better or worse.

However, and unfortunately, at this decisive moment in history, one of the *Sternstunden der Menschheit*, to borrow Stefan Schweig's expression, the following fatal mishap transpired: As a Roman soldier stormed into Archimedes' study with sword in hand, the latter sat immersed in geometrical considerations. He had drawn geometric figures, circles, in the *sand*, as some say. Probably it was *fine dust of glass*, spread out over his drawing-board. This was the scratch paper of antiquity and almost up to our own age, used for writings not to be preserved. Permanent writing material was far too precious to be wasted as scratch paper. Anyway, the soldier demanded Archimedes' name, as he did not know his appearance. But Archimedes had been so engulfed in his geometry, that he was quite unable to say his name. All he could do, was to stotter: – *Please do not touch these* –, as he pointed to his drawing-board. The Roman soldier saw in this remark a lack of due respect for the mighty Roman power, and responded by striking him down with his sword. Thus ended the most remarkable scientific genius of antiquity, perhaps even of the entire human history as we know it.

Marcellus, who was very much saddened and disappointed by the loss of Archimedes, gave orders that he should be buried with full honors. Apologizing to his family, he also carried out Archimedes' wish, stated in his testament, that his most important geometrical theorem be engraved on his tombstone.

Eutocius of Ascalon (see section 4.14) wrote commentaries, among others on Archimedes' work *On the Sphere and the Cylinder*. Based on two of the best Greek manuscripts from Archimedes, with Eutocius' commentaries, William of Moerbeke made a translation into Latin around 1270. The original manuscripts have since disappeared. Archimedes regarded the theorem proved here to be his most profound discovery.

That this inscription really stood on Archimedes' grave, is attested to by the Roman *Marcus Tullius Cicero*, in his *"Tusculian Dialogues"*. He found the grave 140 years after Archimedes' death.

In the year 75 B.C. he was questor at Sicily. In "Tusculian Dialogues" he asserts that virtue is more important for human happiness than power and wealth.

"– Only the wise is really happy," Cicero writes. As proof for this claim he invokes the memory of the tyrant Dionysus, who he compares to Plato and Archimedes. He continues as follows:[5].

" *– But from Dionysus' own city of Syracuse I will summon up to life a humble man in a modest position, from his dust and drawing-board: Archimedes. How can anyone who has ever had the slightest contact with humanity and scholarship not wish to be like this mathematician, rather than the afore mentioned tyrant?*

"While I was questor in Sicily I sought out his grave, which was unknown to the people of Syracuse. Indeed, they denied that any such thing existed. As it happened, the grave was hidden, on all sides overgrown by thorny shrubs. However, I remembered some simple lines of verse, which I knew were inscribed on his monument. They relate that upon his grave there was set a sphere with a cylinder, modeled in stone.

"By the Agrientine Gate there is a large number of old graves. And as I looked everywhere, my attention was caught by a small pillar, which did not reach much above the thorny shrubs. On it there were rendered a sphere and a cylinder.

"The very noblest of the men from the city were with me, and I said to them at once that this was what I sought. Workers with scythes were summoned, and cleared away the scrubs, opening up access to the place. As now entrance had been made possible, we proceeded to the front of the tombstone. And there the inscription became visible. But the last part of the verse was gone, about half of it.

---

[5] As the author is unfortunately unable to read Cicero in the original language, which is of course Latin, the quote from his writings is based upon two translations. Mainly I rely on a translation into Norwegian made for *Viggo Brun*, and used in his excellent book [2]. But I have also supplemented the translation with elements from the one rendered in *Michael Grant*, [9]

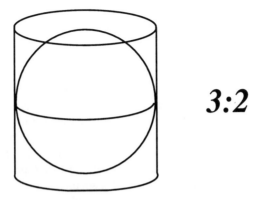

3:2

Fig. 4.4. The monument on Archimedes' grave bore this inscription. It renders what he considered his most profound mathematical theorem, concerning a sphere and its circumscribed cylinder: The proportion of the volumes of the circumscribed cylinder to that of the sphere equals the proportion of the surface areas of the same bodies, counting of course top and bottom of the cylinder. This common proportion is $1\frac{1}{2}$. Or, actually, it would have been written as 3 : 2, since the Greek regarded this as a *ratio*, not a *number* as we do today.

"Thus the noblest of Greek cities, once the most enlightened of them all, would have remained ignorant of the tomb of the most brilliant citizen it had ever produced, had it not been for a man from Arpinum."

Cicero gave orders that the grave should be preserved from then on, but for how long this was done we do not know. The grave was forgotten once more, and although there have been rumors from time to time that the grave has been found, no authoritative photo of the monument seems to exist.

We now turn to the description of some of Archimedes' geometry. A full account is of course impossible to give here, but we refer to Heath's [15] for a fuller and deeper treatment.

In 1906 the Danish classical philologist *Johan Ludwig Heiberg* visited Constantinople to study a parchment from the Saint Sepulchi monastery in Jerusalem. This was a so called *palimpsest*, where an original Greek text had been scraped off and replaced by a text of religious contents. Heiberg realized that the text which had been scraped off, contained among other things the priceless book by Archimedes entitled *The Method*, which had been presumed lost. Fortunately the attempted destruction of the works by Archimedes had served to preserve and protect them through the centuries of darkness. For not only did it turn out to be possible to restore Archimedes' original text – or rather, the text copied by a scribe from earlier copies of copies of... Archimedes' text – but the sacral status of the replacement text had protected the parchment from being destroyed! It turned out that the palimpsest contained *On the Sphere and the Cylinder*, almost all of *On Spirals*, and fragments of some other works which are preserved elsewhere. And then it contained *The Method*.

The book is now being studied and restored at the *Walters Art Museum* in Baltimore, [1].

In the introduction Archimedes explains how he discovered these theorems using mechanics. He studied certain elements in equilibrium, and concluded from that relations between surface areas or volumes. But he emphasized that he would not consider this *as proofs*, only that *"it is easier, when we have found by this method some knowledge about the problem, to find the proof than it would have been to find it without such prior knowledge.*

Archimedes finds the area of a segment of a parabola and the volume of a sphere. We shall treat the latter result.

Archimedes actually leaves it as open whether these methods may be developed into fully valid *proofs*. But he writes: *"I am convinced that this method is not less useful in also proving these statements."* Today it is not difficult to accept his view, since we find ideas which later were developed into the concepts of modern calculus in the form of definite integration.

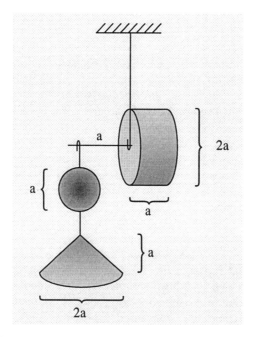

Fig. 4.5. The Sphere, the Cone and the Cylinder at equilibrium.

The ideas in Archimedes' derivation of the formula for the volume of a sphere, is to write the following equality: On one side of the equal sign there stands the volume of a sphere and a cone, the diameter of the sphere is equal to the height of the cone and to the radius of its base.

We denote this radius by a. On the other side of the equal sign stands half of the volume of a cylinder, its height as well as radius of the base is also a. Archimedes proved this equality by first showing that we have the equilibrium displayed in Figure 4.5.

We assume that the sphere, the cone and the cylinder all have density 1. Let us first assume that the three bodies shown in the figure are in equilibrium. Then the equality of momentum yields that if V denotes the volume of the sphere, S the volume of the cylinder and K the volume of the cone, then

$$a(V + K) = \frac{a}{2}S$$

since we may place the total mass of the cylinder in its center of gravity, $\frac{a}{2}$ from the point of suspension.

This yields

$$V = \frac{1}{2}S - K$$

and from this Archimedes could find the volume of a sphere of *diameter* a, since the volume of a cone and a cylinder was known.

This relation yields:

Sphere and Cone The volume of a sphere is equal to the volume of a cone with base four times as big as the area of a great circle in the sphere and height equal to the radius of the sphere. So the cone has base with radius equal to the diameter of the sphere.

We easily deduce this from the relation $V = \frac{1}{2}S - K$. In fact, we have $S = 3K$, so $V = \frac{3}{2}K - K = \frac{1}{2}K$. If K' is the volume of a cone with *half* the height of K and the same base, then $K' = \frac{1}{2}K$, thus $S = K'$.

In modern notation this relation implies $S = \frac{4}{3}\pi r^3$, where r is the *radius* of the sphere.

To prove the equilibrium of Figure 4.5, we consider three circular slices, of a very small thickness Δ, as shown in Figure 4.6. Here x is a number between 0 and a. We now prove that any such configuration is in equilibrium.

The circular slice to the right, the one which has been cut out of the cylinder, has volume equal to $A = \Delta \pi a^2$, and the slice which is cut out of the sphere has volume $B = \Delta \pi y^2$, so that $B = \Delta \pi(ax - x^2)$. Finally the slice cut out from the cone has volume $C = \Delta \pi x^2$. The three slices are in equilibrium if

$$a(B + C) = xA$$

or

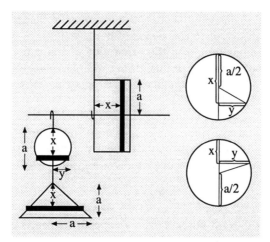

Fig. 4.6. Three circular slices in equilibrium. x is any number between 0 and a, and each slice has a very small thickness Δ. The slice inside the sphere has radius y, the one inside the cone has radius x and the slice inside the cylinder has radius a. To the right we have deduced that $y^2 = ax - x^2$: If $x \leq \frac{a}{2}$ we use the lower circle, and get by Pythagoras that $y^2 = \left(\frac{a}{2}\right)^2 - \left(\frac{a}{2} - x\right)^2 = ax - x^2$, and if $x > \frac{a}{2}$ we use the upper one which yields $y^2 = \left(\frac{a}{2}\right)^2 - \left(x - \frac{a}{2}\right)^2 = ax - x^2$.

$$a\left(\Delta\pi(ax - x^2) + \Delta\pi x^2\right) = x\Delta\pi a^2$$

which clearly holds.

We finally show that this implies the equilibrium of Figure 4.5. In fact, we now let x grow from 0 to a in steps of Δ. Then the corresponding "*slice-configurations*", when taken together, will constitute the configuration shown in Figure 4.7.

By making Δ smaller and smaller, and the number of parts correspondingly bigger, we finally get the equilibrium of the sphere, cylinder and cone as claimed.

From his discovery that the volume of a sphere is equal to the volume of a cone with base of area four times the area of a great circle of the sphere, and with height equal to the radius of the sphere, Archimedes found this result:

Area of the Sphere The surface area of a sphere is four times the area enclosed by a great circle.

Archimedes gives the following argument for this conclusion: "*In the same way that any circle encloses an area equal to the area of a triangle with base equal to the circumference of the circle and height equal to the radius, it is reasonable to conclude that the volume of a sphere is equal to the volume of a cone with base equal to the surface area of the sphere and height equal to the radius.*"

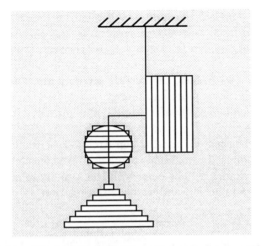

Fig. 4.7. Eight "slice-configurations", which all are in equilibrium, implies equilibrium for the composite configuration.

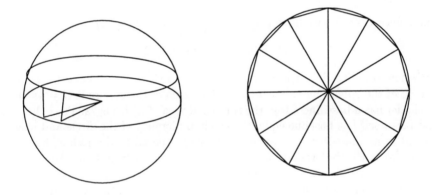

Fig. 4.8. Archimedes' idea to find the surface area of a sphere.

In Figure 4.8 there are two sketches. To the right there is a regular 12-gon, inscribed in a circle. By adding the areas of the twelve triangles, we find that the area of the 12-gon is the circumference of the 12-gon multiplied by the height of the triangles, divided by 2. If we divide all the central angles in two and proceed to the regular 24-gon, we get the same: The area of the 24-gon is the distance from the sides to the center, multiplied with the circumference and divide by 2. Repeating this, we approach the circle, and the claim follows. In order to get the analogous result for the sphere, we divide the surface of the sphere by circles, as shown to the right. We get a net on the surface and an inscribed (non-regular) polyhedron. The volume of this polyhedron will be approximately equal to the radius multiplied by the surface area of the polyhedron divided by 3, as it is composed of several *pyramids*. Using a finer and finer net of circles, we get the claim in the limit. But apart from the anachronistic term *in the limit*, this is only a way of making the claim *plausible*, not of proving it. In fact, the heights involved are not equal, even if they all approach the radius more and more.

In *On the Sphere and the Cylinder* Archimedes provides full proofs of these claims, and more.

We finally deduce the theorem inscribed on Archimedes' tombstone, using the methods which he himself would have employed. We consider a sphere V and a circumscribed cylinder S. The cone with the same base and height as the cylinder will have volume K_1, and we have the relation

$$S = 3K_1$$

We consider the cone K_2 which have base with radius equal to the diameter of the sphere and height equal to the radius of the sphere. Then

$$K_2 = 2K_1$$

and since $K_2 = V$, we have

$$S : V = 3K_1 : 2K_1 = 3 : 2$$

We now come to the surface areas. We denote the area enclosed by a great circle of the sphere by s. Then the surface area of the sphere is $O = 4s$. The cylinder has a base and a top, the combined area of which is $2s$. The cylinders surface itself has baseline equal to the circumference of the circle and height equal to its diameter. Half of this, with height equal to the radius, has area equal to twice the triangle with the same base and height, in other words $2s$, by Archimedes' remark quoted above. The surface area of the cylinder proper is therefore equal to $4s$, so the surface area counting base and top is $A = 2s + 4s = 6s$. We thus have

$$A : O = 6s : 4s = 3s : 2s = 3 : 2$$

Archimedes described the *semi regular polyhedra*. There are altogether 13 of them, they are referred to as the *Archimedian Polyhedra* or as the *Archimedian Solids*. They are convex polyhedra where the sides are regular polygons, but not of the same kind.

In Figure 4.9 two of the Archimedian Solids are presented. The two simplest ones are shown in Figure 4.10, they are obtained by a process which is known as *truncation* from the tetrahedron and the cube, respectively. This process yields new semi regular polyhedra when applied to all five Platonic Solids, as well as to the cubeoctahedron and the icosidodecahedron.

The numbers uniquely identify the semi regular polyhedra For example, $(3, 6, 6)$ for the truncated tetrahedron signifies that at each vertex there are

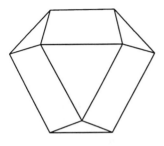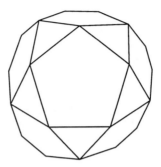

Fig. 4.9. Two of the 13 Archimedian Solids. To the left the cubeoctahedron, designated (3,3,4,4). To the right we see the icosidodecahedron, designated (3,3,5,5).

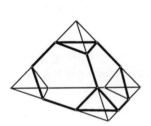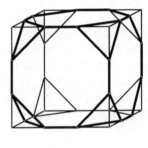

Fig. 4.10. The process of truncation. To the left we show the truncation of the tetrahedron, to the right the truncated cube.

one equilateral triangle and two regular hexagons meeting, while for the so-called *rhombisocahedron* has the numbers $(3, 4, 4, 5)$ which indicate that at each vertex one equilateral triangle, two squares and one regular pentagon meet. For more on these fascinating objects, as well as beautiful pictures and the code allowing you to create your own images, we refer to [7].

In his book *On Spirals* Archimedes studies spirals of the type which today is known as *Archimedian Spirals*. He defines a spiral as follows:

Archimedian Spiral. If a straight line where a point is kept fixed, is made to rotate about the fixed point with even velocity until it comes back to the point of departure, and at the same time a point, starting from the fixed point moves with even velocity outwards along the line, then the point describes a spiral in the plane.

One of the reasons for Archimedes' interest in the spiral came from his work on *squaring the circle*. As we have seen above, Archimedes had drawn certain conclusions from the fact that the area enclosed by a circle is equal to the area of a triangle with base equal to the circumference and height equal to the radius. Another conclusion to be drawn from this fact is the following: *The problem of squaring the circle is equivalent to the problem of rectifying the circle.* In other words, if we are given a circle with center O through P, then we should try to construct a point T on the normal to OP at O, such that the length of OT is equal to the circumference of the circle. Then all we have to do is to construct a square with area equal to the area of $\triangle OTP$. The latter one was a very well known construction to Greek geometers, of course. *Archimedes performed the rectification of the circle by means of the spiral.*

In Figure 4.11 we see to the left how to draw such a spiral, the line rotating counter clockwise: We have drawn 12 straight lines through a point, spaced at 30°. Along the lines we have marked 1,2,3,... ,12 times a certain distance Δ. Now we connect these points with a curved line. To the right we have drawn the spiral in a usual coordinate system, we let the line start from the x-axis, the fixed point is the origin O. When the spiral crosses the x-axis after one rotation, we find the point P. At this point we construct the tangent to the spiral. The tangent intersects the y-axis in the point T. Now Archimedes proved that *the length of OT is equal to the circumference of the circle about O through P*, in our notation it is equal to $2\pi r$, where r is the length of OP.

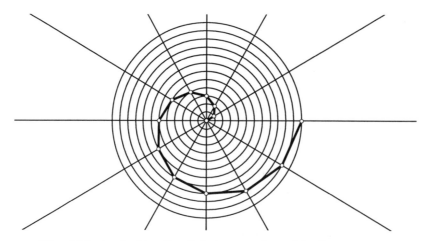

Fig. 4.11. An Archimedian Spiral, approximated by line segments.

We now prove Archimedes' result by modern methods. A point (x, y) on the spiral is given by

$$x = \frac{r}{2\pi} v \cos(v)$$

$$y = \frac{r}{2\pi} v \sin(v)$$

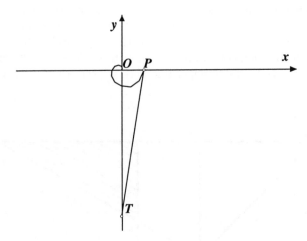

Fig. 4.12. The rectification, hence the squaring, of the circle.

where v is the angle which the rotating line forms with the x-axis. This is the spiral in polar coordinates. We further have that

$$dx = \frac{r}{2\pi}(-v \sin(v) + \cos(v))dv$$

$$dy = \frac{r}{2\pi}(v \cos(v) + \sin(v))dv$$

At the point P we then find the slope of the tangent by substituting $v = 2\pi$ in

$$\frac{dy}{dx} = \frac{\frac{r}{2\pi}(v \cos(v) + \sin(v))}{\frac{r}{2\pi}(-v \sin(v) + \cos(v))}$$

which yields

$$[\frac{dy}{dx}]_{v=2\pi} = 2\pi$$

The equation of the tangent at P therefore becomes

$$y = 2\pi(x - r)$$

and letting $x = 0$, we find the point of intersection between this line and the y-axis: It is the point $(0, -2\pi r)$, and the claim is proven. □

Archimedes gave a construction of the *regular heptagon*, the regular 7-gon. We know this construction through an Arabic translation of an original from Archimedes, which is lost. In fact, according to [35] *Tabit Ben Qurra*, who wrote the Arabic text, complains about the poor condition of the Greek original, he barely could make out the proofs. The construction is an example of what the Greek geometers called a *"Neusis"*-construction, it solves a *verging-problem*: A line is made to slide so that it passes in a certain direction and realizes a specified condition. Typically, this happens with constructions involving a marked straightedge, but here we see a different version. We consider Figure 4.13.

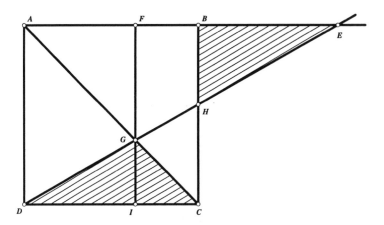

Fig. 4.13. A mechanical instrument for constructing a regular heptagon.

We see a square $ABCD$. The diagonal AC is fixed, but then the line DE produced is made to rotate about D. The point E is where this line intersects AB produced. Now the line is rotated about D until the following happens: *The areas of $\triangle DGC$ and $\triangle BEH$ are equal.* We may think of Figure 4.13 as displaying some kind of mechanical instrument. For example, the lines are metallic rods, $\angle DAB$ and $\angle BCD$ being fixed right angles, so that C may slide up and down along AC produced, the latter rod being fixed in the diagonal position, at 45°. As C moves and DE produced rotates, we may arrange the instrument so that in addition to the area-condition referred to above, the distance between B and E is some distance s, given *à priori*.

When this arrangement is achieved, then *AE will be the largest diagonal in the regular heptagon with side s.* Let us assume this for the time being, and proceed to construct the regular heptagon with sides s on the basis of this information.

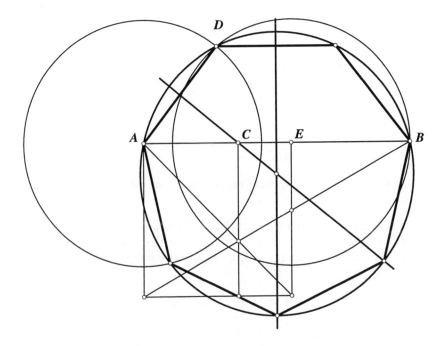

Fig. 4.14. Construction of the Regular Heptagon.

We start by marking off AE from the instrument as AB in the construction on Figure 4.14. F and B from the instrument becomes, respectively, C and E in the construction. Now EB is the length of the side of the heptagon, according to our claim above. We therefore draw a circular arc about A of this radius, EB. We also draw a circular arc of this radius about E, and claim that then the point of intersection of the two arcs, namely D, is a vertex of the regular heptagon. This claim will follow from the analysis to be given below. But now we have three points on our regular heptagon, A, B and D. We therefore find the center O of the circumscribed circle as the point of intersection of the two mid normals, draw the circle and complete the heptagon.

We now refer to Figure 4.15. Since the circumference is subdivided in 7 equal parts, the angles u at the periphery, the *inscribed angles* of the 7 arcs into which the circumference is subdivided, are all $\frac{\pi}{7}$ radians, or $\frac{180°}{7}$, being half the measure of their intercepted arcs, the *central angles*. We thus find that $\angle DAB = 2u$, $\angle DAG = 3u$, and so on. The angular sum of a triangle being π we find $\angle ACD = 3u$, and AFG being a straight line we find $\angle EFG = 3u$.

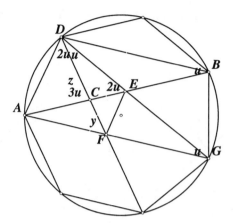

Fig. 4.15. Notations used in the explanation of the construction.

By construction $EB = s$, and we denote AC by z and CE by y. Now $\triangle BED$ being isosceles, we have $ED = s$, this was used in the construction of the heptagon above. All angles being pairwise equal, we have

$$\triangle CED \sim \triangle CDB \sim \triangle EFD$$

This yields

$$\frac{z}{y} = \frac{s+y}{z} \quad \text{and} \quad \frac{z}{s} = \frac{s}{z+y}$$

We now return to Figure 4.13, where we have $BE = s$, and claim that $AF = z$ and $FB = y$. Denote $\angle GDC$ by v, put $FB = IC = a$, $DI = AF = FG = b$ and $BH = c$. Since $\angle GDC = \angle HEB$, we have

$$\tan(v) = \frac{b}{a+s} = \frac{a}{b} = \frac{c}{s}$$

and the equality of the areas of $\triangle GDC$ and $\triangle BEH$ gives

$$a(a+b) = cs$$

Substituting $c = \frac{a(a+b)}{s}$ into the last equality given by $\tan(v)$ we find $\frac{b}{s} = \frac{s}{a+b}$, and as $a+b+s = y+z+s$ we get $a+b = y+z$, thus $\frac{b}{s} = \frac{s}{y+z}$ and hence $b = z$ by the relations from the regular heptagon. Therefore we also have $a = y$, and the claim is proven.

If we set the length $s+y+z$ of the diagonal to 1, then straightforward elimination yields

$$s^3 - s^2 - 2s + 1 = 0.$$

This relation of degree 3 is used to prove that the construction is impossible by *legal use* of compass and straightedge. We return to this in Section 16.7. But the relation also explains that the construction is possible by *a marked straightedge*. We return to this in Section 4.6. An actual construction using a marked straightedge was found by the great French geometer and algebraist *Viète* (1540 – 1603). For the details we refer to [14].

4.5 Erathostenes and the Duplication of the Cube

Erathostenes was born in 276 in Cyrene and died in 194 B.C. in Alexandria.

He was the director of the school at Alexandria, and was an outstanding representative for the refined culture which flowered at the Royal Court there. Some see a certain disdain for Erathostenes in some of Archimedes' writings, others disagree in this interpretations. At any rate, while Archimedes and his fellow citizen in Syracuse were fighting off the Romans, their backs against the wall, Erathostenes and his contemporaries in Alexandria were able to pursue the refined art of dialectics, poetry and rhetoric as well as classical literature and mathematics.

Here we shall only treat a small part of the scientific work of Erathostenes, one of direct relation to Greek geometry: He invented a mechanical instrument for the construction of the continued mean proportionality, to which the problem of the *Duplication of the Cube* had been reduced.

He appears to have been rather pleased with his own invention. Dedicating the device to the king Ptolemy, he had a description of it engraved on a monument which was erected in the kings temple. Eloquently praising the king, a model of the apparatus, in bronze, was put on top of the monument. In van der Waerden's words, in [35]: *"The subtle complement to the king and his son, which occurs in the epigram, betrays the well-versed courtier."*

A horizontal straightedge was equipped with a grove, in which three or more rectangular frames could slide. On these frames a diagonal had been engraved. In Figure 4.17 we show three such frames, then a double proportionality may be determined.

The frames were slid along the grove, so that the one marked I would be the outer end, then II and the innermost III. A second straightedge was fixed at K and could be rotated about this point. The frames were slid between the two straightedges as indicated in the figure.

If the points A, B, C and D are on a straight line, then we get four similar triangles, namely $\triangle AEK$, $\triangle BFK$, $\triangle CGK$ and $\triangle DHK$. This yields

$$AE : BF = BF : CG = CG : DH$$

and we find that BF is the first and CG the second mean proportionality between AE and DH.

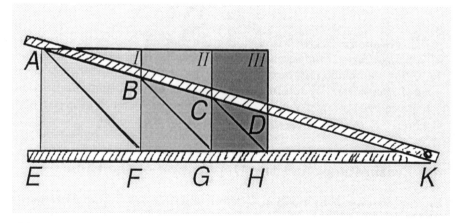

Fig. 4.16. Erathostenes' instrument for the construction of continued proportionalities.

Fig. 4.17. Three of the frames in Erathostenes' instrument.

If we wish to find the mean proportionalities between a and b, we may
assume that $a > b$. If not, then we interchange a and b, and interchange the
two mean proportionals as well: Indeed, if

$$\frac{a}{x} = \frac{x}{y} = \frac{y}{b}$$

then

$$\frac{b}{y} = \frac{y}{x} = \frac{x}{a}$$

Assume first that $a = AE$. We mark off b as HD on frame III. Then slide
frames II and III and rotate the second straightedge, so that the points A,
B, C and D fall along the second straightedge. Then $x = BF$ and $y = CG$ are
the wanted mean proportionals.

For the general case, take $\alpha = AE$, in other words the height of the frame.
We first find β such that

$$a : \alpha = b : \beta$$

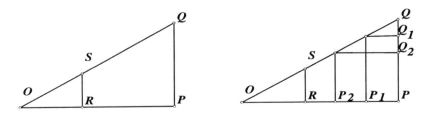

Fig. 4.18. The construction with arbitrary line segments a and b.

This is done by a right triangle as shown to the left on Figure 4.18, where we have $OP = a$, $OR = b$ and $PQ = \alpha$, the height of the frames in Erathostenes' instrument. The parallel to PQ through R intersects OQ in S, we take $RS = \beta$. From the instrument we then find the first and the second continued proportionals ξ and η, respectively. We now look at the figure to the right, where we have marked off $PQ_1 = \xi$ and $PQ_2 = \eta$. We then find the first mean proportional x as OP_1, and the second as $y = OP_2$.

4.6 Nicomedes and his Conchoid

Essentially nothing is known on Nicomedes' life. We estimate that he was born about 280, and died approximately 210 B.C. His main work is *On conchoid lines*, where his one and only great discovery is described, namely the *conchoid curve*. We shall return to it in Section 14.7, when we have sufficient geometric machinery to complete the study of this very interesting curve.

The curve, or rather any finite number of points on it, is constructed as follows: A line ℓ and a point P at a distance a from ℓ is fixed. Then a circle with center P is drawn, and the circumference subdivided in n equal pieces, say with n some power of 2, in which case the subdivision is always possible with compass and straightedge.[6] Corresponding to the subdivision, lines are drawn through P. These lines all intersect ℓ, and on each line points are marked at a fixed distance b from the points of intersection with ℓ. These points all lie on the conchoid, and when $b > a$, as is the case in Figure 4.20, the two sets of points on each line will trace out two branches of the curve.

Note that the conchoid, as with all higher curves considered by Greek geometers, are constructed by compass and straightedge, the Euclidian tools,

---

[6] Of course an exact subdivision is not essential, but mainly serves to make the construction esthetically appealing.

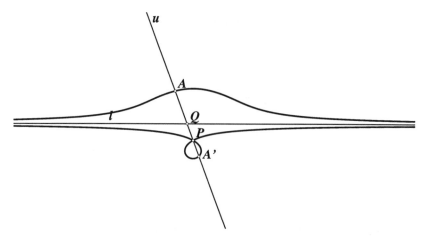

Fig. 4.19. The Conchoid of Nicomedes and how it is defined: Two lengths, or numbers as we would say today a and b are given. In the illustration here $a < b$. A line ℓ and a point P at a distance a from ℓ is fixed. We then draw a line u through P, intersecting ℓ in a point Q. On this line, at the distance b from Q we mark the points A and A'. They trace out the two branches of the conchoid when the line u rotates about P.

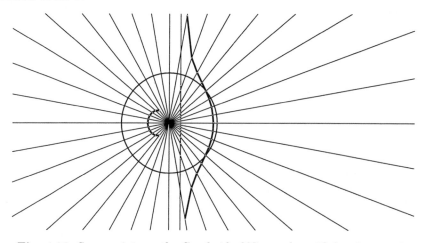

Fig. 4.20. Some points on the Conchoid of Nicomedes with $b = 3 > a = 1$.

but *with an infinite number of steps* in the construction. By this kind of constructions, known as *asymptotic Euclidian constructions*, the classical problems may be "solved". That is to say, if we allow a sufficiently large number of steps in the construction, the problems may be solved approximately with any prescribed degree of accuracy. The constructions we refer to below are of this nature. Indeed, by a large but finite number of steps we may approximate the curves used by a chain of small line segments, and using these approximations the classical problems may be solved approximately.

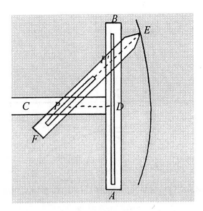

Fig. 4.21. Nicomedes' mechanical device for drawing the conchoid.

However, Nicomedes used a mechanical instrument to draw this curve, which is described by Heath in [15], Vol. I, p. 239. We follow Heath's explanation:

AB is a ruler, with a slot along the middle, and CD is a second ruler at right angle to AB. It has a peg P fixed in it. A third ruler EF has a slot along the middle which fits the peg P, and has a pointed end at E. We could, anachronistically, think of E being fitted with a pencil, to draw the curve on a piece of paper. Finally P' is a fixed peg on EF in a straight line with the slot, P' can move freely along the slot in AB. If now the ruler EF moves so that the peg P' traverses the total length of the slot in AB, then E will describe a segment of one branch of the conchoid. See Figure 4.21.

Nicomedes used the conchoid to solve the problem of *trisecting an angle* and the problem of *doubling a cube*.

Nicomedes was very proud of his invention, and wrote extensively comparing it to Erathostenes' mechanical instrument for constructing the double mean proportional, thus in particular, for doubling the cube. He argued that such a mechanical procedure was alien to geometry, and in every way inferior to the method of the conchoid!

Indeed, the technique of the conchoid is more general. It provides a device to solve a large class of *verging problems*. In the present case it is the problem of inserting a line segment of fixed length between two given curves, usually lines or circles, so that the fixed line segment when produced will pass through a given point.

Doubling the cube, as well as trisecting any angle, may be reduced to two such verging problems.

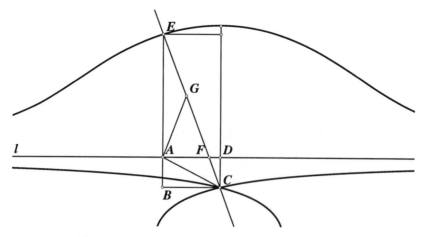

Fig. 4.22. Trisecting any angle using the conchoid.

We shall first show this for the *trisection*-problem, and also show how the conchoid is used to solve it.

The construction is rendered in Figure 4.22, which is pretty much self explanatory. It suffices, of course, to be able to trisect any angle less than a right angle. Such an angle u may be realized as the angle formed by a diagonal and a side in a rectangle $ABCD$, so let $u = \angle ACD$. Now produce DA to the line ℓ, and take C as the point P in the definition of the conchoid. b is taken to be twice the length of the diagonal AC. The corresponding conchoid intersects BA produced in E. The line CE intersects AD in F, and letting G denote the mid point of EF, we find $EG = GF = AG = AC$. Therefore, putting $\angle CEA = v$, we have $\angle AGF = 2v = \angle ACG$, and hence $v + 2v = 3v = \angle ACD = u$.

To double the cube with side a, we set of the line segment AB of length a, and at B erect a line normal to it. We also draw a line forming an angle of $120°$ with AB at B. See Figure 4.23. The normal to AB at B is taken as the line ℓ in the definition of the conchoid, and we take $a = b$ in the construction of the conchoid (note that b has a different meaning in Figure 4.23). We find the point D as shown in the figure, which is such that $\angle DBA = 120°$ and $CD = a$. Put $AC = b$, $BC = c$. Denote $\angle ADB$ by φ. We then have

$$\frac{a}{\sin(30°)} = \frac{c}{\sin(\varphi)}$$

and

$$\frac{a}{\sin(\varphi)} = \frac{a+b}{\sin(120°)}$$

from which we infer that

$$\frac{c}{a} = \sqrt{3}\frac{a}{a+b}$$

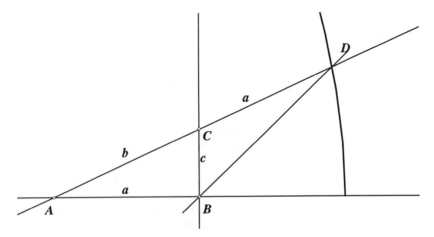

Fig. 4.23. Doubling the cube with side $a = AB$ using the conchoid.

This yields

$$\sqrt{3} = \frac{c(a+b)}{a^2}$$

which when squared implies

$$3a^4 = c^2(a+b)^2 = (b^2 - a^2)(a+b)^2$$

Thus

$$3a^4 = b^4 - a^4 + 2ab^3 - 2a^3b$$

hence

$$2a^3(2a+b) = b^3(2a+b)$$

thus

$$2a^3 = b^3$$

Note that the construction obtained in Figure 4.23 by means of the conchoid, may also be achieved with compass and a straightedge which is capable of moving a distance. To perform the construction, we set off A and B, with the length of AB equal to the side a of our cube. As before we erect the normal n at B and construct another line m forming an angle at B of 120°. Then the distance a is marked by two points C' and D' on the straightedge, which is subsequently slid into a position where C' falls on the normal n and D' falls on m while the straightedge passes through A. This is a typical *Verging Problem*, being solved here by the so called *Insertion Principle*. We may now complete the construction by drawing the line AD, which intersects n at C, such that the length of CD is a.

The construction shown in Figure 4.22 may also be carried out in a similar manner. But this is not all: It suffices to be able to insert, by means of the straightedge, a *single, fixed, distance*. That is to say, construction by a *marked straightedge*. In fact, suppose we wish to insert a line segment of length a into a construction by means of a straightedge. We have a straightedge on which the distance b is marked off. Then scaling the unfinished construction by the ratio $b : a$, the distance b will correspond to a, which is inserted by the marked straightedge, after which the construction is scaled back by the ratio $a : b$. By this, admittedly cumbersome, procedure we may insert any distance by means of the fixed one.

Nicomedes also solved the third classical problem, the *squaring of the circle*, by means of a curve known as the *Quadratrix of Hippias*.

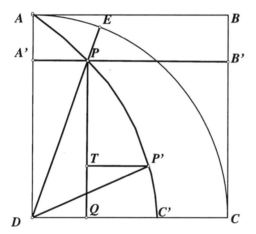

Fig. 4.24. The Quadratrix of Hippias.

Hippias of Elis was born about 460 and died around 400 B.C. He traveled from place to place and earned his living from lecturing on a variety of subjects and providing other services. Plato did not think much of him, but regarded him as boastful and arrogant, with a wide but superficial knowledge of the subjects of his times. He was also known as *Hippias the Sophist*, and here we may find an explanation for a possible prejudice against him on Plato's part. But although covering a wide range of subjects, he *did* specialize in mathematics, and Heath gives a rather favorable assessment of him in [15].

The points P on Hippias' curve are defined as follows, with notation as in Figure 4.24: Construct a square $ABCD$ with side AB. With D as center draw the quarter circular arc AC of radius DA. Let the point E move along this arc at uniform speed, and let at the same time A' move from A to D

also at uniform speed, reaching D at the same time as E reaches C. Let P be the point of intersection between the lines DE and $A'B'$. P then traces out a curve, indicated in Figure 4.24.

We then have

$$\angle EDC : \angle ADC = A'D : AD, \text{ thus } \angle EDC = \frac{A'D}{AD}\frac{\pi}{2}$$

using modern notation.

This curve may also be used to solve the trisection problem. Namely, if we are given an angle $u = \angle EDC$ we may trisect it by trisecting PQ by the point T, drawing the line through T parallel with DC, which intersects the quadratrix in the point P', then $\angle P'DC = \frac{1}{3}u$.

Again, we may find as many points on this curve as we want by subdividing the arc AC in n equal pieces, n being a power of 2, say, and dividing the line segment AD in n equal pieces as well.

The reason why Nicomedes was able to use this curve to square the circle, lies in the observation that, with modern notation,

$$\frac{DC}{DC'} = \frac{\pi}{2}$$

To show this, let $\frac{A'D}{AD} = x$, then $\angle EDC = \frac{\pi}{2}x$, thus $\frac{DQ}{DC} = \frac{DA'}{DC}\frac{DQ}{DA'} = x\cot(\frac{\pi}{2}x)$. So

$$\frac{DQ}{DC} = \frac{x}{\tan(\frac{\pi}{2}x)}$$

i.e.

$$\frac{DC}{DQ} = \frac{\tan(\frac{\pi}{2}x)}{x}$$

$Q = C'$ corresponds to $x = 0$, thus we need to use *l'Hospitale's Rule*, which yields

$$\frac{DC}{DC'} = \frac{\frac{\pi}{2}}{1} = \frac{\pi}{2}$$

proving the claim.

To complete the squaring of the circle, we construct the rectangle and the diagonal in Figure 4.25. We have

$$\frac{DC}{DC'} = \frac{EF}{DC'} = \frac{AE}{AD} = \frac{\pi}{2}$$

thus

$$2AE \cdot AD = \pi AD^2$$

and hence the circle of the given radius AD has the same area as the rectangle of sides $2AE$ and AD.

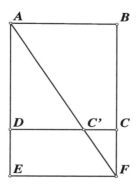

Fig. 4.25. The squaring of the circle according to Nicomedes.

4.7 Apollonius of Perga and the Conic Sections

Apollonius of Perga is the last of the great Greek geometers. He was born in Perga about 262 and died in Alexandria about 190 B.C. As a young man he came to Alexandria, where he studied under Euclid's successors. He also stayed at *Pergamon*. Pergamon and Alexandria were the most important scientific and cultural centers in the Hellenistic world.

Apollonius' most important work is *Conica*, on Conic Sections. This work is in eight books, of which seven are preserved either in the Greek original or in Arabic translation.

Before Apollonius Euclid had written a book on conics, which is lost. Also Archimedes and others had studied them, in particular in the work which took place on the *classical problems.*

The reader who is not familiar with the conic sections, that is to say curves in the plane of degree 2, may want to move to Section 12.5 now for a review from a modern point of view.

But the full fledged algebraic description found there came much later. The Greek geometers understood them as the curves of intersection between a *cone* and a *plane*. Prior to Apollonius conic sections were understood as the intersections between *three different kinds of cones* and a plane at right angle to one of the *generators* of the cone. The cones were right circular cones, of the three kinds *right-angled, obtuse-angled* or *acute-angled.* But both Euclid and Archimedes were well aware that conic sections could be produced in other ways.

Thus an *ellipse* would be produced by an acute-angled cone, with an angle *falling short* of a right angle, as the word *ellipse* reflects. A parabola corresponds to a cone with the top angle *equal to* a right angle, while a *hyperbola* is produced by a cone with top angle *greater* than a right angle. See Figure 4.27. There is a remark to be made at this point, however. Namely,

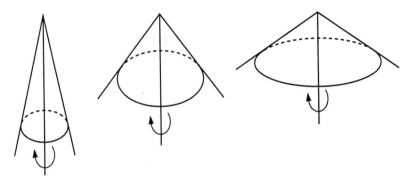

Fig. 4.26. Three different kinds of cones. To the left the top angle is less than 90°, in the middle it is equal to 90° and to the right it is greater than 90°.

we regard a circle as a special ellipse. Thus it would be produced by an acute angled cone as described above. But the angle would have to be equal to zero, in other words the cone would be a cylinder. And the Greek geometers did indeed consider this case as well.

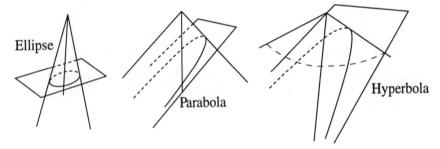

Fig. 4.27. Conic sections prior to Apollonius, cut out in a fixed plane by a varying cone.

Apollonius defines a cone as we do today, by rotating a line. In this way we get the *double cone*, and the hyperbola acquires its two branches. See Figure 4.28.

Also, it was Apollonius who introduced the names ellipse, parabola and hyperbola. These words had been in use in a slightly different way with the Pythagoreans, but the point was the same: The word *ellipse* indicates that *something is left out, it is too little.* The word *parabola* indicates to *liken*, set side by side. The plane cutting the cone runs parallel to the other branch of the cone. And finally, the word *hyperbola* stands for *too much, exaggeration.*

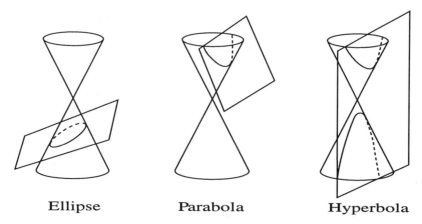

Ellipse   Parabola   Hyperbola

Fig. 4.28. Conic Sections with a varying plane and a fixed cone.

Apollonius' theory of conic sections is a high point of Greek geometry. This theory was essential when *Isaac Newton* very much later deduced the laws of gravity.

Apollonius proves a result known as *The Circle of Apollonius.* In modern notation it may be formulated as follows:

> **The Circle of Apollonius.** Let A and B be fixed points and let k be a constant. Then the set of all points P such that $AP : BP = k$ is either a circle, when $k \neq 1$, or a straight line, when $k = 1$.

This is easy to show with modern algebraic techniques. See Figure 4.29. There we have chosen a coordinate system such that $A = (-1, 0)$ and $B = (1, 0)$. We then have that $AP = kBP$, and by the Pythagorean Theorem we find that

$$AP^2 = (1 + x)^2 + y^2$$

and

$$BP^2 = (1 - x)^2 + y^2$$

Since now $AP^2 = k^2 BP^2$, we get

$$(1 + x)^2 + y^2 = k^2((1 - x)^2 + y^2)$$

The remaining part of the proof is left to the reader.

Another result ascribed to Apollonius is this:

> **Generalized Pythagorean Theorem.** Let ABCD be a parallelogram with sides AB = CD = a, BC = DA = b, and with the diagonals AC = m, BC = n. Then

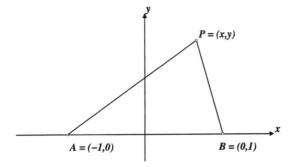

Fig. 4.29. Proof that we get a circle.

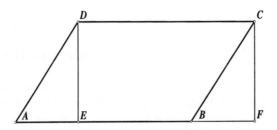

Fig. 4.30. Generalized Pythagorean Theorem.

$$a^2 + b^2 = \frac{1}{2}(m^2 + n^2)$$

We give a modern proof: In Figure 4.30 we denote AE by x. Let DE $= y$. Then BF $= x$ and FC $= y$, since $\triangle AED$ is congruent with $\triangle BFC$. By the Pythagorean Theorem we have

$$b^2 = x^2 + y^2$$

$$m^2 = (a + x)^2 + y^2 = a^2 + 2ax + x^2 + y^2$$

$$n^2 = (a - x)^2 + y^2 = a^2 - 2ax + x^2 + y^2$$

Adding the two last equalities and substituting $b^2 = x^2 + y^2$, we get the claim.

Apollonius worked with an algebraic notation which at least makes him a forerunner for the algebraization of geometry which *Descartes*[7] is credited with.

---

[7] René Descartes, 1596–1650, was an important French philosopher and mathe-matician. He is given credit for having initiated the introduction of algebra into geometry. The name *Cartesian* coordinate system is after him.

Finally Apollonius posed a problem, which in its most general form is very much on the agenda of modern algebraic geometry. Loosely and generally formulated it runs like this:

Apollonius' Problem on Tangency. Given a set of geometric objects, like points, lines or conic sections. Find the objects tangent to the given ones. (Tangency for a point is understood to mean that the curve passes through the point.)

In Apollonius' case there were *three* such fixed geometric objects. For details on the Problem of Apollonius we refer to [15], volume II, page 181, or to [14] page 346.

Apollonius developed geometric algebra a long way towards the complete algebraization of geometry, which came in full much later through the work of Descartes and Fermat.

4.8 Caesar and the End of the Republic in Rome

We have already encountered *Cicero*, in Section 4.4. Cicero represents the finest of the Roman intellectual tradition. Educated in Rome and Athens, he became recognized as the greatest orator of his times. As a politician his words went unheeded, but in his writings his ideas have been preserved up to the 21st. Century. Cicero is credited for having contributed substantially to the continuation of Greek thinking and Greek science through Rome to our days.

He had been skeptical to the young and ambitious *Julius Caesar*. Corrupt as it had become, the political system in Rome was still democratic. Caesar started out as a minor political figure in Rome, plagued by political strife and private scandals. He had been made questor to the province of Iberia (roughly speaking Spain) in 68 B.C., serving under the praetor *Vestus* who he honored ever after. Indeed, when he himself was promoted to praetor of Iberia, he appointed Vestus' son to his old position as questor. Returning to Rome, his political maneuvering led to the famous comment by *Lutatius Catulus*: *"Caesar no longer attempts to subvert the constitution, he now is starting a frontal attack on it."* Caesar rose in importance when he succeeded in being elected to the office of High Priest, *Pontifex Maximus*, but this and other efforts required considerable sums of money, which put him in deep debt. His personal finances were brought in order, however, thanks to a term as *propretor* in Iberia, after which he returned to Rome again, to seek an even more elevated office. He now formed an informal alliance, an agreement known as the *First Triumvirate*, with the successful general *Pompey*, called *Magnus* (The Great), and *Crassus*, called *Dives* (The Wealthy), who had been consul jointly with Pompey in 70 B.C. Crassus had become very wealthy from buying up and speculating with land owned by condemned men, and in 71 B.C. he had together with Pompey brutally suppressed the slave revolt led by Spartacus. Caesar's alliance, later fortified by Pompey's marriage to his daughter *Julia*, got him the highest office in the Roman Republic, that of *Consul*. He was to have served jointly with a certain *Lucius Calpurnius Bibulus*. Although a political opponent of Caesar, Bibulus was in no position to prevail. When Caesar together with his powerful allies pushed through controversial and far reaching land reforms, thereby overstepping his constitutional powers, Bibulus could not prevent it. To save his life while not sharing responsibility for his partner's actions, he finally locked himself up in his house, where he remained for the rest of his term in office.

After his one year term as consul, Caesar received the province of *Gaul*, Gallia, essentially consisting of all of western and northern Europe including Iberia and Britain, as proconsul for five years. This was pushed through by the forceful efforts of his new son in law Pompey, who had filled the Forum with soldiers. Guardians of the constitution like Cicero and *Cato*, later called *Uticenis* (Who Died in Utica), protested but were powerless. Cato was led away to prison on Caesar's orders, but when Caesar, an able political animal, saw the negative reactions to this outrage both from the nobility and the people, he had him secretly released. Cato took up the resistance against Caesar, and when alliances later shifted, he found himself allied with Pompey (Pompey's wife had died in childbirth).

But we are getting ahead of the story. Caesar had his term prolonged by another five years, and spent almost ten years in Gaul, 58 – 49 B.C. This is where he developed into an extremely successful general, paying his soldiers well. He subdued the "barbarians" who had grown increasingly active against the Romans. The rivalry between Caesar and Pompey had been kept in check by their mutual fear of the third man, Crassus. But Crassus finally met his fate in Persia. He had become consul again with Pompey in 55 B.C., after which he got the province of Syria as proconsul, or governor, as the custom was. There he was defeated by the Parthians (Persians) and slain in 53 B.C.

Anarchy and confusion grew worse in Rome, and Pompey was viewed as the only man who could save the republic. The Senate and influential citizen appealed to him not to leave Italy for his provinces which he had been assigned as proconsul, but stay in Rome and preserve the order. In 52 B.C. he was elected consul, some say sole consul, others that he was elected together with *Scipio*, Pompey's new father in law. Scipio belonged to a distinguished family of Roman generals, some of them had played a key role in the destruction of Carthage. Scipio was a staunch opponent of Caesar.

Pompey also had his provincial command extended another five years. Now even Cicero strongly espoused the idea of making Pompey the strong man, the savior of society. But Pompey thus became a close ally of the Senate, and there an active group of senators were urging him to head a final showdown with Caesar. Caesar's command would expire in March 49 B.C., but he would not be replaced until 48 B.C. He wanted to run for a second term as consul without having to come in person to Rome. But his opponents were determined to bring his command to en end, and force him to disband his troops and run for consul as a private citizen without an army to support him. Their apprehension would seem quite justified.

After some haggling with the Senate however, Caesar made up his mind. He uttered his proverbial *"The dice is cast"*, crossed the Rubicon and marched towards Rome. Pompey panicked, and together with the Senate and a large number of nobility he fled into Greece. This was the beginning of a full fledged civil war, which lasted off and on until 31 B.C.

The final battle with Pompey was fought at Pharsalia in Thessaly, in 48 B.C. Here Pompey was completely defeated, and fled to Egypt. Caesar followed him, but arrived too late. In Egypt the political situation was murky. When King Ptolemy Auletes died, he had left his kingdom to his daughter *Cleopatra*, only 17 years old, who was to have ruled jointly with her even younger brother also named Ptolemy, whose wife she was to become according to the customs. A few years later their guardians drove Cleopatra away, and ruled on behalf of the young Ptolemy XIII. The ruling clique consisted of *Pothinius the Eunuch,* who was the head of the royal guard, and *Theodotus of Chios,* a rhetoric master who was the young kings tutor, and finally *Achillas the Egyptian,* who was the commander in chief of the Egyptian army, and technically the regent until the king would come of age. This was the situation as Pompey arrived, defenseless on a single ship, a Seleucian ship not even his own, with his young wife, Scipio's daughter, and their infant son to seek refuge from his *former* father in law. The ruling junta deliberated on how to treat the fallen Roman general: Granting him asylum would certainly infuriate Caesar, while driving him away would create a potentially dangerous enemy, should the fortunes of war change. It was Theodotus of Chios who gave the deciding argument. Rhetorically he declared: *"Receive Caesar's enemy Pompey, but at once do away with him! For in doing so we ingratiate ourselves with the former, and have no reason to fear the latter. A dead man cannot bite!"*

When Caesar arrived in Alexandria after the murder of Pompey, Theodotus wanted to ingratiate himself by presenting Caesar with Pompey's severed head. The effort backfired. Caesar turned away in horror, as from a murderer. Taking Pompey's signet ring, he wept bitterly. Pompey had been too proud to accept Caesar's mercy. Caesar now made an effort to find those of Pompey's men who were in hiding, wandered about or languished in the dungeons of Egypt. He later said that the greatest pleasure his victory had given him, had been to save and offer his friendship to those fellow citizens who had stood against him.

In Alexandria Caesar and his men had received a somewhat disappointing reception. The indefatigable Pothinius the Eunuch was secretly scheming to destroy Caesar, the Roman soldiers were mistreated with inadequate and low quality food, and Pothinius repeatedly offered his unsolicited advice that Caesar should now leave Egypt to attend to his other important matters, like dealing with Cato and Scipio who were still at large on the African coast with a sizable army. But Caesar stayed on, declaring that he did not need Egyptians as his councillors, and secretly sent for Cleopatra.

Cleopatra must have been a remarkable person. Moralistic historians have judged her harshly, as did many Romans of her own time. But she used all means at her disposal to preserve a measure of independence for her country, in the face of Roman imperialism and expansion. Caesar almost immediately fell in love with her, and became a loyal ally as well as her lover. Cleopatra named her son with Caesar *Caesarion*, he was born soon after Caesar had left Egypt. But before that he had helped her defeat her brother, Pothinius having been killed when he was caught plotting to assassinate Caesar. Achillas, who was Pothinius' co-conspirator had escaped to the army, which he commanded against Caesar and Cleopatra.

During this war in Egypt, a *"minor incident"* happened which the soldiers and the generals paid little attention to. Caesar at one point was in the danger of having his communication at sea cut off by the enemy. He averted the danger by setting fire to his ships, the fire spread in the city and the great library was destroyed.

Cato and Scipio had formed an alliance with the king of Numidia, Juba I. But when Caesar attacked, they were defeated at Thapsus in North Africa in 46 B.C. Cato then took refuge at Utica a little north of Carthage. When all hope was gone, he fell upon his sword and killed himself, since *"he would only live in a free state, not under the rule of one man"*. Caesar, who arrived too late, is said to have bitterly uttered the words, *"Cato, I must grudge you your death, as you grudged me the honor of saving your life"*.

Upon his return to Rome Caesar led a triumph for the victory over king Juba, whose infant son was carried in the procession. The little African boy who was brought to Rome in this manner, later became king Juba II. He also became one of the most learned historians of his time.

Caesar was made consul a forth time, and went into Spain where he defeated Pompey's sons. Even if they were quite young, they commanded a large army and fought well. After the battle Caesar remarked that he had fought for victory many times before, but this was the first time that he had fought for his life. The triumph Caesar celebrated offended the Romans immensely. For he had not defeated a foreign enemy, he had instead destroyed the children and family of one of the greatest men of Rome. Nevertheless the Romans elected him *dictator for life*.

Although he had made an effort to attain reconciliation with his opponents from the civil war, and spared many lives, the bitterness must have been deeply rooted against him. When he was murdered on March 15, 44 B.C., the plot was led by one of those whose lives he had spared, *Marcus Brutus*.

Brutus and his co-leader *Cassius* had thought that the republic would be restored, but the struggle which followed turned out to be about who should succeed Caesar as monarch, as the *new Caesar*. In spite of the threatening attitudes of Caesar's former associate *Anthony*, the elderly Cicero worked to enlist the Senate, the people as well as the provincial governors in an effort to restore the republic. His eloquence was, as always, persuasive, his logic convincing. But the time for reason was past, civil war broke out again. Antony and his associate *Lepidus* got together with a young relative and heir of Caesar by the name *Octavian*, called *the young Caesar*, and formed the *Second Triumvirate* in 43 B.C. It was given a legal cover as a *commission for reorganization of the commonwealth*. Appointed for five years, reappointed in 37 B.C. for another five years, the commission pursued a harsh policy against all opposition. One of the first victims was Cicero, who was driven out of Italy at the urging of Antony, and finally murdered. How different this man had been from Caesar and the other generals and warlords: Cicero tried to persuade with words, rather than to compel by violence. And he had understood how much Rome stood to loose by abandoning democracy.

Brutus and Cassius were defeated at Philippi the year after and killed themselves, and all hope seemed to be out for the republican party. The alliance between Antony and Octavian was reinforced by the marriage of Antony to Octavian's sister *Octavia*, and Anthony took command of the eastern part of the Empire while Octavian established his control in the West. Lepidus was given North Africa and thus sidelined.

At this point we have to introduce *Cleopatra* into the narrative once more. Caesar had brought her to Rome, something the Romans strongly disapproved of. She had married a younger brother, remained the mistress of Caesar, and according to some, murdered her husband-brother by poison. After Caesar had been killed, she left and went back to Egypt.

Antony now took up residence in Alexandria. Cleopatra's personality soon won her a total dominance over Anthony, at least this is how Plutarchus tells it in [27]. Anthony's liason with Cleopatra eventually resulted in three children. During this time the library of Alexandria was resupplied with books through an expedition to Pergamon, led by Antony, where the library of Pergamon was raided. At least this is what some sources claim. Unfortunately the library was now in part kept at the *Temple of Serapis*. This had dire consequences later, when zealous Christians decided to burn all pagan literature in Alexandria.

Thus it is fair to say that Caesar's overthrow of the Republic in Rome not only set in motion the events leading to the decline and fall of the Roman Empire, but also started the process which would eventually lead to the end of Alexandria's Great Library and to the end of the classic civilization in Europe, both still half a millennium into the future.

In Alexandria Cleopatra and Antony lived in profuse and wanton luxury, calling themselves *Isis and Osiris*, claiming to be divinities. Octavian did not like Antony's lifestyle in Alexandria, understandably if for no other reason considering the treatment of his own sister, and finally declared war on

Cleopatra. In the sea fight at Actium Cleopatra and Antony were defeated, and in the end they both committed suicide. This happened in the year 31 B.C., and marks the end of the civil war. The Eastern Provinces surrendered in 29 B.C.

This marks the definite end of the Republic, and the beginning of *the Empire*. By general consent Octavian was called upon to be the ruler, and in 27 B.C. he was endowed with the name of honor *Augustus*, meaning *The Just One*. Formally the republic was restored, but it was only in form. The ideas laid out by Cicero in *De Republica* were apparently realized: A constitutional president of a free people. But in reality Octavian now had become the *Emperor Augustus*.

4.9 Heron of Alexandria

Heron lived and worked in Alexandria. He was born about 10 A.D. and died around year 75. These dates were confirmed by Neugebauer, who found that Heron refers to a recent eclipse in one of his books, which he could identify as having occured in Alexandria at 11 p.m. on March 13, in the year 62 A.D.

According to this he lived during a difficult period, at least life was difficult in Rome and probably in the rest of Italy as well. How far the effects of mismanagement under Gaius (Caligula), Claudius or Nero extended out into the provinces and effected life in Alexandria, may be another matter. At any rate Heron was born during the last years of the reign of Emperor Augustus, and died a few years after *Vespasian* had restored the order.

It should be added that especially the reign of Claudius, 41 – 54 A.D., may be cast in an unfair light by the comment above. Under his rule the Claudian aqueduct, as well as the harbor of Ostia were completed. Great engineering enterprises, which happened in the middle of Heron's career.

Heron taught at the Alexandrian Academy, and his preserved books are most probably notes for his lectures, either written by himself or by some of his students. Heron must, without question, have been a brilliant teacher as well as an illustrious engineer and applied mathematician. His lecture notes consist to a very large degree of worked examples, such as his explanation in *Dioptra* of how the Samians used geometry to dig their tunnel through the mountain Castro about 600 years before his time, see Section 3.2. He also explains the construction of mechanical instruments, machines and gadgets intended as toys or for amusement. Also included are war-machines, a wind-organ as well as a steam powered engine working on the same principle as a *jet engine*! It has occasioned some comments that Heron would include the construction of "mere toys for children" in his lectures. Perhaps this was intended to enliven the exposition of otherwise dull principles of mechanics and physics.

From the point of view of geometry, the following mirror-constructions are of interest. They combine nice geometry with some real fun. We quote from [15]:

- To construct a right-handed mirror, i.e. a mirror which makes the right side right and the left side left instead of the opposite,
- to construct the mirror called the *polytheron*, "with many images",
- to construct a mirror inside the window of a house, so that you can see in it, while you are inside the house, everything that passes in the street,
- to arrange mirrors in a given place so that a person who approaches cannot actually see himself or anyone else but can see any image desired, a *"ghost seer"*.

Heron's lecture notes proved so useful, that they were copied and recopied, and used for centuries in Byzantine, Roman and Arab science. As Neugebauer demonstrates definitively in [25], Heron gives expositions of mathematics from the Old Babylonian epoch. He also evidently builds on the mathematics and geometry of Archimedes. Therefore the legacy of Heron is an important segment in the chain tying us, via the Arabs, the Byzantine scholars and the Greek, to the ancient wisdom of the Babylonians and the Sumerians.

The formula for the area of a triangle, known as *Heron's Formula*, very probably comes from Archimedes. It says that if a, b and c are the sides and s is half the circumference, then the area is

$$A = \sqrt{s(s-a)(s-b)(s-c)}$$

A more general formula asserts that if a, b, c and d are the sides of a cyclic quadrilateral, i.e. a quadrilateral which may be circumscribed a circle and s is half the circumference, then the area is given by

$$A = \sqrt{(s-a)(s-b)(s-c)(s-d)}$$

This formula is used by the Indian mathematician *Brahmagupta*, but he does not seem to mention the important assumption that the quadrilateral be cyclic. Thus the formula may well come from another source. By the way, there *is* another formula, valid for any quadrilateral. Its derivation is given in [21].

In our days, when mathematics is so frequently presented to students as a dry and barren symbolic song and dance, it is tempting to conclude this section about Heron by his recipe for finding approximate cube roots. No symbolics, just numbers. We quote from [15]. To find the "cube side" of 100:

Take the nearest cube numbers to 100 both above and below, these are 125 and 64. Then

$$125 - 100 = 25,$$

and

$$100 - 64 = 36.$$

Multiply 5 into 36, this gives 180. Add 100, making 280. Divide 180 by 280, this gives $\frac{9}{14}$. Add this to the side of the smaller cube: this gives $4\frac{9}{14}$. This is as nearly as possible the cube root of 100 units.

Very presumably, Heron had a number of other cube-root examples as well, but this appears to be the only one to have come down to us. Several historians of mathematics have been involved in trying to reconstruct what Heron's formula might have looked like, and from a careful consideration of what one gets from elementary methods, the following amazing formula emerges: Assume that $a^3 < A < (a+1)^3$, and put $d_1 = A - a^3$, $d_2 = (a+1)^3 - A$. Then

$$\sqrt[3]{A} \approx a + \frac{(a+1)d_1}{(a+1)d_1 + ad_2}$$

But there is a simpler interpretation, which however is less elementary to deduce, namely

$$\sqrt[3]{A} \approx a + \frac{d_1\sqrt{d_2}}{A + d_1\sqrt{d_2}}$$

For $A = 100$ this yields the same number as the first formula, of course, a good approximation to $\sqrt[3]{100}$. But for $A = 90$, the first formula still yields a good approximation, but the second one is quite far off the mark.

Finally a more general formula is conjectured by some investigators, the implication being that Heron, or the person taking notes or later copying them, did not quite understand what was going on. For details we refer to [37].

But in teaching young engineers in Alexandria or elsewhere, Heron's examples are undoubtedly more effective than the formulae of his commentators.

4.10 Menelaus of Alexandria

Menelaus of Alexandria was born around 70 and died 130 A.D. Thus his birth coincides with the end of a problematic period in Rome, with emperors like Gaius (Caligula), Claudius and Nero, but others who were able administrators. After Nero's death in 68 A.D. there had followed a period of wars between rival emperors, this was ended by Vespasian, who assumed power in 69 A.D., and marked the beginning of a good and stable period which lasted until 180 A.D., when the capricious and depraved *Commodus* became emperor.

Menelaus spent some time in Rome, where he did astronomical work, and he is mentioned in one of Plutarchus' books. Menelaus wrote extensively, but only one of his works is extant, namely the *Sphaericae*. Here he studies geometry on a sphere, in particular he establishes the properties of spherical triangles in the same way as Euclid treats plane triangles. In this sense one might say that Menelaus' work is a precursor for non-Euclidian geometry.

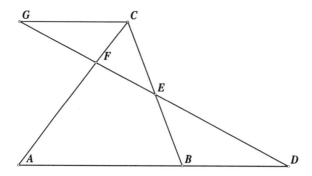

Fig. 4.31. The theorem of Menelaus.

Menelaus' Theorem says the following, in modern notation:

Theorem 2 (Menelaus). *If a triangle ABC be cut by a line ℓ, which cuts the side (possibly produced) AB in the point D, and similarly BC in the point E and CA in F, then the following relation holds*

$$\frac{AD}{DB}\frac{BE}{EC}\frac{CF}{FA} = 1$$

where AD, DB, etc., are the lengths of the line segments AD, DB, etc., all considered as positive numbers.

Proof. Let G be the point on ℓ such that $CG//AB$. The $\triangle FCG \sim \triangle FAD$, from which follows

$$\frac{FC}{CG} = \frac{FA}{AD}$$

and $\triangle ECG \sim \triangle EBD$, from which follows

$$\frac{EC}{CG} = \frac{BE}{BD}$$

The former yields

$$\frac{FC}{CG}\frac{AD}{FA} = 1$$

while the latter yields

$$\frac{CG}{EC}\frac{BE}{BD} = 1$$

and when multiplied these relations yield the claim.   □

Menelaus also proved statements for *spherical triangles*, analogous to theorems for plane ones, including the theorem given above. For details on Menelaus' theorem for spherical triangles as well as related material, we refer to [15].

4.11 Claudius Ptolemy

Claudius Ptolemy was born about the year 85 A.D., probably in Alexandria but possibly in Hermiou in Upper Egypt. He died around 165, in Alexandria. According to his name he would be of an Egyptian family with Greek background, who had been made Roman citizen. This was a usual practice at this time in the history of the Roman Empire, for provincials who had rendered valuable services and thus attained an elevated status.

He lived during a stable period of internal tranquility and good government, and his main work appears to have been done during the reign of the Emperor *Titus Antonius Pius*, which lasted from 138 A.D. to 161.

Ptolemy's main work represents the definitive state of Greek astronomy. Consisting of 13 books, it bore the title *Mathematical Collection*. Later Pappus wrote an introduction to this work, which came to be called the *Little Astronomy*, Ptolemy's original being referred to as the *Great Collection*. When still later translated into the Arabic, the title became something like *"The Greatest"*, or *Al-majisti*, which in turn ended up as *Almagest*. Thus Ptolemy's Mathematical Collection acquired the name under which it was handed down to posterity.

Together with Euclid's Elements the Almagest of Ptolemy is the scientific text longest in use, up to the Renaissance. The idea of the earth-centered universe on which it is based, made necessary intricate mathematical explanations.

Ptolemy develops extensive trigonometric methods, and in particular introduces the *chord-function*, which is essentially equivalent to our trigonometric functions $\sin, \cos, \tan$ etc. The chord function of the angle v may be defined, anachronistically in modern notation, as

$$\mathrm{crd}(v) = 120\sin(\frac{v}{2})$$

In Neugebauer's very interesting book [25] the computations of this trigonometric function is described. One such table carries the title *Table of straight lines in the circle*, it is a table of chords. The computations is to the base 60, and uses a symbol for the number *zero*, used in a fully modern way. Neugebauer asserts the following on page 13, which corrects a very common misconception about Greek mathematics:

> *"According to the prevailing doctrine that Greek mathematics is essentially geometry, the historians of mathematics have badly neglected the enormous amount of numerical computations which are readily accessible in works like Ptolemy's "Almagest" or Theon's "Handy Tables". But long before these classics were written, Greek astronomical papyri were covered with computations. While Ptolemy or Theon are today only preserved in Byzantine manuscripts, we do have papyri from the Ptolemaic period* [the last centuries B.C.] *onwards. In these papyri we can find, e.g., the zero sign as it was actually written.*

Ptolemy starts out in Book I, as a preliminary to the Table of Chords, by dividing the circle into 360 equal parts, or *degrees*, and the diameter into 120 equal parts. It then follows that the chord subtending an arc of $v°$ will have length $120\sin(\frac{v}{2})$, and that $\mathrm{crd}(180° - v) = 120\cos(\frac{v}{2})$.

We now follow the explanation provided by Heath in [15].

First, to find the chords subtending arcs of 72° and 36°, i.e. the sides of the regular pentagon and 10-gon (decagon), Propositions 9 and 10 of Book XIII of Euclid's Elements are used. XIII.9 is equivalent to the formula for s_{10}, the side of a regular decagon inscribed in a circle of radius r,

$$s_{10} = \frac{r}{2}(-1 + \sqrt{5})$$

while XIII.10 is equivalent to the relation for the side of the inscribed regular pentagon

$$s_5^2 = r^2 + s_{10}^2 \text{ or } s_5 = \frac{r}{2}\sqrt{10 - 2\sqrt{5}}$$

We show these relations in Section 16.7. Thus $\mathrm{crd}(72°)$ and $\mathrm{crd}(36°)$ may be computed, the diameter being 120 one gets

$$\mathrm{crd}(72°) = 30\sqrt{10 - 2\sqrt{5}} \text{ and } \mathrm{crd}(36°) = 30(\sqrt{5} - 1)$$

Ptolemy extracts the square root computing to the base 60, by a method later explained by Theon of Alexandria. The answer is rendered in a mixed notation, with the fractional part to the base 60, as $\mathrm{crd}(72°) = 70^p32'3''$, which checks with our calculator which gives the answer as ≈ 70.534236 while $70 + \frac{32}{60} + \frac{3}{3600} \approx 70.5341666$.

Ptolemy utilized the immediate observation that the chords subtending v and $180° - v$ form a right triangle with the respective chords containing the right angle and the diameter as the hypothenuse, thus

$$\mathrm{crd}(v)^2 + \mathrm{crd}(180° - v)^2 = 120^2$$

equivalent to the relation $\sin^2 v + \cos^2 v = 1$.

Now $\operatorname{crd}(60°) = 120$, and $\operatorname{crd}(90°) = \sqrt{2 \cdot 60^2} \approx 84^p51'10''$. To proceed, it is now necessary to have formulas expressing $\operatorname{crd}(\alpha \pm \beta)$ in terms of $\operatorname{crd}(\alpha)$ and $\operatorname{crd}(\beta)$, they are equivalent to the familiar formulas

$$\cos(\alpha + \beta) = \cos(\alpha)\cos(\beta) - \sin(\alpha)\sin(\beta)$$
$$\sin(\alpha - \beta) = \sin(\alpha)\cos(\beta) - \cos(\alpha)\sin(\beta)$$

Of the two formulas, we shall start by deriving the latter, and this is where the famous *Ptolemy's Theorem* enters the scene. The theorem asserts the following:

Theorem 3 (Ptolemy's Theorem). *In a cyclic quadrilateral the sum of the products of opposite sides is equal to the product of the diagonals.*

Proof. We refer to Figure 4.32, where the quadrilateral is $ABCD$, and the claim is that

$$AB \cdot DC + BC \cdot DA = AC \cdot BD$$

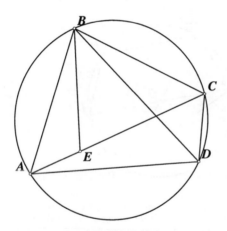

Fig. 4.32. The cyclic quadrilateral is $ABCD$. On AC the point E is marked so that $\angle ABE = \angle DBC$.

On AC the point E is marked so that $\angle ABE = \angle DBC$. It then follows that $\triangle EAB \sim \triangle CDB$ and that $\triangle DAB \sim \triangle CEB$. The former relation yields

$$\frac{AB}{AE} = \frac{BD}{DC} \text{ hence } AB \cdot DC = AE \cdot BD$$

while the latter yields

$$\frac{BC}{CE} = \frac{BD}{DA} \quad \text{hence } BC \cdot DA = CE \cdot BD$$

and adding the two we get

$$AB \cdot DC + BC \cdot DA = AE \cdot BD + CE \cdot BD = AC \cdot BD$$

as claimed. □

We apply the theorem to the cyclic quadrilateral where AD is a *diameter*. Let α be the arc AC, and β be the arc AB. Thus $\mathrm{crd}(\alpha) = AC$, while $\mathrm{crd}(\beta) = AB$. Then the formula of the theorem yields

$$\mathrm{crd}(\alpha - \beta)\mathrm{crd}(180°) + \mathrm{crd}(\beta)\mathrm{crd}(180° - \alpha) = \mathrm{crd}(\alpha)\mathrm{crd}(180° - \beta)$$

or

$$\mathrm{crd}(\alpha - \beta)\mathrm{crd}(180°) = \mathrm{crd}(\alpha)\mathrm{crd}(180° - \beta) - \mathrm{crd}(180° - \alpha)\mathrm{crd}(\beta)$$

To derive a formula for $\mathrm{crd}(\alpha + \beta)$, we use the same cyclic quadrilateral, i.e., we let AD be a diameter. The arc AB is α, but now the arc BC is β. We draw the diameter through B, and get the point E as indicated in Figure 4.33.

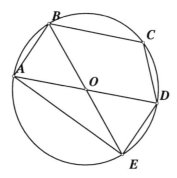

Fig. 4.33. The cyclic quadrilateral $ABCD$ where AD is a diameter and the diameter through B is drawn, yielding the second quadrilateral $BCDE$.

Here $\mathrm{crd}(\alpha + \beta)$ is the length of the chord AC, which is known once we know $CD = \mathrm{crd}(180° - (\alpha + \beta))$. AB as well as BC are known. As $AB = DE$, the latter is known, as is BD, $\triangle ABD$ being right, and $AD = 120$. Thus applying Ptolemy's Theorem to the cyclic quadrilateral $BCDE$ we find the one unknown entity CD, which solves our problem: We get

$$CD \cdot BE + BC \cdot DE = BD \cdot EC$$

thus

$$\mathrm{crd}(180° - (\alpha + \beta))\mathrm{crd}(180°)$$

$$= \mathrm{crd}(180° - \alpha)\mathrm{crd}(180° - \beta) - \mathrm{crd}(\alpha)\mathrm{crd}(\beta)$$

Taking $\alpha = \beta$ this last formula yields

$$\mathrm{crd}(\alpha)^2 = \frac{1}{2}\mathrm{crd}(180°)(\mathrm{crd}(180°) - \mathrm{crd}(180° - 2\alpha))$$

equivalent to the familiar $\sin^2 \frac{v}{2} = \frac{1}{2}(1 - \cos v)$.

Thus Ptolemy obtains $\mathrm{crd}(12°) = \mathrm{crd}(72° - 60°)$, and from the last formula, starting with $\mathrm{crd}(36°)$ he computes $\mathrm{crd}(18°)$ and $\mathrm{crd}(9°)$. Then he also captures $\mathrm{crd}(6°)$ as well as $\mathrm{crd}(3°)$.

To find $\mathrm{crd}(1°)$ is more tricky: Ptolemy readily finds $\mathrm{crd}(1\frac{1}{2}°)$ as well as $\mathrm{crd}(\frac{3}{4}°)$. He then determines the value $\mathrm{crd}(1°) \approx 1^p2'15''$ by an ingenious method of interpolation, using the fact which in modern notation says that the function $f(v) = \frac{\sin(v)}{v}$ is monotonously decreasing in the interval $< 0, \frac{\pi}{2} >$.

The method was not new, in fact it is due to an earlier great Greek mathematician, namely *Aristarchus of Samos*, 310 – 230 B.C. Heath writes about him as follows:

> Historians of mathematics have, as a rule, given too little attention to Aristarchus of Samos. The reason is no doubt that he was an astronomer, and therefore it might be supposed that his work would have no sufficient interest for the mathematician. The Greeks knew better; they called him "Aristarchus the mathematician".

Aristarchus was a precursor for *Copernicus*, in that he was the first to propose a sun-centered universe. He is also remembered for his attempt to determine the sizes and distances of the sun and moon. It is ironical, perhaps, that Ptolemy who so carefully had studied Aristarchus' mathematics, did not know or make use of the work Aristarchus had left behind concerning a sun-centered universe! And that he used Aristarchus' mathematics for the computations when he compiled his tables of chords, partly intended to explain the doctrine of an earth-based universe.

For details we refer to [15].

4.12 Pappus of Alexandria

As is frequently the case with the ancient mathematicians, there has been some disagreement on his dates. Thus in [15] it is asserted that Pappus lived at the end of the third century AD. However, it can be deduced from Pappus' commentary on the Almagest that he observed an eclipse of the sun in

Alexandria October 18, 320. For this and other reason one now fixes his date of birth to about 290 A.D., and his year of death to 350.

This puts his birth to the first years of the reign of *Diocletian*, who strengthened and reformed the government of the Roman Empire after a dismal century of civil war and disorder. But already when Pappus was 15 years old, Diocletian and his co-emperor *Maximian* abdicated, after which there followed 18 years of fighting between rival emperors. We give more details on this period in Section 4.15. During Pappus' life *Christianity* became the state religion.

Pappus was born in Alexandria, where he lived all his life. Proclus writes that he headed the school in Alexandria, which certainly stands to reason.

Pappus' major work in geometry is entitled *Synagoge* or the *Mathematical Collection*. It is a collection consisting of eight books, probably written around 340. In [15] Heath describes the Mathematical Collection as follows:

> Obviously written with the object of reviving the classical Greek geometry, it covers practically the whole field. It is, however, a handbook or guide to Greek geometry rather than an encyclopaedia; it was intended, that is, to be read with the original works (where still extant) rather than to enable them to be dispensed with. ... Without pretending to great originality, the whole work shows, on the part of the author, a thorough grasp of all the subjects treated, independence of judgement, mastery of technique; the style is terse and clear; in short, Pappus stands out as an accomplished and versatile mathematician, a worthy representative of the classical Greek geometry.

The golden age of Greek geometry had ended with Apollonius of Perga, but for a while there still were competent mathematicians who kept the field alive, although not producing any significant original research. But as time passed even such scholars became rare, and Heron of Alexandria was one of the last great expositors among them. The last one was Pappus of Alexandria.

But Pappus did indeed leave behind original research of considerable interest as well. We return to his celebrated theorem in Section 12.9.

4.13 The Murder of Hypatia

We have taken a long step from the time of Euclid. 600 years has passed. Alexandria has developed into a magnificent metropolis. Mathematics and geometry, science and the humanities have been nurtured. The greatest minds of the known world have spent time there as students or as visitors. Now, 600 years after Euclid, we find a considerably lesser geometer in Alexandria. Lesser than Euclid, but in no way insignificant.

His name was Theon, Theon of Alexandria. He was active in the fourth century A.D. He wrote commentaries to the unquestioned mathematical masterpiece, *Euclid's Elements*.

Some historians of mathematics slightly contemptuously refer to Theon and others as *The Commentators*. This might imply a certain lack of originality, and a period of decline in Alexandria. In general the mathematical research in Alexandria was indeed in a state of decline at Theon's time. But for Theon that judgment is not an altogether just one. This is best illustrated by moving even further ahead in time, in fact a staggering leap of 16 centuries. In 1899 one of the greatest minds of modern mathematics, *David Hilbert* (1862-1943), published his work [18]. Is this fundamentally important work "merely commentaries" to Euclid? Commentaries, yes. But "merely", absolutely not!

David Hilbert was much reverenced in his time, also by the present rulers at the time of his death. But he never concealed his contempt for the Nazis in power. It is told that just before his death, he was asked by one of the leading figures: *"– Now, Herr Professor Hilbert, how is your Institute now that we have gotten rid of all the Jews?"* Hilbert looked at his questioner and answered coldly: *"– The question is easy to answer. My institute does not exist any more."*

At this time the population of Alexandria had grown quite cosmopolitan. Besides Egyptians, it consisted of Greeks and Jews. Many among the Greeks had converted to Christianity, thereby causing the translation of the Old Testament into Greek. The Alexandrian Museum, or the *Academy of Alexandria* was a true temple for learning, scientific pursuits and culture. The foremost thinkers, philosophers, mathematicians and scientists of the world lived here. At this unique institution they were still free to carry out their spiritual activities according to their own wishes. The library in Alexandria still contained some of the finest works produced by mankind.

Theon had a daughter, who became one of the greatest names within philosophy, mathematics and other sciences. Hypatia was born around the year 355 A.D.[8] She is the first woman mathematician we know with absolute certainty, although undoubtedly there were others before her. She worked with her father, Theon, on commentaries to the great geographer, astronomer and mathematician *Claudius Ptolemy's* work, and revisions of *Euclid's Elements*. She also wrote commentaries to the works of the great classical mathematicians *Apollonius and Diophantus*. Here we have to understand, as already mentioned, that *"Commentators"* in the antique really were original scientists, working from knowledge obtained by past scientific generations. How much of todays science, even research at the Nobel Price level will not fit the description of "commentaries"?

Much of her work, as well as that of numerous other mathematicians from antiquity, is lost in its original form. It is only known to us through copies, translations, summaries and as rendered by "Commentators".

---

[8] Some put her year of birth at 370 A.D., but M. Dzielska argues persuasively for 355 in [4].

Hypatia gave lectures and did research, and around the year 400 she became the leading philosopher of the *Neoplatonic Academy in Alexandria*. She also became the Head of the *Museum of Alexandria* and the *Library*. She then had reached the peak within the Alexandrian intellectual élite, as the unquestioned leader of cultural life there.

Hypatia must have been a remarkable person, in more than one way. She was superbly gifted as a scientist and scholar. Eminent thinkers from the entire antique world traveled to Alexandria to hear her speak. She also was a very beautiful woman. Numerous were the offers of marriage she had received from kings and noblemen, but politely turned down: She preferred to devote her life to the pursuit of the eternal truths in philosophy and mathematics.

But Hypatia lived during turbulent times. This period was dominated by the struggle between old ideas which had been forming the core of the antique civilization on one hand, and emerging new ones on the other. The confrontation was hard and merciless. The new religion, *Christianity,* was on the rise. And the old Gods like Enlil and Ishtar, Zeus, Apollo and Athena, Jupiter and Venus, or the most important one in Egypt at this time, *Serapis,*[9] they were *pagan gods*, reprehensible to the Christian zealot. Paganism had to be fought by all means avaliable.

Fanaticism and fundamentalistic narrow-mindedness and bigotry have been present throughout human history. Combined with xenophobia and disdain for the *"aberrant"*, as well as with plain and simple *ignorance*, these negative forces of human life have haunted humanity since time immemorial. Such collective tendencies of human nature may well lie dormant under the surface during shorter or longer intervals, and then suddenly flare up under some contemporary pretext, igniting the flames of a Holocaust, of a Witch Hunt, ultimately an apocalyptic conflagration in which civilizations are reduced to smoking ruins.

One of the features which we keep finding time and again, is the need these fundamentalists have for some groups of *scapegoats*. Groups of people who are anathematized as reprehensible enemies, be it for their race, ethnicity, beliefs or sexual orientation. Women are especially vulnerable, above all if they "do not know their proper place". The zealot will frequently stop at nothing, certainly not physical elimination, murder.

Cyril of Alexandria is the name of a man who was elected Patriarch of Alexandria in the year 412 A.D. He was a very partisan warrior for the young Christian Church. An uncompromising guardian of the *true faith*, who not only fought vigorously and without scruples against the unbending *pagans*, but also used all means at his disposal to go after and fight down the abominable *heretics* within the church itself. His detest for Hypatia and all she stood for was intense. She was the epitome of everything he hated.

---

[9] Serapis, or *Osarapis* was the dead Apis worshiped as Osiris. He was the lord of the Nether world, and the Serapis cult incorporated elements of the Greek Gods into the traditional Egyptian ones.

Cyril regarded the Jews as dangerous enemies. But before he could turn his attention to them, he had to secure his position within his own Church.

Novatianus was one of the early leaders of the Christian Church in Rome. Born about 200 and martyred in 258, he had a high reputation as a learned theologian, but he lost out to a rival named *Cornelius* in the vote for pope. A minority declared itself for Novatianus, who then became the second *antipope* in the history of the Church. His views would in many ways be reprehensible to us today, for example he refused reentry into the Church for those unfortunate Christians who had denounced their faith as a result of persecution. Novatianus and his followers were excommunicated in 251. The schism developed into a sect which spread across the entire Roman Empire. Novatianus managed to build his own church with his own bishops, and the *Novatians* still had some following in Alexandria at this time. Cyril had to deal with the Novatians in Alexandria, which he did. Their churches were stormed, plundered and burned, the unrepentant killed or driven away, but the ones who converted to the true faith were graciously forgiven.

Having eliminated opposition from Christian *heretics*, Cyril was now free to turn to external enemies. He did so with a vengeance. As far as I know this is the first instance of a large scale pogrom, where practically the entire Jewish population of Alexandria was eliminated, driven away or murdered. Their homes were pillaged and then ignited, their property plundered. The synagogues were burned down. This infamous crime rests on the conscience of concerned Christians of today. Adding to this picture is the unbelievable "explanation" offered by some historians that the Jews had "provoked" the Christians first, by attacking them! It sounds all too hauntingly familiar.

Cyril was now at last strong enough to take on *The Pagans*. His success with the Christian heretics and the Jews, made him confident that he would soon have cleansed this sinful city. And he understood full well that in order to win, he would have to aim for the top. Thus Hypatia had to be eliminated. She had stubbornly refused to become a Christian, and unrepentantly stood by her pagan beliefs, worshiping the sorcerer Pythagoras and his followers with their satanic secret rites. She led young people astray with her talk about the old pagans Socrates and Plato. But she had allies in Alexandria, powerful allies. The allies of Hypatia would have to be immobilized first.

Hypatia's most prominent friend in Alexandria was no other than the Roman Prefect there, an enlightened man named Orestes. He had studied at the Academy, under direction of Hypatia. And he harbored a deep sense of admiration and esteem for his former teacher, now a dear friend and close advisor. Even Cyril could not dare to cross the Roman Prefect. Not yet.

Orestes had kept the peace, he had kept the *Pax Romanum*, the Roman Peace of which the Romans prided themselves. After the final conquest of Egypt, when Cleopatra and Anthony were defeated and utterly destroyed in 31 B.C., the Romans had brought peace and, by and large, prosperity to the region. Certainly order and the rule of law. But now the Empire was in decline. Rome had been hard pressed from many sides, and just to hold the outer provinces together had become a heavy burden. At the death of Theodosius the Great i 395 the Roman Empire was formally split into the East and the West Empire. One of his sons, *Honorius*, had become Emperor in the West, but his power rapidly fell apart, the West Roman Empire had now less than a hundred years left before its final fall. See Section 4.15 for more on this. In the East a second son of Theodosius, *Arcadius*, became Emperor. Alexandria and the rest of Egypt had belonged to the eastern part of the Roman Empire, and was ruled from Constantinople after the separation. Arcadius was succeeded by Theodosius II in 408, who originally entertained dreams of reuniting the Roman Empire as a mighty power. But the pressure in the West from aggressive Germanic tribes proved too strong, and the ideas of a reconstructed Empire eventually had to be abandoned.

This was the political situation at the time when Cyril became the leader of the Christian Church in Alexandria.

Orestes had been a good man to have in Alexandria for the emperors in Rome and in Constantinople. He had kept that part of the Empire comparatively quiet. This was an important prerequisite for Theodosius' plans of reuniting the Empire under his own rule.

Cyril must have realized that this was Orestes' weakness. Cyril could muster troops of ardent followers on short notice. And he had organized an echelon of lieutenants, taking their commands from him and carrying them out precisely as ordered.

Orestes soon had serious problems on his hands. Riots flared up all over Alexandria. The mobs chanted their accusations against the ungodly Romans, who showed such respect for the pagan sorcerers, based at the Museum and in the Library where these ideas were preserved. Hypatia was singled out, as having poisoned Orestes' mind against the pious Cyril. Orestes' hands were tied, or so he felt. News of him having sent the Roman legions out onto the streets of Alexandria to put down the disturbances would certainly not have been understandingly received in Constantinople. He had himself become a Christian, out of political convenience more than anything else. Cyril and his followers were not impressed. And as the pogroms were in the making, he had intervened and arrested, at the behest of some of the persecuted Jews, one of Cyril's lieutenants named Hierax. For good measure he had him tortured as well. But Orestes had grossly overplayed his hand, and Hierax had to be freed, emerging as a martyr and hero to Cyril's followers. Not to speak of the unbelievable incident, when he himself was bodily attacked on the street by a mob, led by the monk Ammonius. The same Ammonius had hit him in the head with a stone, causing him to bleed profusely. As the courage of his guards wavered, brave citizen of Alexandria rushed to his assistance, and Ammonius was seized and brought before the enraged Orestes, still bleeding from his wound in the head. Ammonius, far from receiving a fair trial, was sentenced to severe torture, from which he died, thus transformed into another martyr of the Church. Orestes' account with the Emperor in Constantinople was already overdraw. Orestes knew this perfectly well. And he knew all too well who and what where the real targets for this frenzy: The Neo-Platonists at the Museum of Alexandria, and above all, their spiritual leader Hypatia. The writings were on the wall.

But Hypatia felt unable to remain silent in the face of the injustices and atrocities committed by Cyril and his people. And she certainly did not wish to flee Alexandria. So she stayed on, apparently not really being able to believe that Cyril would harm her because of her science and philosophy. She attempted to support Orestes, in his feeble efforts to resist Cyril.

Cyril had put one of his monks named *Peter* in charge. They waited for her as she rode home from her lecture at the Museum in her chariot.

Socrates Scolasticus was born in Constantinople towards the end of the fourth century A.D. He was a historian of the early Church, and wrote the fundamental source *Ekklesiastike historia* (Ecclesiastical History). In a time of turmoil and acrimonious disputes, he is generally credited for striving to avoid the animosities and hatred often engendered by theological disputes in these times. Himself not being a priest, he honored clerics and venerated monks, but also urged the study of works by pagan authors. As a historian he is credited with thorough research and with seeking out the primary sources. See [38] for more details. He has related what happened next in Alexandria, which we summarize as follows omitting the graphic details:

In front of the Caesarum Church her chariot was stopped, and she was pulled down. They dragged her into the church, where she was killed. They then cut her body in pieces, carried it to a place called Cinaron where her remains were burned.

Orestes resigned and left Alexandria. The City Council reported her murder to the court in Constantinople, and demanded an investigation. But no investigation ever took place, since no witnesses could be found and no evidence seemed to exist.

Cyril went on to new victories in forming the dogmas of the Church, which was still united up until 1054. His writings on ethical and theological questions are extensive, and after his death he was sainted by the Church. St. Cyril's day is June 27.[10]

The Museum and the Library were burned down. Much of what remained of the Library was burned already in 392 A.D., when the Christians destroyed the Temple of Serapis, which had also been a center of learning and culture. In any case, the year 415 A.D. marks the beginning of the end of antique civilization, and the end of the beginning of the dark Middle Age.

Part of Theon's mathematical and astronomical work has survived. This include a student edition of Euclid's Elements *The Data* and *The Optics*,

---

[10] There are four more saints by the name Cyril, among them the monk who designed the Russian alphabet.

which were used by Byzantine scholars in their effort to reconstruct Euclid's work. Also preserved are his commentaries on work by Claudius Ptolemy. Theon also comments on work by the geometer *Menelaus of Alexandria*. Theon's and Hypatia's mathematical and astronomical work also relied on work by another geometer of Alexandria at that time, *Pappus*. Hypatia's mathematical work has been presumed lost. But recent research indicate that we may be able to piece together her contributions. As it probably happened, her work did not get lost, she "just" did not get credit for it. Not an infrequent occurrence in the history of mathematics. Thus efforts to reconstruct her work on Apollonius' *The Conic Sections* indicate that she made substantial contributions. Also, it is now believed that the survival of Diophantus' *Arithmetica* is due in large part to Hypatia's elucidation. Theon of Alexandria, Hypatia's father, was very much engulfed in astrology and Babylonian mysticism, through his strong involvement with Pythagorean doctrine and philosophical thinking. Thus it is not unlikely that some of this material formed part of the ancient insights which flowed to Alexandria from what had been the Babylonian Empire.

For more on these questions we refer to M. Dzielska, [4]. This reference also recounts the events in Alexandria. I have included some of the details from the narrative given there, but omitted others. I admit that my position is a personal one, Dzielska views Cyril of Alexandria in a somewhat more favorable light than the present author does.

4.14 Preservation of a Heritage

For a while antique mathematics and philosophy lingered on, in Alexandria and elsewhere. Longest, perhaps, lasted the Academy of Constantinople, where many works were preserved. *Proclus* (410-485) headed a Neoplatonic Academy in Athens, where he wrote *Commentaries on the First Book of Euclid*. This work is, as already explained, our main source for the history of early Greek mathematics, as so many of the originals went lost by the conflagration in Alexandria, and for other reasons as well.

Ammonius is reported to have been a student of Proclus. It would be consistent with this to put his year of birth to around 450 A.D. He wrote commentaries on Aristotle, and Ammonius was appointed head of the Alexandrian school, which still existed in his time.

Ammonius had two students in Alexandria, who both contributed significantly to the preservation of the classical heritage. *Eutocius of Ascalon* was born around 480 A.D. and died about the year 540. Ascalon, now named Ashqelon in Israel, was the city of *Herod the Great*. It had an old history when it was conquered by Alexander the Great in 332 B.C., then it became a Roman city in 104 B.C. and finally was destroyed completely during the Crusades. Excavations at the cite reveal what a magnificent city this was.

Eutocius wrote commentaries on Archimedes and Apollonius. As we have seen in Section 4.4, his commentaries on Archimedes' work *On the Sphere and the Cylinder* served to preserve this important work for posterity.

Another student of Ammonius was *Simplicius of Cilicia.* Cilicia is located in southern Anatolia in present day Turkey. After completing his studies under Ammonius of Alexandria, he went to Athens where the Academy of Plato was still in existence. There he studied under the Neoplatonist Damacius, who had become head of the Academy around 520 A.D.

Simplicius is given credit for preserving numerous classical works for posterity, through his comments and writings. Another important teacher of mathematics around this time was the architect and mechanical engineer *Isidorus of Miletus*, who directed the building of the Hagia Sophia (*Holy Wisdom*) in 537.

In 529 A.D. the Academy in Athens was closed by the (East Roman) Emperor Justinian as being pagan. Damacius, who was still the director of the Academy, together with Simplicius and others had to flee, and were well received by the Persian King Khrosrow I, an enlightened patron of philosophy and culture. Although the exiles were able to return to Athens under a peace agreement worked out between Justinian and Khrosrow in 532 A.D., their freedom of expression was now severely constrained.

The Arabs conquered Alexandria in 630, and are generally blamed for having burned the Library and the Museum at that time. But this is almost certainly not the case, as in fact very little, if anything, remained of the Library nor of the Academy at this late date. On the contrary the Arabs should be credited for having saved ancient mathematics for posterity. Several fundamental texts, among them Euclid's Elements, were actually reintroduced into the Christian world as translations from the Arabic to Latin, at a much later time.

Indeed, as the Arabs expanded into southern Spain, we westerners generally view this as a grave threat to the civilized world. But the Arabian Muslims founded Academies for mathematics, science and medicine there, and as Europe later began its reawakening, Christian scholars traveled south, and disguised as Arabs managed to attend these centers of culture and learning.

4.15 The Decline and Fall of the Roman Empire

Historians have written extensively on how such a mighty power as the Roman Empire would crumble and disintegrate. Towering among these works is the six-volume tome of Gibbon, available as a one-volume edition in [8]. Not that it was the first time such a thing happened in human history, nor has it been the last. But the beginning of the process which ended in the decline and fall, probably lie about 500 years before the imperial city of Rome fell to the Ostrogoths in 476 A.D. When Caesar had brought down the Republic, no rational and stable procedure existed for the transfer of power from one ruler to his successor.

For about 150 years after *Augustus* became emperor, or *Imperator*, Commander, in Rome his policies were by and large continued and secured peace as well as prosperity. But already emperors like Gaius, known as *Caligula* (meaning "Little Boot"), Claudius and above all, Nero, showed how vulnerable the system of government had become, mainly through the capricious system of succession.

Caligula succeeded his grand uncle Tiberius in 37 A.D. Already in 38 he fell ill, apparently with serious consequences for his mental condition. He started to lead a reckless and boozy personal life, ruthlessly doing away with people who opposed him. He wanted to be worshiped as a good, thus emulating oriental practices, and among all the bizarre tales told about him, is the appointment if his horse to the office of consul. After two unsuccessful attempts on his life, he was killed in his palace on the third. Claudius had been viewed as a less than gifted politician, but was a learned scholar. He got his first public office under Caligula, and succeeded him as emperor in 41, thanks to large sums of money changing hands. But he seems to have been an able administrator, who completed projects like the Claudian aqueduct, built a new harbor at Ostia and improved Roman jurisprudence in several ways. He is said to have been largely led by his wives, especially the third, Messalina. His fourth wife had him poisoned to make her son by a previous marriage, Nero, the new emperor.

Nero was a real disaster as emperor. The first five years were not so bad, thanks to good advisers. But then he had his mother murdered, as well as his wife *Octavia*, the daughter of his predecessor. He also had the son of his predecessor, the rightful heir to the throne so to speak, done away with. The great fire in Rome in 64, which lasted for nine days, was probably set on his orders, and used as a pretext to persecute Christians and Jews in Rome. After his fall in 68 there ensued a turbulent year or so, after which Vespasian in 69 established a firm rule and repaired some of the damage inflicted by his immediate predecessors.

Now followed a good period for the Roman Empire, with able and competent rulers. This ended in 180 at the death of *Marcus Aurelius* He was followed by a century of war and disorder. During the period 235 – 284 there were more than 20 emperors, warlords with basis in various parts of the empire, and engaged in bloody fights over the power. In the time span 237 – 38 a total of six "emperors" were killed within a few months.

With the accession of *Diocletian* in 284 the government was strengthened and reformed, and in the second year of his reign he associated *Maximian* with himself as co-emperor. They both abdicated in 305, and during the following 18 years six rival emperors competed for power, in the end *Constantine* united the whole empire under his own rule. During his reign, which lasted 14 years, two important changes took place. The first being the recognition of Christianity as the religion of the state, and the second being the establishment at Byzantium of a new capital city, *the City of Constantine,* Constantinople. There it was established a second Senate, a prefect of the city, and so on, duplicating the institutions in Rome. This contributed to paving the way for the final separation of the Empire. Again, when Constantine died in 337 several competing caesars fought among themselves, and in 364 some stability was reestablished but now with one emperor in Rome and another in Constantinople. From 383 to 395 the Roman Empire was led by *Theodosius the Great,* emperor of the East but engrossed by the duty of upholding the feeble authority of his colleges in the West. The latter were under an increasing pressure from the "barbarians", who now threatened to overrun the western part of the empire. When he fell ill and died in 395, his two sons took over the East and the West, and the separation into the West Roman Empire and the East Roman Empire was formally effectuated.

When the emperor in the West, Theodosius' son Honorius, died in 423 his authority had been seriously eroded. Valentian III ruled for a relatively long time, from 423 to 455 when he was murdered. During his reign the province of Africa was lost, and Atilla invaded Gaul and Italy. He was repelled thanks to the aid of the Christened and half Romanized "barbarians", the Visigoths. In 455 *Maximus* was emperor for three months, during which time Rome was overrun, plundered and partly burned by the Vandals. From now on the emperors in Rome ruled on the mercy of the *barbarian mercenaries.* When the Emperor *Romulus Augustulus* tried to assert his authority, or more precisely his father *Orestes,* who was the man behind him tried to do so, he was ousted as emperor and the father killed. Exit the Emperor, enter *Odoacer the Rugian,* King of Italy.

This was in the year 476, which we call the year of the fall of the Roman Empire. But the East Empire lived on, however slowly lost its power and glory. The East Empire came to be named *Byzantium,* it finally fell to the Tucks in 1453.

Sicily in many ways fared better than Italy. First plundered by the Vandals, it was conquered in 493 by the Ostrogoths. But in 535 the East Roman Empire under Emperor *Justinian I* opened an offensive westwards, with the intention of restoring the old Roman Empire. His generals *Belisarius* and *Narses* conquered Italy as well as a large portion of the North African coast, as well as southern Spain. Again the shores of the Mediterranean were essentially Roman land. The largest extension of the Byzantine Empire was reached in 565, after which it started to fall apart. But by the middle of the ninth century it still included Sardinia, Sicily and southern Italy. But the conquest by Justinian did not carry with it a full revitalization of the culture and mathematics of the old days, in fact Justinian was the one who closed the Academy in Athens, as being *pagan.*

5 The Geometry of Yesterday and Today

5.1 The Dark Middle Ages

The Roman Empire had become the carrier of a common civilization of philosophy, learning and mathematics throughout *"The Known World"*, the *The Oikumene*. Strangely, the Romans themselves were not particularly interested in it. It has been said that the only contribution the Romans ever made to mathematics is due to *Cicero*, when he rediscovered Archimedes' grave. But the fall of the Roman Empire marks the end of this common civilization, at least as a web encompassing the entire "known world". When the Roman Empire ceased to exist, this cultural web shrunk, and went into a kind of hibernation. In *Constantinople*, among the Arabs with the Caliphs at Baghdad and elsewhere, and to some extent on Sicily the seeds of culture and learning were preserved, as well as among individual thinkers, many of them monks in the monasteries, within the Christian Church.

The Church itself was very powerful, and may well be viewed as a successor to the Roman Empire. To the constantly changing kingdoms, countries and alliances, the Pope in Rome with his administration and wide network was the single stable institution. More than a millennium after the fall of the West Roman Empire, the Pope in Rome crowned the "Holy Roman Emperors" of Europe.

The legacy of the Roman Empire lasted a long time, in many ways it is still with us today. In fact, some of the most dangerous conflicts in Europe and the Middle East may be traced back to events during the final 500 years of the empire. And we may speculate what the situation in the world would have been today, if the Romans then had listened to Cicero and rejected Caesar.

The *Dark Ages* commences with the fall of the West Roman Empire and lasts until the middle of the eleventh century. During this time practically no mathematical activity took place in Europe. The exceptions to this claim are so meagre as only to strengthen the assertion.

Boethius was born around 480 and died 524. He belonged to a distinguished family, which counted several important senators from the old days of the Empire. He is said to have been *"the last of the Romans which Cato and Cicero could have acknowledged for their countrymen"*. He spent his boyhood in Rome when Odoacer was monarch. His father, *Flavius Manlius Boetius* had been consul in 487, and probably died soon after that year. According to Boetius himself, when he lost his parents *"men of the highest rank"* took him under their charge. He received an excellent education, as such scholarship still existed in Rome. He soon became known as a promising and able young man. But these were turbulent times, and Odoacer was by no means firmly seated on the throne of Italy. A dangerous rival was the Ostrogoth *Theodoric*, with his army of formidable warriors. A vicious enemy as well as an unreliable ally, he had received an extraordinary education for a barbarian: At the age of 7 he had been sent to the court in Constantinople as a hostage, and he remained there for 10 years. Returning to his father he established himself as an able leader, who greatly increased his father's domain. After he had succeeded his father as king, he invaded Italy in 489, and pushed Odoacer further and further back until he was able to lure him into a trap in the form of a grandiose banquet at the palace of Laurentum. There Theodoric killed Odoacer with his own hands. Now Theodoric became the new king, and his eyes fell on Boetius who became a favorite with the new ruler.

In 510 Boetius became consul, and in 522 his two sons, who were still young, became consuls together. Boetius now stood on the top of his career, when he was placed between his two sons in the Circus and received the ovations from the people.

But the fall from the summit was near. Intrigues at the palace resulted in charges of treason. He was supposed to have written letters to the Emperor in Constantinople, in order to restore the Empire in Rome. He unequivocally denied the accusations, but of course the resentment against the barbarians in power was deep among the old Roman families. So such accusations would not be altogether preposterous. At any rate he was brought before the king, and found guilty. He was thrown in jail, where he wrote his most famous work *De Consolatione Philosophiae*. He was executed in 524. The king is said to have later very much repented his rashness in putting Boetius to this mistreatment, so much so that he passed away soon after 524.

Boetius' mathematical works became standard texts throughout the Middle Ages. But his *Geometry* consist only of the propositions in Book I and parts of books III and IV of Euclid's Elements. It also contains some elementary applications to mensuration. His *Arithmetic* is based on the work by *Nicomachus*, four centuries earlier.

Nevertheless, he appeared at a time when contempt for intellectual pursuits was widespread, and his fame and influence increased after his death. A central theme in his philosophical writing was to reconcile the ideas of Plato with those of Aristotle, and his grand idea was to revive the spirit of his countrymen by filling them with the thoughts of the ancient Greek writers.

Now Theodoric had been an Arian, that is to say a follower of the heretic *Arius*. He was the origin of the first great heresy-struggle in the Christian Church. His teachings, dealing with the relation between the *Father* and the *Son*, had been repudiated at the council of Nica in 325, but the controversy raged on. In fact Arius himself would have been readmitted into the Church in Constantinople, from which he had been excommunicated, had he not died suddenly while walking with a friend one evening. Since Boetius could be viewed as having lost his life while trying to depose the heretic ruler Theodoric, by aiding and abetting the pious orthodox emperor in Constantinople, his star rose even higher. So now Boetius was canonized as *Saint Severinus*, precisely on the merit of the false accusations once made against him.

Bede, called *the Venerable*, was the most learned Englishman of his age. He was born in 673 in Northumberland and died in 735. He is considered the father of English history, but his writings were on a broad range of subjects, making a total of about forty different treatises. These essentially amount to an entire encyclopedia. He also included some mathematics, a calendar and a treatment of finger reckoning. But there is no geometry, and Bede who was brought up in a monastery from childhood and remained there his entire life, was first and foremost a pious religious thinker.

Alcuin, who gave his full name in Latin as *Flaccus Albinus*, was another learned Englishman. He was born in Yorkshire in 735 and died in 804. He was educated in Yorkshire under the direction of *Archbishop Egbert*, whom he succeeded as director of the seminary. The word of the learned Englishman reached the Emperor *Charlemagne*, and Alcuin was called upon to instruct the Emperor and his family in the subjects of *rhetoric, logic, mathematics* as well as *divinity*, the study of Christianity. He assisted Charlemagne in building up a seminary, or university, in Tours, which had a strong influence on higher education in France. But Alquin forbade the reading of the classical poets.

Gerbert was born about 950 in Auvergne in France and died i 1003 as *Pope Sylvester II*. It is generally assumed that he came from a rather poor background, but as a young boy he entered the Benedictine cloister of St. Gerald at Aurillac, where he received a good education, showing himself as exceptionally gifted. In 967 Count Borrell of Barcelona visited the monastery, and the abbot asked the count to take Gerbert back to Spain with him so that the promising student could study mathematics there. The count did so, and put the young man under the protection of the bishop of Vic in Catalunya, where there was a cathedral school. Here there was extensive contact with the Arab culture and civilization of al-Andalus to the south. Al-Andalus was much more advanced that Christian Europe, the library in the Islamic capital of Cordoba was overwhelming by contemporary European standards. The libraries of the cathedral of Vic and the nearby monastery of Ripoll were among the best in Europe.

Here he came into contact with learned Arabian Islamic scholars. His studies earned him profound insights, by European standards at this time, into the subjects of *mathematics, astronomy* and *music*. Then in 970 he came to Rome, where the unusual sagacity of this young man caught the attention of the Pope himself. The Pope recommended him to the Emperor, *Otto the Great*, and for a while he served as the tutor and advisor of the latter. Later he went to Reims, where he had a group of students, and wrote on geometry as well as arithmetic. Gerbert was active in politics as well, worldly and ecclestiastical. In 999 this led to his election to the elevated office as Pope. As Pope he is not considered to be particularly outstanding, while as a mathematician and astronomer he is regarded the foremost for this period, around the previous turn of millennium. This is the more remarkable considering the rather meager nature of his scientific findings.

In his work on geometry, Gerbert solves the following problem, considered to be very difficult: In a right triangle the area and the hypothenuse are given. Find the remaining two sides. Of course this problem would have been handled easily by the Babylonians, and certainly as well by the Greek. It is a trivial consequence of the Pythagorean theorem with the use of some algebra, as the Babylonians would done it, or by geometric algebra, as the Greek would have proceeded. With our modern notation we proceed as follows: Let the two sides in question be of lengths x and y, the known diagonal be d and the area be A. Then we have

$$x^2 + y^2 = d^2$$
$$xy = 2A$$

We then find

$$(x + y)^2 = d^2 + 4A$$
$$(x - y)^2 = d^2 - 4A$$

from which x and y are easily found.

Gerbert also expresses the area of an equilateral triangle of side s as

$$A = \frac{s}{2}\left(s - \frac{s}{7}\right)$$

which corresponds to $\sqrt{3} \approx 1.714$, not a very good approximation by any means.

Gerbert is credited by some with being the first to have used the Indian-Arabic numerals in Europe. However, this is not a unanimously accepted view among historians of mathematics. Such were the superstitions and general ignorance in Europe at this time, that rumors of him having sold himself to Satan started to gain acceptance after his death.

5.2 Geometry Reawakening: A New Dawn in Europe

The classical mathematical books had been lost to Europe, but were preserved by the Arabs. In the twelfth century they were reintroduced into Europe, being translated from Arabic to Latin. Among the first was *Euclid's Elements*.

The translation of Euclid's Elements from the Arabic to Latin was done by the English monk *Adelard of Bath*. He visited Spain between 1126 and 1129, and traveled widely in Greece, Syria and Egypt.

The Jewish mathematician *Abraham bar Hiyya* played an important role during this time. He is also known under the name *Savasoda*. He lived and worked in Barcelona, being born there in 1070, and died in Provence in France in 1136.

He has written a book entitled *Treatise on Measurement and Calculation*, covering a broad range of subjects. It is the first book in Europe covering Arab algebra, containing the complete solution of the general quadratic equation. Abraham was familiar with the works of Greek geometers such as Euclid, Theodosius, Apollonius and Heron.

Gherardo of Cremona was active in the twelfth century, and oversaw the work of a group of translators who worked with material obtained from *Toledo*, which had been captured by the Christians in 1085. They translated more than ninety works from Arabic into Latin.

As we have already told in Section 4.15, Sicily was part of the Byzantine Empire until the middle of the ninth century. Then for about fifty years it was captured by the Arabs, so recaptured by the Byzantine Empire, and then captured by the Normans. The period of Norman rule was very good for science and culture. Sicily was then a melting pot for the Greek, Arab and Latin cultures, the contact with Constantinople and Baghdad was good and Greek and Arab manuscripts were translated into Latin.

Merchants from Italian and Spanish cities established ties with the East, so in this way as well the undeveloped and barbaric Europe was little by little gaining access to the cultural heritage of which it had been so long deprived.

In conclusion it should be emphasized that the Arabs not only served as preservers and messengers of the ancient civilization, they developed their science vigorously and originally in their own right.

5.3 Elementary Geometry and Higher Geometry

Following Euclid, it is has been customary to distinguish between *Elementary Geometry and Higher Geometry*. According to this tradition, elementary geometry deals with configurations built up from *points and lines*, as well as *circles*. Higher geometry is concerned with *general conic sections* as well as curves of *higher degrees*, or even *transcendental curves*. The line between these two parts has not always been sharply drawn, and the distinction is today rendered obsolete, at least within the main stream of research in geometry. It has, however, to some extent survived in some didactical treatments of the subject.

From a historical point of view, however, it is important to keep these two faces of geometry in mind. Indeed, let us consider the three famous classical problems, namely *Squaring the Circle, Doubling the Cube* and *Trisecting an*

Angle. In their original and enigmatic form, these three constructions should be performed by straightedge and compass only, used in the *legal fashion* as explained in Section 3.6. That is to say, the problems should be solved with tools from *Elementary Geometry*, with the *Euclidian Tools*. As we shall prove in Chapter 16, in this form they are insoluble, all three of them. However, with tools from *higher geometry* they do have solutions, some of them very beautiful and deep ones, as we have seen in Chapters 3 and 4.

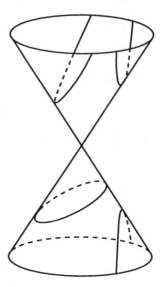

Fig. 5.1. In this figure we have cut a fixed cone with a plane in several different ways, and thus obtained one from each of the classes of conic sections: We see an ellipses, a parabola and a hyperbola.

Conic Sections is a class of curves in the plane, which are obtainable by *intersecting a circular cone with a plane.* This is indicated in Figure 5.1.

Of course a *circle* is a conic section, but traditionally it was regarded as *"more elementary"* than the others. This is undoubtedly due to the fact that a circle may be produced by the Euclidian tool *compass*, while a general ellipse, for example, may not.

Up until the 15 th. Century the two parts of geometry, elementary and higher, were pursued by methods which were basically the same, by the so-called *synthetic* methods. A circle may be drawn with a compass, a line using a straightedge. Points are found as intersections of lines, lines and circles or two circles. As for *higher curves* or conic sections, they could not be drawn in one piece by straightedge and compass, but were given in terms of definite constructions or procedures, as the *loci of points satisfying certain defining properties*, allowing the construction, by straightedge and compass, of a finite but arbitrarily large number of points on them. In many cases, as we have seen in the previous chapters, this made possible the manufacture of certain mechanical tools, which like a straightedge or compass made it possible to produce the curve in question. The simplest case is the string-rule for producing an ellipse, explained in Figure 5.2.

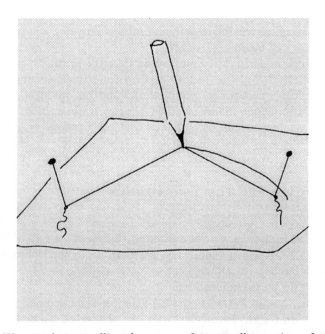

Fig. 5.2. We may draw an ellipse by means of two needles, a piece of sting and a pencil as shown here.

In the 15 th. Century the methods from Greek geometry are augmented by powerful new techniques through work by *Roberval, Torricelli, Pascal, Desargues* and others. Roberval studied tangency for curves, considering a curve as being the locus of a point whose movement is the result of two separate movements. The vector sum of the two corresponding velocities would then define the tangent direction at any given point. We understand his approach today in terms of curves on parametric form. Torricelli had similar ideas, and the two mathematicians locked horns in priority disputes over their work, a disease endemic to the entire field of mathematics, past and present.

Descartes introduced *algebra* into geometry in a decisive manner. One might perhaps, with some justification, say that algebra was *reintroduced* into geometry, since the *Babylonians* had a well developed geometry based on their superb mastery of algebra. To a somewhat lesser extent this may also be said of the ancient Egyptians. And of course Greek geometers had used geometric algebra, above all Euclid and Apollonius. However, these observations do not belittle the importance of Descartes' contribution. Babylonian mathematics had become completely forgotten, and its rediscovery is an accomplishment of the 20th. Century.

Moreover, Descartes' credit for the algebraization of geometry should be shared with another great French mathematician, *Pierre de Fermat*. His work was outlined in a letter to Roberval in 1636, it was then already seven years old. His work on these matters was not published until after his death.

The line which we will follow is that of *higher geometry*, which has become the main stream of modern geometry.

But it certainly should be emphasized that the synthetic methods persist in *Axiomatic Geometry*. Today this is a comparatively small field, but tied to interesting questions in combinatorics and general algebraic systems. However, it falls outside the scope of the present book.

Another area which split off is that of *General Topology*, again giving rise to *Algebraic Topology*. This will also not be treated here.

5.4 Desargues and the Two Pascals

The French engineer and architect *Gérard Desargues (1591 – 1662)* lived in Lyon, which at that time was the second most important city of France. He belonged to a very wealthy family. He participated in the campaign against the city of la Rochelle as an engineer, spent some time in Paris and later in life retired to his estate in Condrieu.

During his time in Paris he pursued his interest in geometry in an environment including great mathematicians like *Descartes, Étienne Pascal* and his son, the young *Blaise Pascal.*

Desargues published a treatise on conic sections in Paris in 1639, without question his most important work. The work had a very limited circulation, according to some it was mostly ignored, although it is said by others to have influenced Desargues' student, the young Blaise Pascal. At any rate, all copies disappeared, and it was not until 1845 when the important French Geometer *Chasles* found a copy of it by one of Desargues' students, that the significance of Desargues' work was recognized. Then, in 1951 an original copy of the book resurfaced, so today the record has been straightened out as far as this part of Desargues' work is concerned.

The book is short and densely written. He gives a unified treatment of *conic sections*, and proves his *Theorem of Perspective*. This theorem asserts that if two triangles are such that the three lines through corresponding vertices pass through a common point, then the three points of intersection of the (prolongations of) corresponding sides will lie on a common line. This important result will be treated in detail in section 12.1.

Blaise Pascal's father Étienne Pascal also made contributions to mathematics in general and geometry in particular, see Section 14.7. But his son was by far a more important mathematician. Already at the age of 16 he wrote a treatise on conic sections, where he proved the theorems which we treat in sections 12.9 and 14.10. His contribution to projective geometry and conic sections is of fundamental importance. He also wrote on the *cycloids*, using methods which essentially amount to integration. And he laid the foundation for probability theory, corresponding with Fermat. His work was not confined to mathematics alone, indeed he profoundly influenced physics by bold and controversial ideas, like the possibility of a *vacuum*.

Then he turned to religious problems, and finally spent the last eight years of his life in a monastery.

5.5 Descartes and Analytic Geometry

René du Perron Descartes (1596 – 1650) was a French mathematician and philosopher. In 1637 he published a book which contained an appendix, containing some pathbreaking ideas. The appendix had the title *La Géométrie*, while the book itself was entitled *"Discours de la méthode pour bien conduire sa raison et cherche la vérité dans les sciences"*. In translation it should be something like "A discussion of the method for correct reasoning and seeking the truth in science".

Descartes main contribution to science is contained in this appendix. It consists in about hundred pages, is divided in three parts, and presents the important ideas it contains in a rather obscure style. Some say that this was intentional on the part of Descartes. Be this as it may, subsequent translators and commentators have contributed to a more lucid exposition, and later mathematicians like the eminent *Gottfried Wilhelm Leibniz*, eventually provided insights which finally resulted in the main framework of what is known as *Analytic Geometry* to all students of introductory calculus.

Today Descartes' main idea would strike us as so obvious that it is hard to see how this would not be self evident to anyone doing geometry. The set of real numbers are identified with the points on a line. A point in the plane, correspondingly, is represented by a pair (x, y) of real numbers. The set of points $(x, 0)$ then form the x-axis, and the y-axis is the set of all points $(0, y)$. x and y are called the coordinates of the point $P = (x, y)$, x being referred to as the *abscissa* and y as the *ordinate* of P.

Now a line is defined by a relation, where a, b and c are fixed real numbers

$$ax + by + c = 0,$$

which holds if and only if the point (x, y) is on the line, and similarly any planar curve is given by a corresponding equation: Thus for example,

$$(x - 1)^2 + (y - 5)^2 - 25 = 0$$

represents a circle of radius 5 about the point (1,5).

In this way geometric considerations are translated into algebraic or analytic ones, depending on what kinds of equations we are dealing with: A line or a circle are algebraic curves, being defined by polynomial equations, while something like

$$y = \sin(x) \text{ or } y = e^x$$

are *analytic* curves which are not algebraic, we call such curves *transcendental curves*.

Of course geometry in 3-space is done similarly, a point now being $P = (x, y, z)$.

5.6 Geometry in the 18th. Century

In the 18th. Century calculus as we know it today took shape, amidst strong controversies. *Michel Rolle,* a noted French mathematician, found Calculus totally inferior to Geometry with the rigorous standard of proofs inherited from Euclid. He went so far as to assert that Calculus was merely *"a collection of ingenious fallacies"*. He nevertheless made important contributions to the subject himself, and in later years became more approving of the new ideas of Calculus! The Scottish mathematician *Collin Maclaurin* is best known by students of Calculus for his series-expansion, but also made contributions to Geometry. Indeed, he proposed an ingenious construction for *trisecting an angle in three equal parts* by means of a curve of degree 3, today known as *the Trisectrix of Maclaurin*. Maclaurin also published a theorem which is known today as *Cramer's Rule*. *Gabriel Cramer* found this result independently a little later than Maclaurin, and got all the credit for it. Probably this is due to a better notation in his treatment than the one employed by Maclaurin.

For the benefit of readers who are not familiar with linear algebra, we recall Cramer's theorem, which is important for geometry. This will be used later on. Actually Cramer published the result in a treatise on the geometry of lines, precisely a special case of the way we shall also benefit from the theorem here.

We start out with the simplest case, namely a system of two linear equations:

$$a_{1,1}x_1 + a_{1,2}x_2 = b_1$$

$$a_{2,1}x_1 + a_{2,2}x_2 = b_1$$

In this case Cramer's Theorem has the following simple form:

Theorem 4. *The system above has a unique solution if and only if*

$$a_{1,1}a_{2,2} - a_{1,2}a_{2,1} \neq 0$$

If so, then the solution is

$$x_1 = \frac{b_1 a_{2,2} - b_2 a_{1,2}}{a_{1,1} a_{2,2} - a_{1,2} a_{2,1}}, x_2 = \frac{a_{1,1} b_2 - a_{2,1} b_1}{a_{1,1} a_{2,2} - a_{1,2} a_{2,1}}$$

If $a_{1,1} a_{2,2} - a_{1,2} a_{2,1} = 0$ then the system may have infinitely many solutions, or none. If $b_1 = b_2 = 0$, then there are infinitely many solutions in this case.

We shall not give a separate proof for this special case. The conventional proof becomes quite messy even in this very simple special case. Instead, we give a proof valid in complete generality, which is very simple and conceptual. Strangely it does not seem to appear in the standard textbooks, but we have used it in [20].

The number $\Delta = a_{1,1} a_{2,2} - a_{1,2} a_{2,1}$ is called *the determinant* of the *matrix* of the system of equations. The number is denoted by

$$\begin{vmatrix} a_{1,1} & a_{1,2} \\ a_{2,1} & a_{2,2} \end{vmatrix}$$

In general any such $n \times n$ arrangement of numbers, which is called an $n \times n$ *matrix*, may be assigned a number called its *determinant*:

$$\det \begin{Bmatrix} a_{1,1} & \cdots & a_{1,n} \\ & \cdots & \\ a_{n,1} & \cdots & a_{n,n} \end{Bmatrix} = \begin{vmatrix} a_{1,1} & \cdots & a_{1,n} \\ & \cdots & \\ a_{n,1} & \cdots & a_{n,n} \end{vmatrix}$$

An important feature of the determinant is its behavior under elementary row operations in the matrix:

1. If all the numbers on one of the rows are multiplied by the same number r, then the determinant gets multiplied by that number.
2. The determinant changes sign when two rows are interchanged.
3. If a row is multiplied by some number and added to another row, the determinant is unchanged.

In addition we need another rule:

Rule of the diagonal. If all the numbers under the diagonal are zero, then the determinant is the product of all the numbers on the diagonal.

Elementary row operations may be performed in any $m \times n$ matrix

$$A = \begin{Bmatrix} a_{1,1} & \cdots & a_{1,n} \\ a_{2,1} & \cdots & a_{2,n} \\ & \cdots & \\ a_{m,1} & \cdots & a_{m,n} \end{Bmatrix}$$

and by a finite number of such steps, where the number r in 1. is assumed to be non-zero, a matrix which does not consist of exclusively zeroes, may be brought on the form shown in Figure 5.3. We then say the matrix A has been brought on *reduced row echelon form.*

The procedure is to first select the first column in A which does not consist of all zeroes, by interchanging rows we may assume that this non zero number lies in the first row. Dividing the first row by that number, we may assume that it is a 1, usually referred to as a "leading 1".

We then modify the matrix according to Rule 3, subtracting suitable multiples of the first row from the second, third and so on. We thus finally get zeroes under the leading 1 of the first row.

Looking away from the first row, we then treat the remaining part of the matrix in the same way, getting the next box, also containing a leading 1. Subtracting a multiple of the second row from the first, we ensure that there is a zero directly above it. Repeating this a finite number of times, we get the result shown in Figure 5.3.

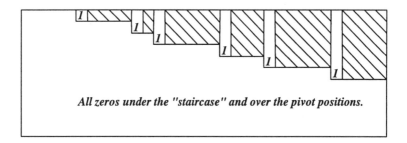

All zeros under the "staircase" and over the pivot positions.

Fig. 5.3. A matrix on reduced row echelon form. The white area consists of only zeroes, and the shaded area may contain any numbers, depending on the original matrix A, of course. The so called leading 1's are shown, their positions are called the pivot positions.

Now consider a general system of equations

$$
\begin{aligned}
a_{1,1}x_1 + a_{1,2}x_2 + \ldots + a_{1,n}x_n &= b_1 \\
a_{2,1}x_1 + a_{2,2}x_2 + \ldots + a_{2,n}x_n &= b_2 \\
&\ldots \\
a_{m,1}x_1 + a_{m,2}x_2 + \ldots + a_{m,n}x_n &= b_m
\end{aligned}
$$

It is clear that if we perform elementary row operations in the matrix

$$T = \left\{ \begin{array}{cccc} a_{1,1} & \cdots & a_{1,n} & b_1 \\ a_{2,1} & \cdots & a_{2,n} & b_2 \\ & \cdots & & \\ a_{m,1} & \cdots & a_{m,n} & b_m \end{array} \right\}$$

then we get a system of equations which is *equivalent to the original one*. Thus we may bring the above matrix on reduced row echelon form, and then it is much simpler to analyze the system. This method is referred to as *Gaussian elimination*. We use this important principle to prove the following

Theorem 5 (Cramer's theorem). *Given a system of n linear equations in n unknowns*

$$\begin{array}{c} a_{1,1}x_1 + a_{1,2}x_2 + \ldots + a_{1,n}x_n = b_1 \\ a_{2,1}x_1 + a_{2,2}x_2 + \ldots + a_{2,n}x_n = b_2 \\ \cdots \\ a_{n,1}x_1 + a_{n,2}x_2 + \ldots + a_{n,n}x_n = b_n \end{array}$$

and put

$$A = \left\{ \begin{array}{ccc} a_{1,1} & \cdots & a_{1,n} \\ a_{2,1} & \cdots & a_{2,n} \\ & \cdots & \\ a_{n,1} & \cdots & a_{n,n} \end{array} \right\}$$

Moreover, let A_i be the matrix obtained by replacing column number i in A with the $b_1, b_2, \ldots, b_n$. Let $a = \det(A)$ and $a_i = \det(A_i)$. Then
i) The system has a unique solution if and only if $a \neq 0$.
ii) In that case, the solution is

$$x_1 = \frac{a_1}{a}, x_2 = \frac{a_2}{a}, \ldots, x_n = \frac{a_n}{a}$$

Proof. i): The system has a unique solution if and only if the reduced row echelon form of

$$T = \left\{ \begin{array}{cccc} a_{1,1} & \cdots & a_{1,n} & b_1 \\ a_{2,1} & \cdots & a_{2,n} & b_2 \\ & \cdots & & \\ a_{n,1} & \cdots & a_{n,n} & b_n \end{array} \right\}$$

is

$$T' = \left\{ \begin{array}{ccccc} 1 & 0 & \cdots & 0 & b_1' \\ 0 & 1 & \cdots & 0 & b_2' \\ & & \cdots & & \\ 0 & 0 & \cdots & 1 & b_n' \end{array} \right\}$$

since otherwise there would be an equation saying $0 = b'_n$, either being a contradiction or an empty condition leading to an infinite number of solutions. But T' is of this form if and only if $\det(A) \neq 0$, by the rules for the determinant quoted above.

ii): Since each elementary row operation in T gives the same operation in A, and all the A_i's, all $\frac{a_i}{a}$ are unchanged at each step which transforms T into T'. So to check the formulas we may assume that the system is such that T is on reduced row echelon form. But then the formula is trivial to verify. □

In general the number r of 1's in the reduced row echelon form of a matrix A is called its *rank*.

5.7 Some Features of Modern Geometry

In modern times the higher geometry has developed rapidly, and split up into several different subfields. The difference between them may be in terms of the geometric contents, or it may be in terms of the methods employed. One of the features they share, is the central position occupied by geometric objects of *higher dimensions*, in addition to classical and reasonably familiar *curves* and *surfaces*.

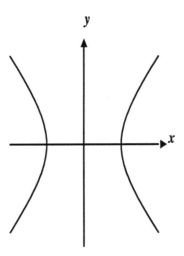

Fig. 5.4. A familiar geometric object: The curve given by the equation $x^2 - y^2 = 1$, which is a hyperbola.

In Figure 5.4 we show the familiar *hyperbola*, given as the set of all points in the plane $\mathbb{R}^2$ satisfying the equation

$$x^2 - y^2 = 1$$

In Figure 5.5 we look at a somewhat less familiar case, namely a hyperbola where the equation has a *cross term*, where the product xy occurs in the equation. This corresponds to the curve being *tilted* relative to the coordinate system.

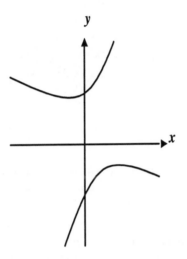

Fig. 5.5. A somewhat less familiar geometric object: The curve given by the equation $x^2 + 2xy - y^2 - x + 1 = 0$, which is also a hyperbola.

Algebraic surfaces in the Euclidian space, in $\mathbb{R}^3$, are given by equations in x, y and z. In Figure 5.6 we show the surface defined by the equation

$$x^2 + y^2 - z^2 - 1 = 0,$$

which belongs to the family of *conic surfaces* in $\mathbb{R}^3$, that is to say surfaces given in $\mathbb{R}^3$ by an equation of degree 2.

The higher dimensional geometric objects are referred to by names like *manifolds, varieties* or *schemes*, and are the objects of study within the *Differential Geometry*, the *Analytic Geometry* or the *Algebraic Geometry*. The difference between these three branches of geometry basically lies in the *functions*, in several variables, used to describe the geometric objects in question. We illustrate the point by loosely explaining what an $\mathbb{R}$-variety or as we also say, a *real algebraic variety*, is.

The simplest case is that of an *affine algebraic variety*. This is the zero locus in $\mathbb{R}^n$ of a finite set of polynomials in n variables, $f_i(X_1, \ldots, X_n), i = 1, \ldots, r$. For the complex numbers instead of the reals, affine algebraic $\mathbb{C}$-varieties are defined similarly as subsets of $\mathbb{C}^n$. We talk about affine algebraic

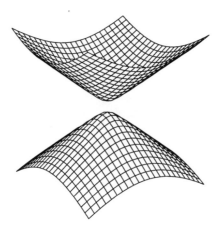

Fig. 5.6. Another geometric object, familiar to some readers: The surface given by the equation $x^2 + y^2 - z^2 - 1 = 0$, which is a rotational hyperboloid. We have omitted the coordinate axes here.

varieties over $\mathbb{R}$ or over $\mathbb{C}$. For such varieties we have a good concept of *dimension*. It will take us too far to provide a detailed discussion, but the dimension then is an integer ≥ 0, dimension equal to zero corresponding to points, dimension 1 to curves and dimension two to surfaces.

In general and roughly speaking, a real algebraic variety X is glued together by pieces which are affine varieties, $V_\alpha, \alpha \in A$, A denoting some indexing set. So the pieces V are as given below:

$$V = \{(a_1, \ldots, a_n) \in \mathbb{R}^n |\ f_i(a_1, \ldots, a_n) = 0, i = 1, \ldots, r\}$$

There are more technical points in the definition, but the main idea is captured by the above. The individual pieces referred to above are called *affine open subsets* of X. The local dimension of X at a point x is the dimension of a sufficiently small affine piece of X containing x, and if the local dimension is constant throughout X, this number is called the *dimension* of the algebraic variety X.

In *Differential Geometry* the objects under study are defined similarly, but the functions are no longer polynomials but functions in n variables for which *all partial derivatives*, mixed and to any order, exist.

In *Analytic Geometry* this condition is replaced by the condition that all the f_i's possess power series expansions.

Note that any variety or manifold over $\mathbb{C}$ also is one over $\mathbb{R}$, but the dimension as an $\mathbb{R}$ - variety is twice that as a $\mathbb{C}$ - variety.

We leave the matter of explaining what varieties are at this point. But general as it is, we are only at a timid beginning. The theory of *Grothendieck schemes* is even more general, the definition of these objects will not be given here. But with this theory we have come full circle. The basic reference for this is [11]. Now it is no longer a matter of having introduced algebra into geometry, but perhaps of introducing geometry into algebra: The theory of Grothendieck schemes represents a single, powerful common generalization of *algebra and geometry*. And then, beyond the schemes of Grothendieck there is *non-commutative* algebraic geometry...

The obvious question of *why we want to study such objects* may be given several answers. Basically, one answer would be that we study them for the same reason as the Greeks studied circles and straight lines: They encountered them when dealing with the tasks undertaken in their science and technology. So do we, with these objects. In fact, to take a comparatively recent development: Physicists consider *models* for the physical universe where the space in which we live is not 3-dimensional, but of a considerably higher dimension. Only three of these dimensions are noticeable in our life, the remaining ones being *curled up* in *very small* cylinders, or something like that. The diameter of these cylinders would be so small as to fit well inside the nucleus of the atoms. Thus a *point* would actually be a *very small* but higher dimensional, *subspace* of the space in which we live. The idea is not as preposterous as one might think at first encounter: A point is a *zero-dimensional* subspace of the usual Euclidian space, in our conventional way of thinking, a line is a one-dimensional subspace and a plane a two-dimensional one. *"Fattening"* the usual space by some extra, curled up, dimensions, would fatten up the points, lines and planes correspondingly. A point in our usual space would then have to be understood as the manifestation, the section, of a higher-dimensional *"point-manifold"*.

Modern geometry studies geometric objects which are very difficult to visualize. Even capturing them as some sort of subspaces of simple spaces like $\mathbb{R}^n$, $\mathbb{C}^n$, or the corresponding projective spaces which will be explained in Chapter 11 may not be possible. But it should be noted that if we have r polynomials $f_1(X_1, \ldots, X_n), \ldots, f_r(X_1, \ldots, X_n)$ with real coefficients, then we obtain an algebraic variety consisting of only one affine piece, namely the one given by the zero-locus of the given polynomials in $\mathbb{R}^n$. Such algebraic varieties are, as we have explained, called *affine* algebraic varieties, and as we shall explain in Chapter 11, we may similarly define *projective algebraic varieties* as well, contained in projective spaces.

An important phenomenon studied extensively in modern geometry is that of *singular points*. If X is an $\mathbb{R}$-variety in the sense loosely explained above, then it turns out that for most points $x \in X$ there is a small neighborhood of x which may be identified with the unit ball in $\mathbb{R}^n$, by modifications without breaking or merging. These points are called *smooth points*. Points which are not smooth are called *singular points*. A simple example of a singular point is the origin in the plot of the curve in $\mathbb{R}^2$ with equation

$$y^3 + y^2 - x^2 = 0,$$

shown in Figure 5.7.

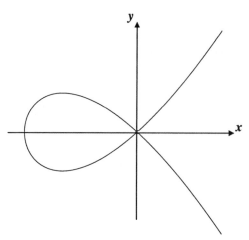

Fig. 5.7. The origin is a singular point, since any small neighborhood would look like two crossed line segments, which can not be deformed into a single line segment without breaking or merging.

Another example is provided by the *vertex of a cone*, shown in Figure 5.8. Here we see that around any other point of the cone than the vertex, we may take a small disk which can be identified with the unit disc around the origin in the plane. For the vertex, however, this is not possible.

We also have surfaces where there is an entire curve of singularities, as in the example shown in Figure 5.9.

Today physicists study some interesting objects called *black holes*. Black holes started out as being highly hypothetical, but are becoming more and more part of what we know to be the reality we live in. A black hole is created as the density of a collapsing star becomes so immense that the generated gravitational force near its surface makes it impossible for light to escape: The object then becomes invisible, and anything coming within a certain boundary, including light, will inevitably be sucked into it. Black holes may be understood as singularities of space itself. Matter has been merged with geometry to become part of the very fabric of space.

Another modern application of geometry is the so-called *Catastrophe Theory*. We treat this theme in some detail in Section 6.2. Catastrophe Theory was very much on the public agenda some years ago, but the interest in it has since abated somewhat. However, the field still exists and very much also

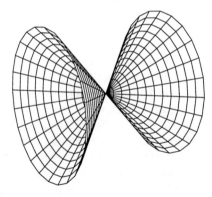

Fig. 5.8. The vertex of a cone is a singular point. If the vertex is the origin, and the axis is along the x-axis, the equation for it as a subset of affine 3-space is $y^2 + z^2 - R^2x^2 = 0$, where R is a constant.

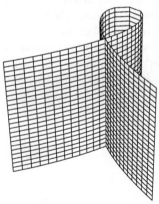

Fig. 5.9. The cylinder over the curve, shown in Figure 5.7, with a singular point at the origin yields a surface, with an entire curve of singular points.

still holds the promise of becoming an effective tool in predicting dramatic changes in important systems we depend on. New insights have already been gained, one such insight being a renewed awareness of how misleading an old *dogma* of natural science really is: Namely the assertion that *Nature non facit saltus*: Nature does not make jumps, natural phenomena always change continuously. At the micro-level this dogma has been out for some time, contradicting as it does the very fundament of quantum theory. But a chilling realization is that at the macroscopic level nature may indeed perform jumps as well. Thus in particular, the *global warming* we are now experiencing, or so some claim, might not just smoothly move our global climate from where we have been to some new slightly different state, but the climate may be thrown suddenly into a new mode. And not only that: Even if we succeed in reducing the greenhouse effect again, the climate may remain in the new state for a long, long time. We shall return to these ideas in the light of the geometry of Catastrophe Theory in Section 6.2.

The concept of dimension is more interesting and more subtle than our definition above would suggest. In fact, there are geometric objects where the original definition of dimension, due to Felix Hausdorff, yields a number which is not an integer. Such objects, having a *fractional dimension,* are referred to as *fractals.* The *Theory of fractals* has received considerable attention during recent years, and one might say captured the limelight from Catastrophe Theory. Fractals are, according to some of the proponents of this field, to be found everywhere in nature. A shoreline could be viewed as, or *modeled as* a fractal: Looking at it from high above, it would give one appearance, but as the observer descends and moves closer to the surface, its features change. In the end the observer is looking at small pebbles and grains of sand, and the "curve" which was the shoreline is no longer that, instead it has dissolved into something between a curve and a strip of surface: It suggests a dimension less than 2 but more than 1. Some fractal theorists assert that the typical shoreline would have a dimension of about 1.2. Similarly *clouds* would be objects with an estimated dimension of more than 2 but less than 3, most being deemed to be of dimension around 2.3. Computing or estimating dimensions in this way is somewhat controversial. Certain estimates on the degree of *"self similarity"* of the object under examination have to be made, and these estimates are certainly not above discussion. We give details in Chapter 17.

6 Geometry and the Real World

6.1 Mathematics and Predicting Catastrophes

Mathematics is important in understanding nature. By mathematics we create *models*, which provide explanations for the phenomena we observe. If a mathematical model yields a result which is in contradiction to the observations we make, then the theory will have to be scrapped, no matter how beautiful the mathematics in it should be. But can mathematics guide us in finding new knowledge about nature itself? In other words, not just arrange the knowledge we already have in a nice and orderly model, but actually *predict observations* which we have not yet made? The answer is, of course, affirmative. In fact this phenomenon is the reason for the usefulness of working with models in the first place.

A classic example is the discovery of the planet Neptune. In 1843 *John Adams* used certain irregularities in the orbits of the known planets to predict the existence of an unknown planet outside the orbit of the planet Uranus. He even gave the exact coordinates where the telescope should be aimed to find this new planet. But his computations were not taken seriously, and it was not until 1846 when *Urbaine le Verrier* arrived at the same result that the astronomers bothered to look. And there it was!

Later the planet Pluto has been found in a similar manner. Some years ago new claims were made concerning a further planet, far beyond the orbit of Pluto. Some even claim that the Sun may have a dark companion-star, tentatively named *"Nemesis"*. This ominous name is due to speculations concerning occurences of periodic *mass extinctions*, documented in the fissile record.

During one such mass extinction, the *Cretaceous-Tertiary mass extinction*, the dinosaurs died out in a, geologically speaking, very short time-interval. This happened 65 million years ago. The leading theory explaining that catastrophic event, is that the Earth was hit by an astroid of about 10 kilometers in diameter. An impact crater, presumably being the result of this killer-astroid, is located partly off the coast of Yucatan. Other theories are also being offered, but as of today the astroid-theory seems to be the most plausible and best supported by evidence.

At any rate, the connection with the hypothetical "Nemesis" is that it would have a very eccentric orbit around the Sun, and therefore at intervals of about 60 million years pass through a belt of comets or asteroids. Deflecting some of these objects, the inner solar system would, at such intervals, be exposed to showers of comets or asteroids.

The Nemesis-theory has been loosing ground, however. Today there seems be general consensus among astronomers that the culprit is the planet *Jupiter*. In fact, Jupiter influences the orbits of the asteroids inhabiting the astroid belt beyond Mars. The influence happens in a way which is perfectly determined by the mathematics of planetary orbits, but which is nevertheless of such complexity that the phenomenon appears *chaotic*. Moreover, small changes from a stable situation will lead to abrupt and large changes in the orbital structures. These *chaotic instabilities* caused by the proximity of this giant planet occasionally sends an astroid off course and into the inner solar system. Such an event could happen tomorrow, or in one thousand years, or in 65 million years. It could be a very unpleasant surprise for us Earthlings!

But as it has happened many times in the past, we already have numerous asteroids with odd orbits, some of them coming near Earth from time to time.

An impact of an astroid of a diameter of 10 kilometers would lead to gigantic sunamis, flooding large areas, dust would be ejected into the air blocking sunlight, and leading to a global winter lasting for several years. Impact by an astroid with a diameter of 1 kilometer would also lead to a catastrophe on a global scale. But even impact of an object with a diameter of 100 meters could, under unfortunate circumstances, devastate a large modern city. To get an impression of the force involved here, it can be recalled that on June 30, 1908, 1000 square kilometers of Siberian pine forest was flattened by the blast caused by an impacting astroid. The estimate is that the explosion, which happened above ground, was equivalent to the detonation of a 10 megaton hydrogen bomb. The astroid is estimated to have had a diameter of merely 70 meters.

This scenario is taken seriously. Enough so as to having the watch for near-Earth asteroids being one of the activities of NASA. But the program is to map all asteroids in near-Earth orbit – NEOs as they are known – of more than 1 kilometer diameter. The mapping moves along very slowly, one estimate being that it will take 20 years to have mapped 90% of all NEOs of this size.

These insights contributed to alerting the public to the possibility of a *nuclear winter* which could follow an all-out nuclear war, and so has served a usefully purpose already. In relatively recent times a volcanic eruption is known to have caused conditions with a taste of this kind of calamity. It happened in 1815, when the volcano *Gunung Tanbore* at the island of Sumbawa in present day Indonesia erupted in the period between April 7 and April 12, the largest volcanic eruption known in historic times. More than 30 cubic kilometers of dust, ashes and stone was flung into the atmosphere, the plume extending 44 kilometers up in the air. An estimated 90000 people died in the area, and that as well as the following year the summer was extremely bad over the entire northern hemisphere. Famine resulted in Europe, and in New England 1816 was known as *"the year without summer"*.

Climatic change is not only threatening in the form of the "nuclear winter" scenario. The insidious process known as *global warming*, caused by the increased emission of CO_2 as well as other *greenhouse gasses* might be leading us into a situation where the ice is starting to melt in the polar regions, where the oceans are heating up, storing immense additional quantities of energy, giving rise to a variety of troubling and threatening scenarios.

Thus it not surprising that the prediction of volcanic eruptions and earthquakes is an important task for modern science. Again this is a question of having a model for the geological dynamics of the planet Earth, which is good enough to not only explain past events, but which is also capable of yielding reliable predictions. Such models will certainly have to rely heavily on sophisticated mathematics, including the mathematics which goes into the field some call Catastrophe Theory. But names for the same mathematical phenomena may vary, in the present case we find terms like *Bifurcation Theory* or *Stability Theory*.

Even if the chilling prospect of Earth being hit by an astroid captures our imagination, the climatic change due to global warming may constitute a greater danger. The nature of this process is different from the sudden impact, which might be prevented entirely by invoking the sophisticated technology, heroic individuals and the combined resources of a united Humankind. Less heroic is the task of arguing to convince politicians, dependent on contributions from various business groups, that measures limiting the emissions of CO_2 are needed to avoid a more or less drastic climatic change on our planet. But at least the problem and the process may be understood better by means of concepts from Catastrophe Theory. We shall give some indications of this in Section 6.2, and provide a more complete mathematical treatment in Chapter 18.

6.2 Catastrophe Theory

In the early 1960's the distinguished mathematician *René Thom* created the foundation of this field. Among other significant contributors are *V.I. Arnold, B. Malgrange, John Mather* and *Christopher Zeeman*.

In this book we may only touch upon this interesting field, but readers who would like to learn more are referred to the book by *P.T. Saunders* [31].

An instructive introduction to Catastrophe Theory is provided by the famous example of Zeeman dealing with *aggression* in dogs as a function of the emotions *rage* and *fear*. The example is treated in his classic Scientific American article from 1976 [41], and also explained in [31]. The basic assumption on which this example builds is that aggression in dogs is a result of the two emotions *rage* and *fear*. Rage and fear is estimated on a numerical scale by

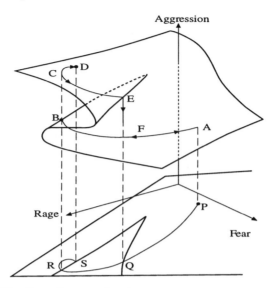

Fig. 6.1. The Cusp Catastrophe: Aggression as a result of rage and fear.

observing by how much the mouth is open (*rage*) and by how much the ears are laid back (*fear*). The aggression displayed by the animal's behavior is then estimated on a numerical scale as well, and an important observation is made: In some situations the same animal may display different levels of aggression when being under the same amount of fear and rage: The aggression produced depends, in some cases, on the previous history.

Zeeman uses the paradigm of Catastrophe Theory to model this situation. He postulates the variable z for *Aggression* being such that the point $(x, y, z) \in \mathbb{R}^3$, where x and y are the numerical estimates for *Rage* and *Fear*, respectively, lies on a certain fixed surface in $\mathbb{R}^3$, see Figure 6.1. Surfaces of this shape may be given by polynomial equations of degree 3, this is explained in Section 18.1. There it is also explained that in the (x, y)-pane there is a wedge-shaped area, inside a certain curve known as a *semi cubical parabola*, see Section 12.3, such that for a point in this area there are three points on this surface directly above it, and for $P = (x, y)$ outside the area there is a unique point on the surface above P. Thus for these values of rage and fear, the dog's behavior is perfectly predictable: The level of aggression is uniquely determined by its rage and fear.

However, for a point P *inside* the wedge shaped area there are in principle three points above P. But only two of them correspond to actual levels of aggression. This may sound strange and appear unmotivated, but it follows from the mathematics of Catastrophe Theory that this is the situation as far as the theory is concerned. Again, this will all be explained in Chapter 18. So with this model for aggression in dogs, there are two possible *modes* corresponding to rage-fear levels inside the wedge. *The middle sheet of the fold of the surface is removed.*

Now suppose that the dog starts out at the rage-fear point P, and a sequence of events moves it along the curve indicated, from P through Q and R to S. At the point Q it passes into the critical area, but nothing much happens, aggression remains about the same. The path on the surface has passed from A, through F, and proceeds to B. This corresponds to the rage-fear point R. This is where the *"Catastrophe"* occurs, in the present case, this is where the apparently quiet dog (who has, however, quietly more and more bared its teeth) suddenly tries to bite you! Aggression jumps, as it has to according to our model, to the level given by the point C. Zeeman calls this event *an attack catastrophe*. You now have to control the dog, inducing fright or reducing its rage by taking various measures. At first this does not help much. Moving back through S the dog still stays at a high level of aggression. If the path is traversed form S and back to P, then the point on the surface follows a different path, namely from D, through C, then onwards to E, where the dog suddenly ceases being aggressive, the point jumping down to F. Zeeman calls this *a flight catastrophe*. The dog's behavior now follows the original path back to A.

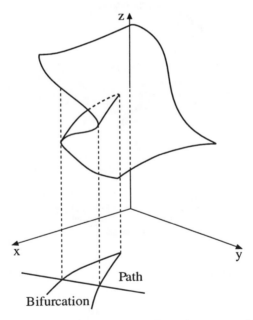

Fig. 6.2. The paradigm of the Cusp Catastrophe: z is average global temperature, x is the level of greenhouse gasses in the atmosphere and y the area covered by carbon dioxide-consuming plants. As we cut down the forests and produce carbon dioxide, we move along the indicated path, from right to left.

The singularity of the wedge shaped curve, the semi cubical parabola, is known as a *cusp*. Correspondingly this situation is referred to as a *Cusp Catastrophe*.

Some feel that Catastrophe Theory has been somewhat oversold during the last two or three decades. Today a common view is that while the theory does not provide explanations for everything, it may provide a point of view which elucidates some of the issues on our agenda. Take for example the issue of *Global Warming*. We represent the mean global temperature, on some scale, by the variable z, we let x be the level of *greenhouse gasses*, and let y denote the *area covered by carbon dioxide-consuming plants* on the Earth. Could it be that the situation may be modeled by a cusp catastrophe? And if so, could it be that we are indeed moving along the path indicated, from right to left, presently cruising smugly inside the bifurcation set? This disturbing scenario is illustrated in Figure 6.2.

6.3 Geometric Shapes in Nature

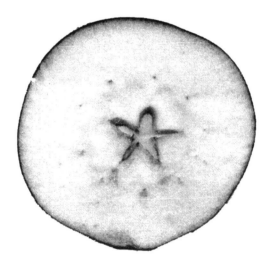

Fig. 6.3. A section through an apple, revealing a pentagram and a regular 10-gon around it!

It is a fascinating aspect of studying geometry to search for and to find geometric shapes in nature. As a source of inspiration to geometers this activity goes back to the very origin of the subject. And closely related to it is the creation of decorative patterns for pottery or mosaics, often reproducing patterns observed in the natural world.

One of the simplest ways to encounter *the Golden Section* is to cut an apple along "the equator". There the kernels are arranged in a chamber, the cross section of which has an outline which is *a pentagram*. We show an example of this in Figure 6.3.

The pentagon and the pentagram, as well as the Golden Section were important to Pythagoras and the Pythagoreans, as explained in Section 3.2. And how the Golden Section determines the pentagram – and conversely – was explained in Section 3.4.

By cutting a Kiwi-fruit, we get a different geometric image. It seems very plausible that such natural works of art have played a part in the beginnings of geometry and mathematics.

Fig. 6.4. A section through a Kiwi-fruit.

In Figure 6.5 we finally cut through an *orange*, and find the assignment of subdividing the circumference of a circle in equal parts. In addition to many other things the reader may find. What do *you* see in this picture?

In Figure 6.6 we show a computer generated, stylizer representation of a tree. The author made this image by means of a simple algorithm implemented in the system MATHEMATICA, see [12]. The image shows an appealing regularity, but looks as much like some type of *fern* than a tree.

In Figure 6.7 the algorithm has been modified in some obvious ways, mimicking the presumed randomness in the details as branches develop: Thickness increases with age, direction and length, to some extent randomly. Here the randomness and the developments are determined by a set of input parameters to the program. Also, foliage is now included. The result is a much more

Fig. 6.5. Section through an orange.

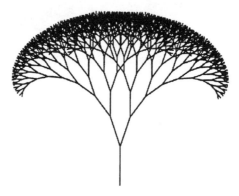

Fig. 6.6. A computer generated, stylizer representation of a tree.

natural-looking tree. But with other input parameters, the same program will yield an object looking more like a pineapple than a tree. And, by the way, the original stylized image above also comes from the general program by setting all input parameters equal to zero.

This has brought us to the realm of computer generated images, we return to this theme when we come to the mathematical theory of *Fractal Geometry*. For now, we briefly indicate how fractals may constitute a realistic way for describing certain natural phenomena.

Fig. 6.7. A more natural-looking, though still computer generated, rendering of a tree.

6.4 Fractal Structures in Nature

For a long time the Euclidian geometry was the basis for understanding space. But this traditional way of describing real-life phenomena is increasingly coming under question: Are *straight line segments or smooth curves* really suitable tools in describing the real world as we experience it? For example, take a *shoreline*. As viewed from above, from an airliner, at an altitude of say 30000 feet, it may look like a smooth curve. A lake with a shoreline, en route from San Francisco to Minneapolis, is shown in Figure 6.8.

Fig. 6.8. A lake with a shoreline, en route from San Francisco to Minneapolis.

As the plane starts to descend, this smooth curve undergoes a transition: It starts to dissolve. If we observe the same stretch of it from various distances, what looked once as a smooth and straight curve will become more like an irregularly wiggling one. The situation is very much different from what we get if we take a photo at the first elevation, and later enlarge a portion of it! This is shown in Figure 6.9.

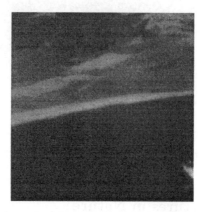

Fig. 6.9. A closer look at the lake and the shoreline is not at all what we get by magnifying a portion of a photo from the original distance, as we have done here!

Coming yet closer, some of the wiggles may turn out to be, actually, small *islands*. Coming even closer, the coastal line may reveal itself as being composed of large rocks, then of somewhat smaller rocks.

In the presence of surf the shoreline eventually will disappear, or more precisely, it will be constantly changing. But assuming a totally calm sea, we may proceed with a closer and closer examination. Being now out of the airplane, we start to examine the shoreline with a microscope, and find that the *shoreline* simply does not exist, as a *curve*.

Is it therefore impossible to describe, to find *a model for* a shoreline using geometry? Well, at least we may find a better model than the naive one we started out with, using pieces of smooth curves.

Pieces of smooth curves have a property which makes them quite unsuitable for describing a shoreline. Namely, if we consider a geometric object build up from such pieces, then zooming in on a point of it the pieces of smooth curves adjacent to the point will look more and more like *pieces of straight lines*. This is practically the definition of smooth curves, as we shall see in Section 12.3.

Euclidian Geometry, which we have studied in Section 4.1, is build on the basic elements of *points* and *lines*, as well as *circles*. That is for the planar case, in space we also consider planes and spheres of course. For giving a description of the physical world, however, these elements fall short of even yielding anything approximately correct. As our example with the shoreline shows, reality does not fit into this framework. Plato explained the situation by proclaiming *geometry* to be a *perfect realm of ideas*, while real life suffered from all kinds of imperfections. This was, and still is, a very fruitful point of view, which has inspired geometers and mathematicians throughout our history. But today scientists have turned the tables, viewing *reality* as a perfect and enigmatic realm, which we may never be able to comprehend fully. We have to content ourselves by constructing *mathematical models* for reality. These models make it possible to predict the end results of various sequences of events which we may chose to set in motion. Or we may be able to predict with some reasonable accuracy, the outcome of some natural processes. But the models, the mathematical systems, are imperfect. Not in the sense of being erroneous, of having faulty mathematics build into them. Rather, it is the question of only being able to include assumptions of limited validity into any mathematical model. And no mathematics, as advanced as it may be, can extract more from a system of assumptions than what is already implicit in the agreed-upon *first principles.*

Now the *shoreline* which we investigated have very little in common with a *straight line. Lines and planes* are completely unsuitable to even approximately describe shorelines, mountains, hillsides or forests. Nevertheless they do share an important property: They have the property known as *self similarity.* We may phrase this roughly as follows:

Self Similar Objects: An informal definition An object is said to be self similar, if any part of it is equal to some strictly smaller part when the latter is magnified up to the size of the former.

Lines and planes certainly have this property. Circles, however, do not. The shoreline do have the property, within certain limits: As always in the real world there are limits to any description we choose to give!

Objects which are *approximately self similar* are ubiquitous in nature. *Fractal Geometry* deals with mathematical objects of this kind, offering fascinating mathematical models for objects encountered in the real world. We shall return to this in Chapter 17, but conclude this section with pictures of self similar objects from nature and from mathematics. It is not easy to tell one type from the other. First we show a variety of the *von Koch snow-flake curve,* which will be explained in more detail in Chapter 17. Here we just observe the appearance, which amply justifies its name.

The von Koch Snowflake Curve is a mathematical object, but it resembles shapes which we encounter in the real world. We cut out two segments, one containing the other. Then when the smaller piece is enlarged up to the size of the bigger one, it will coincide with the bigger piece. The curve is a *self similar* object.

The rendering of the Snowflake Curve shown in Figure 6.10 has been created by a computer program. This curve was constructed by the Swedish mathematician *Helge von Koch,* with the aim of giving an example of a function $y = f(x)$ defined on an interval, say $I = [0, 1]$, which is continuous

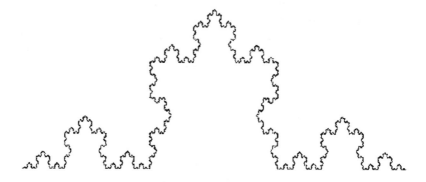

Fig. 6.10. The celebrated von Koch Snowflake Curve.

everywhere in the interval where it is defined, but such that the *derivative* does not exist anywhere. For such a function the resulting graph is a curve, in this case the von Koch Snowflake Curve, but a rather *pathological curve*: It is continuous but has no tangents, it is of finite extension but of infinite length. As if this were not enough, its dimension is not a whole number! We shall return to this fascinating mathematical object in Section 17.2, where we compute the dimension to be approximately 1.262.

The self similarity of an object resembling *a tree* is illustrated in Figure 6.11, where a tree is sketched. In the picture a piece is cut out and enlarged, this procedure is repeated six times, starting top left and ending bottom right. This object is *approximately self similar*.

The *fern* shown in Figure 6.12 is another self similar object, generated by a computer program.

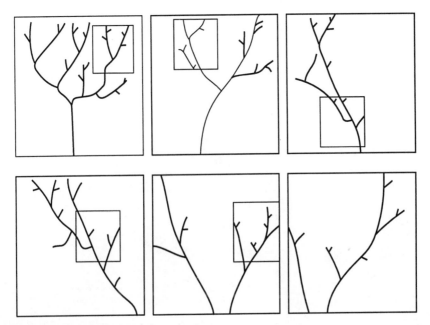

Fig. 6.11. A tree is sketched, and pieces of the picture cut out and enlarged, moving from left to right starting top left.

Fig. 6.12. A computer generated fern.

Part II

Introduction to Geometry

7 Axiomatic Geometry

7.1 The Postulates of Euclid and Hilbert's Explanation

While representing a true watershed in the development of mathematics, in their original formulations the postulates of Euclid for Planar Geometry are not easy to understand. In fact, according to present day standards of rigor, they need to be made more precise as well as to be supplemented by additional statements.

The ideas Hilbert developed in *Grundlagen der Geometrie*, [18] have had a profound influence on subsequent mathematical thinking.

Euclid's five postulates for planar geometry is frequently reformulated as follows, modified to a contemporary mathematical language. The original wording is by no means inferior, we have presented Euclid's own words, in Heaths vivid translation, in Section 4.1. In the statements below, "line" means *straight line*, contrary to what is the case in Euclid's own formulations. We also note that the Fifth Postulate is given in the version due to Proclus.

Postulate 7.1 *Through two different points there passes one and only one line.*

Postulate 7.2 *If two points on a line are in a plane, then the line lies in the plane.*

Postulate 7.3 *There exists right angles.*[1]

Postulate 7.4 *Given two points in a plane. Then there may be drawn a circle with the first point as center, passing through the second point.*

Postulate 7.5 *Given a straight line α and a point P outside it. Then there is one and only one line β passing through P which does not intersect α.*

Hilbert took as starting point three undefined terms, namely *point*, *line* and *plane*. Among these he introduced altogether *six* undefined relations, namely *being on*, *being in*, *being between*, *being congruent*, *being parallel*, *being continuous*. And finally, Euclid's five *Axioms* (common notions) and five

---

[1] In keeping with the spirit of Hilbert's explanation, this ought to be formulated as *all right angles are congruent*

postulates, were replaced by a collection of *21 statements*. This system has since been known as *Hilbert's Axioms*. Since Hilbert's work, other axiomatic systems for planar geometry have been devised. At this point, what is important for us in this exposition is not the details of Hilbert's Axioms, but rather their existence: That essentially Euclid's postulates can be made to work in a rigorous modern axiomatic setting.

A system based on Hilbert's axioms, but without the *Parallel postulate*, is frequently referred to as *Neutral geometry*. The study of neutral geometry is interesting, since it points out all the theorems whose proofs do not require the Parallel postulate. Thus for example it follows that even though the parallel to a line ℓ through a point P outside it is not uniquely determined, there is always a unique *normal* from P to ℓ. In neutral geometry there always exist a line through P parallel to ℓ, but it is not necessarily unique.

In neutral geometry there is an absolute angular measure, namely the *radian*. By contrast, in order to introduce *distance* or *measure of length*, it is necessary to *choose* some line segment AB and declare it to be of length 1. It is a theorem in neutral geometry that the angular sum of any triangle is less than or equal to π radians.

We refer to [10] for details on neutral geometry.

Under the assumptions of Hilbert's axioms, he proves that the Euclidian plane may be identified with $\mathbb{R}^2$, and with the usual definition of distance. Today we may short-circuit the axiomatic approach to Euclidian Geometry altogether, by defining Euclidian n-space simply as $\mathbb{R}^n$, with the distance between the points $P = (p_1, \dots, p_n)$ and $Q = (q_1, \dots, q_n)$ given by

$$d(P,Q) = \sqrt{(p_1 - q_1)^2 + \cdots + (p_n - q_n)^2}$$

The interplay between algebraic properties of $\mathbb{R}$ and geometric assertions is very interesting. The point is that by starting from an initial, weak, set of axioms one can show that a plane subject to these axioms may be parameterized as pairs of elements (x, y), x and y taking their values in some general algebraic system, of which the real numbers would be a special example. Then as axioms are added, each one will be equivalent to some algebraic property of the algebraic system providing the parameterization. In the end, when the axioms are complete, the algebraic system is uniquely determined as $\mathbb{R}$.

We do keep Euclid's postulates in mind, in the version due to Hilbert, when we next discuss the emergence of *non-Euclidian* geometry.

7.2 Non-Euclidian Geometry

The fifth and last statement among Euclid's postulates is the celebrated *Euclid's Fifth Postulate*, also referred to as the *Parallel Postulate*. Euclid himself must somehow have been unhappy with it. While apparently never doubting its truth, he made, according to what we know, numerous attempts of proving it as a consequence of the other four. Not only did Euclid do that, but the pursuit of this elusive goal dominated the lives and careers of many geometers for the next two millennia.

Today we know why it was so difficult. A very important discovery was made to the effect that this postulate can not be deduced from the other ones. It is *independent* of them. Thus we may construct geometries in which the Fifth Postulate and its consequences are not valid, but where otherwise everything functions as in the Euclidian plane.

This discovery was probably first made by the great German mathematician *Carl Friedrich Gauss*, (1777 – 1855). There had been others before him, coming close to the discovery as they relentlessly worked on finding a contradiction from assuming the converse of the Fifth Postulate. But Gauss never published his discovery. At his time it was a difficult issue. Euclidian Geometry had, so to say, been canonized by the Catholic Church. Its truth was considered only second to the Bible itself. *God had created the World, complete with its Euclidian Geometry.* Ludicrous as this appears to us now, then it was dead serious. Literally, only somewhat earlier, when the scientist and philosopher *Giordano Bruno* was burned alive for espousing the opinion that Earth moved in a circular orbit around the Sun, thus not being the Center of the Universe.

It is told that Gauss had his assistants climb various high mountaintops in Germany, measuring large triangles with the mountaintops as their corners. He wanted to check the sum of the angles in these large triangles, to see if it really added up to 180°! Why would a knowledgeable mathematician do such a thing?

Some claim that already Euclid had realized that his Fifth Postulate was equivalent to the assertion that the sum of the angles in any triangle is twice a right angle. But Heath ascribes this version of the Fifth Postulate to Legendre, from Gauss' own times.

The Hungarian mathematician *Jáons Bolyai* and the Russian *Nikolai Ivanovitch Lobachevsky* (1793 – 1856), independent of each other found the existence of non-Euclidian geometry somewhat later. And these two are credited with the discovery. Bolyai's system of axioms for *hyperbolic Geometry* was published in 1832.

The Fifth Postulate may be replaced with either of the two following, yielding geometries as mathematically valid as the Euclidian one:

Postulate 7.6 *Given a line α and a point P outside it. Then all lines β through P will intersect α.*

However, it should be pointed out that we may not just replace the Parallel postulate in *Hilbert's* system by Postulate 7.6, since parallels always exist in neutral geometry, as pointed out in the previous section. Other modifications of the axioms have to be carried out as well.

Postulate 7.7 *Given a line α and a point P outside it. Then there are at least two lines β_1 and β_2 through P which do not intersect α.*

If we use the last Postulate, we obtain the so-called *hyperbolic geometry*. The first yields the *elliptic geometry*. We return to the issue of how to realize these geometries. But before we can do so, we first have to explain how we understand the concepts of *axioms, systems of axioms and models for such systems* in a precise mathematical setting. This requires some preliminaries on *Set Theory*.

7.3 Logic and Intuitive Set Theory

Logic and Set Theory are mathematical fields with a high level of precision, as they represent the very foundation for the edifice of mathematical theory. And if the foundation is not sound, then the total body of knowledge can not be considered secure.

It therefore was considered deeply troubling when *contradictory assertions* could be deduced by the same methods of proofs which were, unquestioningly, used to prove *the theorems*. Some of these contradictions were quite technical, but a very simple one was found in 1902 by the mathematician and philosopher *Bertrand Russell*. We shall treat this *Russell's Paradox* in Section 7.4.

We shall now give a brief summary of some basic notions from logic and set theory. In practice we do not need the intricacies of these theories, however. Indeed, it will suffice to understand the concept of a *statement* as assertions like "$2 + 2 = 4$", $2 + 2 = 3$" or, say " *The moon is made of cheese*". Our statements may be either *true*, in which case they will be assigned the *"truth-value"* T, or *false*, in which case they are assigned the truth-value F. No other alternatives exist.

For a given statement P, we let $\neg P$ denote *the negation* of P: If P is the statement $n = m$ then $\neg P$ is $n \neq m$. Further, for two statements P and Q, we let $P \wedge Q$ and $P \vee Q$ denote the statements "*P and Q*" and "*P or Q*", respectively. Finally, let $P \implies Q$ denote the statement "*P implies Q*", and we let $P \iff Q$ denote the statement "*P implies Q and Q implies P*", thus that P and Q are equivalent.

The composite statements introduced above may be viewed as *boolean functions*, which means functions where the variables P and Q only may take the values T or F, and the functions themselves also may take only the values T or F. Tables like the one given in Figure 7.1 define such functions.

We also recall the following *set notations*: If A is a set and a is an element in it, we express this by writing $a \in A$. Moreover, $A \subset B$ signifies that A is a proper subset of B, while $A \subseteq B$ means that A is a subset, possibly equal

P	Q	$P \wedge Q$	$P \vee Q$	$P \Longrightarrow Q$	$P \Longleftrightarrow Q$	$\neg P$
T	T	T	T	T	T	F
T	F	F	T	F	F	F
F	T	F	T	T	F	T
F	F	F	F	T	T	T

Fig. 7.1. Truth table for some composite statements.

to, B.[2] $A \cup B$ denotes *union* of A and B, i.e., all elements which lie either in A or in B, or both. $A \cap B$ denotes *the intersection* of A and B, thus all elements which both lie in A and in B.

The set of all elements a which satisfy a statement $P(a)$ we write as

$$\{a|\ P(a)\}\,,$$

or expressed in words: *The set of all a such that the statement $P(a)$ is true.*

7.4 Axioms, Axiomatic Theories and Models

An *axiomatic theory* consists of a set *undefined terms*, and a system of *axioms* which these terms satisfy. Throughout modern mathematics one encounters a number of such axiomatic theories, all mathematical disciplines are in one way or another build on such a foundation.

But the foundation under the mathematical edifice was often put in place long after the construction of the building started, indeed frequently long after it had been completed. And furthermore, through the ongoing research on the foundations of mathematics, new levels under the building is being added all the time. Thus the final result is elusive, if even attainable. In fact, the process itself is probably more important than its "goal". Not only does mathematics reach out towards the outer limits of our universe of thinking, it also pierces deep into the microcosmos of its foundations.

At certain points in the history of mathematics it really looked as if the whole edifice might collapse, or at least would have to be thoroughly rebuild.

Mathematics had been build on the so-called *intuitive* – non-axiomatic – set theory. Through the efforts of the brilliant number theorist and set-theorist *Georg Cantor* this foundation appeared to be safe and secure. It therefore came as a considerable shock when several apparently grave inconsistencies surfaced. The first one was discovered in 1897 by the Italian mathematician *C. Bural-Forti*, and two years later Cantor himself found a similar paradox. Annoying as these paradoxes were, they dealt with rather

---

[2] The reader should take note that some authors use the symbol $\subset$ to mean $\subseteq$. There exist other variants of this notation as well, so it may be prudent of a reader to check out the notation in use before drawing any conclusions.

exotic constructions known as *transfinite numbers*. These concepts lay, at the time, at the outer fringes of mathematical knowledge. The paradoxes were, therefore, not as threatening as possible inconsistencies right within the central body of mathematical knowledge would have been. Then, in 1902, the real bombshell struck: Bertrand Russell discovered a paradox which only depended on the basic concept of *a set* and *an element being a member of a set*. Proofs based on similar reasoning had been accepted as fully valid throughout mathematics.

At the time of this discovery, *Gottlob Frege*[3] had just completed a prodigious work in two volumes, on the foundations of arithmetic based on Cantor's Set-Theory. At the end of the last volume he is then obliged to acknowledge, in a note added in print, that due to a communication from Russell he now finds himself in the undesirable position of seeing the very base under his efforts giving way. We shall now explain the famous *Russell's Paradox*.

Let Ω denote the set of *all existing objects*, the *universal set which contains absolutely everything in the universe*. Since this set contains absolutely everything, it must contain itself, thus is *an element in itself*:

$$\Omega \in \Omega$$

Ω has of course many elements which themselves are sets. Some of these are elements in themselves, such as Ω, while others – the most – do not have this somewhat unusual property. We denote *the set of all objects having the property that it is not an element in itself* by Δ:

$$\Delta = \{\Gamma \mid \Gamma \notin \Gamma\}$$

The question now becomes: *Is it true or false that $\Delta \in \Delta$?*

Assume that $\Delta \in \Delta$. Then Δ does not satisfy its defining condition, thus Δ can not be an element in Δ, i.e., $\Delta \notin \Delta$. Assume on the other hand that $\Delta \notin \Delta$. Then *the defining condition is satisfied*. Hence we do get $\Delta \in \Delta$. So both alternatives are impossible. This is clearly a self-contradictory result.

The solution to this and other mathematical debacles stemming from deficiencies in the set-theoretical foundation of mathematics, was to introduce *Axiomatic Set-Theory*. The axioms for set theory prescribe in detail exactly which sets can exist. Thus for instance, the set Ω introduced above *does not exist*, it is "too big".

The Axiomatic Set-Theory takes as its starting point the undefined terms *Set* and *Element*, as well as a relation $\in$ which may exist between an element and a set. The theory does not take any position on "what the sets and elements really are", the interplay between the terms as prescribed by the axioms being the issue of concern.

We shall not give the complete system of axioms for set theory, but confine ourselves to some selected, and even somewhat simplified axioms to give a

---

[3] He is quoted as having asserted that *"Every good mathematician is at least half a philosopher, and every good philosopher is at least half a mathematician"*

flavor of the theory. The axioms behind our treatment here are due to *Ernst Zermelo, Adolf Abraham Halevi Fraenkel* and *Albert Thoralf Skolem*.

As usual we use the simplified notation $\alpha \notin A$ instead of $\neg(\alpha \in A)$. Also, we shall use the term *"statement"* as explained in Section 7.3. In addition to the symbols we have explained above, we also use the so-called *quantifiers*, $\forall$ and $\exists$: When writing $(\forall\, a \in A)(P(a))$ what we mean to say is this: *"For all elements a in the set A the statement P(a) is true"*. Furthermore, the symbols $(\exists\, a \in A)(P(a))$ expresses that there exists an element a in the set A such that the statement $P(a)$ is true. We are now ready to state the first axiom in the *Zermelo-Fraenkel-Skolem* system for Axiomatic set theory:

Axiom 7.1 (ZFS 1) *Given two sets A and B for which the following statement is true:*

$$(\forall\alpha)(\alpha \in A \Longleftrightarrow \alpha \in B)$$

Then $A = B$.

This is certainly a property which is required in such a theory. It asserts that the membership-relation $\in$ and the rules of mathematical logic determine the sets uniquely. The next axiom assures that whenever a set is given, and a statement involving elements in the given set is formulated, then that statement determines a unique *subset* of the given set:

Axiom 7.2 (ZFS 2) *Given a set Ω, and a statement $P(\alpha)$ about elements α. Then there exists a set Δ such that the following statement is true:*

$$(\forall\alpha)(\alpha \in \Delta \Longleftrightarrow (\alpha \in \Omega) \wedge P(\alpha))$$

Almost as in the *intuitive set theory* we write:

$$\Delta = \{\alpha \in \Omega \mid P(\alpha)\}$$

Already these two simple axioms suffice to salvage some of the most elementary rules from the *intuitive* set theory. But note that there is a big difference between the axiom above on the one hand, and to postulate the existence of the following set on the other:

$$\Delta' = \{\alpha \mid P(\alpha)\}$$

Assuming this would lead to a contradiction in the same manner as for Russell's Paradox in the intuitive (more precisely: Cantor's) setting.

So far so good. But we wish to make sure that there does exist sets at all! Therefore we add the

Axiom 7.3 *There exists one and only one set $\emptyset$ such that the statement*

$$\alpha \in \emptyset$$

is false for all elements α.

The existence of the empty set may of course not be taken for granted, it has to be secured by a separate axiom in the *Zermelo-Fraenkel-Skolem System*. Actually as it turns out, the assertion will follow from some stronger existence axioms. We omit these considerations here. But if we choose to include the empty-set axiom in this form, then we need only to postulate *the existence*, as *the uniqueness* follows from the axioms we have already introduced. This proof is left to the reader.

We now should see what the guiding principle in building the system of axioms is: As much as possible of the intuition should be saved, and put on firm and secure ground. A further step in this process comes with the next axiom:

Axiom 7.4 (ZFS 3) *Given two sets A and B. Then there exists a set Γ such that $A \in \Gamma$ and $B \in \Gamma$*

It follows from this, of course, that any *set* also appears as an element in some other set.

We would of course like to have the usual constructions of union, intersection and complement. For this we need the

Axiom 7.5 (ZFS 4) *Given two sets A and B. Then there exists one and only one set, which is denoted by $A \cup B$, such that the following statement is true:*

$$(\forall \alpha)(\alpha \in A \cup B \Longleftrightarrow (\alpha \in A) \vee (\alpha \in B))$$

Using these axiom we may prove the existence of a fairly large collection of sets. As an example we shall see how *the complement* of one set in another set is constructed:

Theorem 6. *Given two sets A and B. Then there exists a uniquely determined set $B - A$ such that the following statement is true:*

$$(\forall \alpha)(\alpha \in B - A \Longleftrightarrow \neg(\alpha \in A) \wedge \alpha \in B))$$

Proof. We use ZFS 2 with $\Omega = B$ and $P(\alpha) = \neg(\alpha \in A)$. $\square$

We have not postulated the existence of *the intersection* of two sets. This may also be deduced from the axioms:

Theorem 7. *Given two sets A and B. Then there exists one and only one set, which is denoted by $A \cap B$, such that the following statement is true:*

$$(\forall \alpha)(\alpha \in A \cap B \iff (\alpha \in A) \wedge (\alpha \in B))$$

Proof. We use ZFS 2 with $\Omega = A$ and $P(\alpha) = (\alpha \in B)$ $\quad\square$

Having reached this stage, it is not difficult to define – by abstract use of the axioms – what we mean by the relation $A \subset B$ between to sets. We also easily give meaning to the usual notation

$$A = \{\alpha_1, \alpha_2, \ldots, \alpha_n\}$$

for a *finite set.*

In the further development of the system we come to an axiom *ZFS 5,* which guarantees the existence of *the set of all subsets* $\mathcal{P}(A)$ of an arbitrary set A. Using this and the earlier axioms one may prove the existence of *the product* of to sets A and B:

$$A \times B = \{(\alpha, \beta) | \ (\alpha \in A) \wedge (\beta \in B)\}$$

A further axiom *ZFS 6* guarantees the existence of sets of *infinitely many elements*: This axiom postulates the existence of a set A such that $\emptyset \in A$ and such that if $\alpha \in A$, then we have $\alpha \cup \{\alpha\} \in A$.

The last axiom which we shall explain here, is the seventh one, *ZF 7.* It is called *The Axiom of Choice*, and may be formulated as follows in plain language: *If Ω is a set whose elements are sets, then we may form a set Δ by choosing one element from each set which occur as an element in Ω.*

This apparently innocuous statement is actually somewhat controversial. The point is that the notion of *a set* is still quite far-reaching, and some mathematicians feel that to just *choose with no specification of constructive procedure* from a *set of sets*, is a rather sweeping acceptance of what kinds of sets we will allow to exist. But in many cases where one proves a result using the axiom of choice, it is actually possible to proceed by more constructive and thus, safer, means.

The author learned set theory from Skolem in the fall of 1962, attending the very last lectures he gave on this subject as *Professor Emeritus* in Oslo, Norway. Skolem was very skeptical to the Axiom of Choice, and spent several lectures deducing seemingly absurd consequences of this axiom. The implied question being: *"Now, do we really want this?"* In the end he summed up his view as follows: *"In mathematics, we can not and indeed, should not, prohibit anyone from the study of any odd system of axioms. However, we must be allowed to question the meaning of results obtained from such a set of assumptions."*

The two last axioms are of a more technical nature, and will not be treated here. The reader may consult [32] for this and further reading.

7.5 General Theory of Axiomatic Systems

Set theory lies at the base for all mathematics, at least in theory. Thus in all other axiomatic theories one may take this as starting point. The undefined terms could then be elements in certain sets, and relations between these. Other axiomatic theories may deal with more general classes of objects than sets, this is the case with *Category Theory,* for instance. Such theories fall outside the scope of the this book.

When it is possible to give a concrete interpretation of the undefined terms, in such a way that all the axioms become true statements, then we have constructed a *model* for the axiomatic theory. If this is possible, then clearly the system of axioms do not contain any contradiction.[4] In this case we say that the system is *consistent.* If no axiom can be deduced from the remaining ones, then we say that the system is *independent.*

Finally, we say that an axiomatic system is *complete* if any statement involving the undefine terms either may be proven or may be disproven, i.e., the *negation* of the statement may be proven, by means of the axioms.

We cannot leave the subject of general axiomatic theories without mentioning a result which is considered deeply troubling by some, but others have found ways to live with: In 1931 the brilliant mathematician and logician *Kurt Gödel* showed that *if an axiomatic system is complete, then it can not be proven consistent by means from inside the system itself.* Thus if we want to be certain that there are no contradictions in our system, then we have to live with some *undecidable* questions. This insight certainly came as a shock to Hilbert and others, who had wanted to treat all of mathematics as a huge axiomatic system, complete and consistent, thus providing a firm and absolutely secure basis for our knowledge. Unfortunately – or fortunately – the reality was not that simple.

---

[4] This statement has to be qualified somewhat. Suppose that the model is constructed in terms of the real numbers. This will be the case for all the models we are going to consider here. Then we may conclude from the existence of the model that if there is no contradiction among the axioms leading up to the construction of the reals, then there is no contradiction in our axiomatic system under consideration. As the former statement is taken for granted, we may say that our new system is free of contradictions. This is *the fine print of math.* We should realize it is there, but not spend our lives worrying about it.

8 Axiomatic Projective Geometry

8.1 Plane Projective Geometry

The axiomatic treatment of *plane projective geometry* has at its starting point three *undefined terms: point, line and incidence*. We are given one *set* $\mathcal{P}$, which we call *the set of points*, and another set $\mathcal{L}$ which we call *the set of lines*. Further, there is given a *relation* between elements from $\mathcal{P}$ and elements from $\mathcal{L}$ which is denoted by I, and referred to as *incidence*. If $PI\alpha$ holds for $P \in \mathcal{P}$ and $\alpha \in \mathcal{L}$, then we say that *the point P is incident with the line α*.

Of course we will tie our intuition to the implementation of the abstract setting where α is a "real" line, P a "real" point and that $PI\alpha$ means that P lies on the line α in the sense that $P \in \alpha$. But as we will see later, it is both possible and indeed some times desirable, to give *other* implementations of these terms, such that the axioms which we shall formulate below hold also in these cases.

Furthermore, we write $\alpha I P$ as meaning the same as $PI\alpha$. It simplifies the exposition, and is in close agreement with our intuition and usual language to extend the meaning of the relation I in this manner.

Thus our point of departure is to have abstractly given two sets, $\mathcal{P}$ of "points" and $\mathcal{L}$ of "lines", and a relation I which may hold between elements $\alpha \in \mathcal{L}$ and $P \in \mathcal{P}$, and which we write as $PI\alpha$ or as $\alpha I P$.

For reasons of practicality only, we will not use this unfamiliar and a little awkward relation I, but instead of "$PI\alpha$" or P is incident by α we simply say that "*the point P lies on the line α*". Similarly we replace the statement $\alpha I P$ or α is incident with P by "*the line α passes through the point P*". We also retain, based on the stringent meaning prescribed above, such formulations as: "*The two lines α and β meet in the point P.*

This is a matter of pure practicality, to get a more natural language in line with our usual geometric intuition. As made precise in this way, the "informal" statements this leads to will not at all be less "mathematically precise" than the more formal statements using the relation I as well as logical and set-theoretical notation. As we have given the intuitive geometric language a *new meaning*, tying it to our new, stringent axiomatic context, there is nothing preventing us from continuing to use the familiar words in our further work. Indeed, as some educators made their ill-fated experiment with the so-called *"New Math"* in Elementary School, one of the reasons why it turned out to be a disappointment may have been an unfounded belief that a high level of formalism is a precondition for a high level of mathematical precision. But that is by no means always the case, and a needlessly high symbolic complexity may indeed hide rather than enhance the mathematical ideas. This is particularly true for the teaching of *geometry*.

Projective Plane Geometry is built on the following system of axioms. As we shall see later on, there are other, equivalent axiom-systems as well.

Axiom 8.1 *Let P and Q be two different points. Then there exists one and only one line α such that P and Q lie on α.*

Axiom 8.2 *Two different lines meet in one and only one point.*

This axiom shows that we are no longer dealing with *Euclidian geometry*: Tow lines will always meet. Intuitively we may adapt the familiar notions from Euclidian plane geometry by adding points at infinity, and then prescribing that parallel lines meet at infinity. Then we still have an axiom which expresses something not alien to our experience.

Three different points are said to be *collinear* if they lie on the same line.

The next axiom is needed to get more than a single line in our projective plane:

Axiom 8.3 *There are at least three non-collinear points.*

We need an axiom to the effect that lines may not merely be pairs of points.

Axiom 8.4 *On every line there are at least three points.*

A model for this axiomatic system is depicted in Figure 8.1. This model is forced on us as the smallest geometry which satisfy the four axioms above. In fact, there must exist a line, and on it there must be at least three points. Then, there are at least three non-collinear points, thus a point outside the line. We then have the "triangle", as well as three points on the " base-line". But by the system of axioms, the other two lines must have an extra point, yielding at least three on each of them. We now have six points but only three lines. But the "middle point" on each of these three lines and the point outside it, together determine a line, yielding three more, so we have six. But so far these new three lines have only two points. If we let them all pass through a seventh point in the middle, we have added new points in the most economical manner. So we do that, wishing as we do, to keep the number of points minimal. The system demands that two points on different "sides" of the triangle, together determine a line. This is accomplished, in the minimal way, by letting that line pass through all the "middle points". Now all of the axioms are satisfied, and there is no need to add more points or more lines.

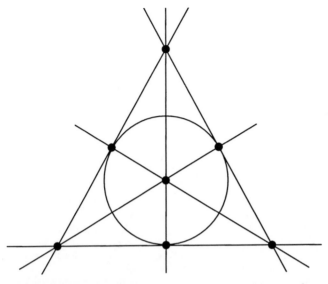

Fig. 8.1. This geometry consists of seven points and seven lines. Each line has three points on it. Three of the lines are drawn as the sides in a triangle, with three of the points appearing as the vertices. On each line the third point is depicted between the two "vertices". The seventh point in the geometry is depicted in the center of the triangle, with three lines passing through it, as shown. We now have six lines, and the seventh is shown as a circle, in the picture it is inscribed in the triangle. Note that there are no other points than the ones marked, thus the lines and the circle are drawn only for the purpose of illustration.

We have thus shown that *this axiom system is consistent* or *that it contains no contradiction*. Indeed, if it contained contradictions, then it could not have a model. We make the following

Definition 1 (PROJECTIVE PLANE). *A model for the above system of axioms is called a projective plane, or a plane projective geometry.*

We now pose the question of whether this axiom-system is *minimal*. Conceivably, for example, the fourth axiom might be deducible from the first three. We now prove that this is not possible.

Theorem 8. *The first three axioms do not imply the fourth.*

Proof. It suffices to exhibit a model in which the first three axioms hold, but not the fourth. Such a model is depicted in Figure 8.2. □

In a similar way we easily prove that none of the four axioms follow by the remaining three. Thus we have the

Theorem 9. *The four axioms for plane projective geometry are independent.*

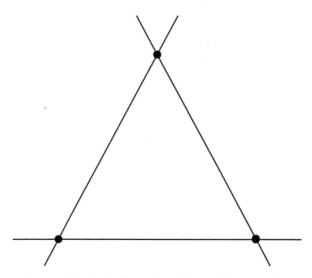

Fig. 8.2. This model, which shows that the fourth axiom does not follow from the first three, consists of only three points, not on the same line. Evidently the three first axioms are satisfied in this model, while the fourth does not hold.

We shall now state and prove the first simple theorems in projective plane geometry. We start by making a definition.

Definition 2. *An Arc of Four is a set of four points such that any choice of three among them is not collinear.*

We may prove the following:

Theorem 10. *There exists at least one Arc of Four.*

Proof. The construction in the proof is shown in Figure 8.3. By Axiom 8.3 there exist three non-collinear points A, B and C. By Axiom 8.1 we have uniquely determined lines AB, AC and BC. By 8.4 the line AC contains one more point, say D. For the same reason, the line BD must contain one more point, which we denote by E. We now claim that $\{A, B, C, E\}$ will constitute an Arc of Four. Namely, by construction A, B, C are not collinear. If A, B, E were collinear then the point D would be on AB, such that C would be on AB, contrary to the above. If A, C, E were collinear, then E and D would coincide because of the uniqueness of the point of intersection of two different lines, Axiom 8.2. It remains to check B, C, E: If these were collinear, then C and D would coincide by Axiom 8.2 as in the previous case. Thus the claim is proven. □

The four axioms for the Projective Plane have a remarkable property: If we interchange the words "point" and "line", then we get *true* statements,

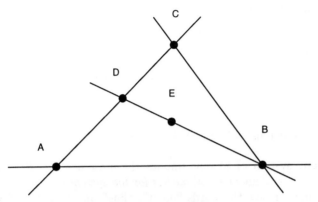

Fig. 8.3. In this illustration the construction of the Arc of Four carried out in the proof is depicted.

in the sense that they may be deduced from the given four axioms. Thus, for instance, Axiom 8.1 is transformed into Axiom 8.2, while Axiom 8.2 is transformed into Axiom 8.1 by interchanging "point" and "line". The axioms 8.3 and 8.4 are transformed into the statements of the following two theorems:

Theorem 11. *There are at least three lines not passing through the same point.*

Proof. Choose the points A, B, C, by Axiom 8.3. Then the lines AB, AC and BC will not pass through the same point. □

Theorem 12. *For all points there exist at least three lines passing through it.*

Proof. Let A be an arbitrary point, and choose two new points B, C such that A, B, C are not collinear. This is possible, as otherwise all points in the geometry would lie on the same line, so that Axiom 8.4 would not hold. We now choose a third point on the line BC, call it D. Then the three lines AB, AC, AD satisfy the claim. □

This simple observation is of fundamental and far reaching significance: It lies at the foundation of the so-called *Principle of Duality*.

We shall say that a statement is a *valid theorem in projective plane geometry*, if it may be deduced from the axiom-system for the projective plane given above.

Then we have the following result:

Theorem 13 (Principle of Duality). *We get a new valid theorem in projective plane geometry whenever we interchange the words "point" and "line" but retain incidence, in a valid theorem for plane projective geometry.*

We have proved a valid theorem in projective plane geometry above, namely the existence of an Arc of Four, Theorem 10. This now implies the

Corollary 1. *There are at least four lines so that no selection of three among them all pass through the same point.*

Proof of the Principle of duality: Let P be a valid theorem in plane projective geometry. Then

$$\text{The axioms } 8.1, \ 8.2, \ 8.3 \ \text{ and } \ 8.4 \Longrightarrow P$$

In other words, our system of axioms implies the statement P. But this means that P is a true statement *in all models for the system of axioms*, i.e., P will be true no matter how the words "point", "line" and "incidence" are interpreted, as long as the statements provided by the four axioms are valid for the interpretation.

We now form a model by interchanging the words "point" and "line", retaining the meaning of incidence. Abstract as it is, this still will be a model, since we have proven that all four axioms hold with this interchange of "points" and "line". In particular the statement P is true for this model, i.e., the dual statement is true. This completes the proof. $\square$

We may replace the axioms 8.3 and 8.4 by the statements in Theorem 10 and Corollary 1. To prove this, it remains to show that these four statements conversely imply the axioms 8.1, 8.2, 8.3 and 8.4. This system of axioms would have the advantage that the Principle of duality now would be completely obvious, as the axiom-system would be *self dual*. The disadvantage is that such an axiom-system would appear less intuitively evident and less natural.

8.2 An Unsolved Geometric Problem

Already at this stage we now may formulate an interesting open geometric problem, which in some form or another goes back to Euler, around 1780. Conforming, perhaps, to the needs of governments at his time, concerning mathematical knowledge, he formulated the problem as different ways of arranging soldiers in rows and columns. A problem which fails to stir our interest nowadays. But mathematically Euler's problem is interesting enough, and deals with what we today call *Latin Squares*.

We shall say that a projective plane is *finite* if the set $\mathcal{P}$ of points is a finite set. Omitting the set of lines from our notation, we will from now on say *"the projective plane $\mathcal{P}$"*, letting the set of lines be understood.

Definition 3. *Given a finite projective plane $\mathcal{P}$. Let M denote the maximal number of points on any line. Then say that the geometry $\mathcal{P}$ is of order $m = M - 1$.*

Thus for example, the 7-point geometry which we have encountered already, is of order 2. Not only that, but the number of points on *all* lines is $M = 3$ in this case.

Indeed, one may prove in general that the number of points on all lines is equal to the maximal number M. Moreover, the number of lines through any point is equal to M as well.

We have, as noted, constructed one geometry of order 2. It also is very easy to construct geometries of order p, where p is a prime number. And it is not difficult to construct geometries of order p^r, where p is a prime and r a natural number > 0. But the following conjecture has turned out to be very difficult.

Conjecture 1. All finite geometries have order equal to p^r, where p is a prime number and r a natural number > 0.

It may appear surprising that this should be so difficult to decide. But this is the way it frequently turns out in mathematics: A problem may be very simple both to state and to understand, but extremely difficult to solve.

Thus the problem is the following: *Assume that m is a natural number which is not a power of a prime. Show that then there is no Plane Projective Geometry of order m.*

As far as I know the best result towards a solution of this conjecture was obtained in 1949 by *R. H. Bruck* and *H. J. Ryser.* They showed the following:

Theorem 14 (Bruck-Ryser). *If the number m is not a power of a prime, and is such that division by by 4 leaves a remainder of 1 or 2, and furthermore m can not be written as a sum of two squares, then m is not the order of some plane projective geometry.*

This remarkable result excludes an infinite number of cases. The first of them are the numbers $6, 14, 21$ and 22. Take for instance 6: Division by 4 yields the remainder 2, and as $6 = 1 + 5 = 2 + 4 = 3 + 3$ are the only ways in which one may write 6 as the sum of two natural numbers, we see that 6 can not be written as the sum of two squares. Thus the result of Theorem 14 shows that 6 is not the order in a plane projective geometry.

The first case which remains open after the Bruck-Ryser Theorem is therefore the case $m = 10$. Division by 4 does give the remainder 2, but $10 = 1 + 9$, so that it is the sum of two squares and the last part of the test fails. So the first case to check comes down to the following problem:

Problem 1. Does there exist a plane projective geometry in which all lines have 11 points?

Other open cases are $m = 12, 15, 18, 20$. There are infinitely many such open cases.

It may appear surprising that the case $n = 10$ is so difficult to decide. But we may rapidly see why one can not just sit down with paper and pencil and check out all cases which may occur. Namely, as mentioned above there are $m = 11$ lines passing through each point. Since two lines will have exactly one point in common, the total number of points in such a projective plane would be

$$11 \times 10 + 1 = 111 :$$

Indeed, choose a point say P_0. Through this point passes 11 lines. Each of them have 10 points in addition to P_0. By Axiom 8.1 all points in the plane will appear on one of these lines, so in addition to P_0 there are altogether 11×10 points. Including P_0 this gives a total 111 of points, as claimed.

We now wish to check all possibilities for prescribing subsets of a set containing 111 elements, or "points", yielding a set of subsets, called "lines", such that our four axioms are satisfied.

A digression at this stage: The alert reader may now be concerned, that we have moved to a concrete interpretation of the undefined term "line". But this is always possible: We may harmlessly identify a "line" ℓ with the *subset* $\mathcal{P}_\ell$ of $\mathcal{P}$ consisting of all $P \in \mathcal{P}$ such that $PI\ell$. Namely, as is easily seen we have that for two lines α and β we have $\alpha = \beta$ if and only if P is incident with α exactly when P is incident with β. The formal proof is left to the reader, the key being that if $\alpha \neq \beta$, then they have exactly one point in common.

If we make this identification, then the incidence relation is interpreted as $\in$, the membership relation between an element and a set.

Thus a "line" will be a subset of $\mathcal{P}$ containing 11 points. Altogether there are a total of

$$\binom{111}{11} = 473239787751081$$

such subsets. Now the number of "lines" in a Plane Projective Geometry is equal to the number of "points", thus in this case 111. Hence the candidates for projective planes of order 10 will be every selection of 111 among these 473239787751081 subsets, altogether a staggering $\binom{473239787751081}{111}$ possibilities to be checked for compliance with our four axioms. The reader is recommended to take a few minutes to compute this number, say by MAPLE or a similar system. Just do not try to do it by hand or calculator.

This is certainly a completely insurmountable task. Even though we do not have to work through absolutely all the possible selections, and even though the checking can be made a lot smarter than just working through all possibilities, even this initial case presents us with a real challenge.

Thus it attracted a great deal of attention when a group of mathematicians and computer scientists at Concordia University in Montreal showed that $m = 10$ can not be the order of any plane projective geometry. Their proof is dependent on massive computer usage, but also involves intricate mathematical considerations. For details we refer to the interesting article by *B. A. Cipra* in Science, [3].

8.3 The Real Projective Plane

We start by recalling some standard analytic geometry. When we introduce coordinates, a point is represented by a *pair* of real numbers, (x, y). A line is represented by its equation,

$$AX + BY + C = 0,$$

where A, B, C are real constants and X, Y are the *variables*. The line given by the equation, then, is the set of all points (x, y) in the plane represented as $\mathbb{R}^2$ consisting of all points (x, y) such that

$$Ax + By + C = 0.$$

This gives us the tool for transforming geometric considerations from the usual Euclidian plane into algebraic computations.

To be precise, we have here constructed a *model for the Euclidian plane*, in an axiomatic setting. We realize, however, that this will not be a model for the axiomatic system defining plane projective geometry. Indeed, all the axioms hold, *except for* Axiom 8.2.

We now show how $\mathbb{R}^2$ may be extended to the *real projective plane*, such that this axiom also holds. We do that by adding some points, which we will call *the points at infinity*.

The resulting model will be the so-called *real projective plane* $\mathbb{P}^2(\mathbb{R})$.

At first the definition may look strange and unnatural:

Definition 4. *The set $\mathcal{P}$ of points in the real projective plane $\mathbb{P}^2(\mathbb{R})$ is the set of all lines through the origin $(0, 0, 0) \in \mathbb{R}^3$.*

The set $\mathcal{L}$ of lines is the set of all planes in $\mathbb{R}^3$ which pass through $(0, 0, 0)$.

We say that the projective point P, that is to say, the line $a \subset \mathbb{R}^3$, is incident with the projective line L, i.e, the plane $p \subset \mathbb{R}^3$, if $a \subset p$. We write then, as in the formal setting, PIa.

We next verify that all four axioms are satisfied.

Verification of Axiom 8.1: Let P and Q be two different "points" in *the real, projective plane*, as defined above. Thus P and Q are two different lines through $(0, 0, 0)$ in $\mathbb{R}^3$. Two such lines span a unique plane through $(0, 0, 0)$. This plane is by our definition a "line" in $\mathbb{P}^2(\mathbb{R})$, which we denote by a in Figure 8.4.

Verification of Axiom 8.2: Let a and b be two different "lines" in the real, projective plane, thus two planes through $(0, 0, 0)$ in $\mathbb{R}^3$. These two planes will intersect in a unique line p in $\mathbb{R}^3$ passing through the origin. This line is, according to our definition, a "point" in the real, projective plane $\mathbb{P}^2(\mathbb{R})$. We change its name, denoting it by P, and thus have that the two lines a and b intersect in the point P. Moreover, P is uniquely determined. The situation is illustrated in Figure 8.5.

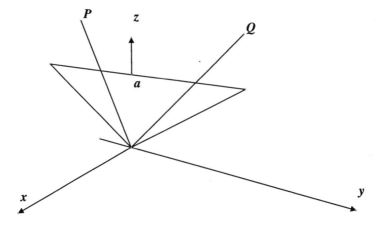

Fig. 8.4. This illustration shows how the two points $P \neq Q$ in the model $\mathbb{P}^2(\mathbb{R})$ determine a unique line a in $\mathbb{P}^2(\mathbb{R})$.

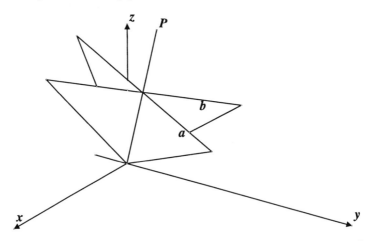

Fig. 8.5. This illustration shows how the two lines $a \neq b$ in the model $\mathbb{P}^2(\mathbb{R})$ determine a unique point P in $\mathbb{P}^2(\mathbb{R})$.

Verification of Axiom 8.3: Let p, q and r be three non-coplanar lines (that is to say, lines not in the same plane) passing through $(0,0,0)$ in $\mathbb{R}^3$, see Figure 8.6. These lines then represent points, which we rename P, Q, R, which are not on the same line in the model $\mathbb{P}^2(\mathbb{R})$.

Verification of Axiom 8.4: Let ℓ be a line in the model $\mathbb{P}^2(\mathbb{R})$. According to our definition, this is a plane passing through the origin in $\mathbb{R}^3$. Clearly there are at least three different lines p, q, r in that plane, passing through the origin. The situation is shown in Figure 8.7.

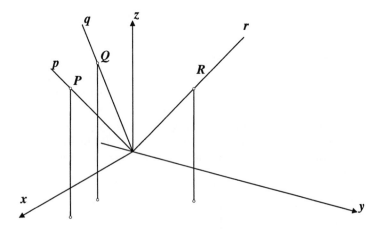

Fig. 8.6. This illustration shows how we prove that there are three non-collinear points in $\mathbb{P}^2(\mathbb{R})$.

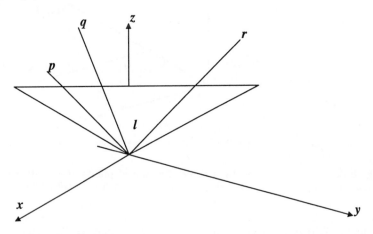

Fig. 8.7. This illustration indicates how we prove that there are at least three points on every line in $\mathbb{P}^2(\mathbb{R})$.

We shall now compare the points in the model $\mathbb{P}^2(\mathbb{R})$ to the usual points of $\mathbb{R}^2$. In fact, we prove that the points in $\mathbb{R}^2$ may be identified with *certain* points in $\mathbb{P}^2(\mathbb{R})$.

We let the point $(x, y) \in \mathbb{R}^2$ correspond to the line which passes through the points $(0, 0, 0)$ and $(x, y, 1)$ in $\mathbb{R}^3$. This line is uniquely determined by the point $(x, y) \in \mathbb{R}^2$, and will be denoted by $a(x, y)$. When we let the point (x, y) run through $\mathbb{R}^2$ then the corresponding line $a(x, y)$ will run through the set of lines through the origin in $\mathbb{R}^3$ which are not contained in the xy-plane. We denote the set of these lines, considered as points in $\mathbb{P}^2(\mathbb{R})$, by $\mathbb{P}^2(\mathbb{R})_0$. We sum up what we have proved as the

Proposition 1. *The mapping which sends the point (x, y) to the line $a(x, y)$ identifies $\mathbb{R}^2$ with $\mathbb{P}^2(\mathbb{R})_0$.*

Denote the complement of the set $\mathbb{P}^2(\mathbb{R})_0$ in $\mathbb{P}^2(\mathbb{R})$ by $\mathbb{P}^2(\mathbb{R})_\infty$. This is the set of points in the model $\mathbb{P}^2(\mathbb{R})$ given by lines which lie in the xy-plane, and those are the "new" points which have been added to $\mathbb{R}^2$. They are thus the *points at infinity*. If we now let (x, y) move outward towards infinity in $\mathbb{R}^2$, then the corresponding line will approach the xy-plane more and more, but never actually quite reach it. The *limiting positions* for these lines will therefore be the lines in the xy-plane, which we view as points at infinity. See Figure 8.8.

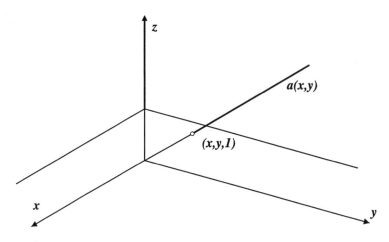

Fig. 8.8. $\mathbb{R}^2$ is augmented by a projective line of points at infinity to form the projective plane $\mathbb{P}^2(\mathbb{R})$: A line through the origin $(0, 0, 0) \in \mathbb{R}^3$ which does not lie in the xy-plane corresponds to the point $(x, y) \in \mathbb{R}^2$, where the coordinates are given by the line meeting the plane $z = 1$ in the point $(x, y, 1)$. The projective line at infinity is then the projective line which corresponds to the plane $z = 0$, the xy-plane.

We take a closer look at the points in $\mathbb{P}^2(\mathbb{R})_\infty$, the set of all lines through the origin and contained in the xy-plane. The basic observation is the following: The *xy-plane is simply a line in the model $\mathbb{P}^2(\mathbb{R})$*, whose points, which we have denoted by $\mathbb{P}^2(\mathbb{R})_\infty$, all lie at infinity. The xy-plane is therefore nothing but *the line at infinity in $\mathbb{P}^2(\mathbb{R})$*. We also denote this line by L_∞. Adding this line to $\mathbb{R}^2$ represents a *compactification* of $\mathbb{R}^2$.

The set of points which lie on a fixed line ℓ in $\mathbb{P}^2(\mathbb{R})$ correspond to lines through the origin $(0, 0, 0) \in \mathbb{R}^3$, contained in the plane p which corresponds to the line ℓ. We may identify the plane p with $\mathbb{R}^2$, and then the points on the projective line ℓ correspond to the lines in $\mathbb{R}^2$ passing through the origin $(0, 0) \in \mathbb{R}^2$. Thus we have a completely analogous situation to the points in

$\mathbb{P}^2(\mathbb{R})$, except for the dimension being reduced from 3 to 2. Indeed, in the previous situation we had a *projective plane*, whereas we now are dealing with a *projective line*.

Definition 5. *The set of lines through* $(0,0)$ *in* $\mathbb{R}^2$ *is called the points on the real, projective line, and is denoted by* $\mathbb{P}^1(\mathbb{R})$.

Again, in the same way as we saw for $\mathbb{P}^2(\mathbb{R})$, $\mathbb{P}^1(\mathbb{R})$ may be viewed as $\mathbb{R}$ augmented by a point at infinity, ∞. In Figure 8.9 this point at infinity corresponds to the x-axis.

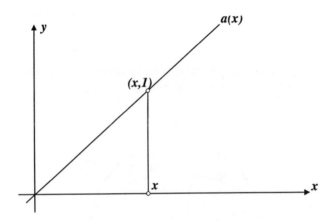

Fig. 8.9. $\mathbb{R}$ is augmented by a point at infinity to form the projective line $\mathbb{P}^1(\mathbb{R})$.

The real, projective plane $\mathbb{P}^2(\mathbb{R})$ may be viewed as a *compactification* of $\mathbb{R}^2$, obtained by adding a *boundary* to the surface $\mathbb{R}^2$ consisting of a *real, projective line* at infinity. This real, projective line in turn consists of $\mathbb{R}$, to which there is added one point at infinity, the *one point compactification* of the reals $\mathbb{R}$. That, of course, may be identified with *a circle*, as illustrated in Figure 8.10.

The lines through $(0,0) \in \mathbb{R}^2$ may be identified with the points of the semicircle shown in Figure 8.10, with the exception of the two points on the x-axis, which corresponds to the two diametrically opposite points A and B. When these two points are identified, then the semicircle is joined to a circle, and the set of lines through $(0,0) \in \mathbb{R}^2$ is identified with it. Thus the real projective line $\mathbb{P}^1(\mathbb{R})$ may be identified with a circle.

The projective plane $\mathbb{P}^2(\mathbb{R})$ may be given a similar interpretation. This is illustrated in Figure 8.11.

Here we represent a line through $(0,0,0) \in \mathbb{R}^3$ by its point of intersection with the northern hemisphere of a fixed sphere with center at the origin. Under this correspondence the points at infinity, that is to say, the ones

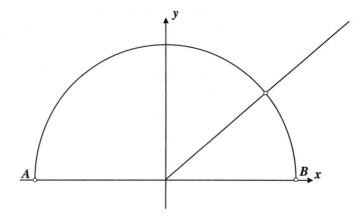

Fig. 8.10. The projective line $\mathbb{P}^1(\mathbb{R})$ may be identified with a circle, as it consists of $\mathbb{R}$ to which is added a single point at infinity.

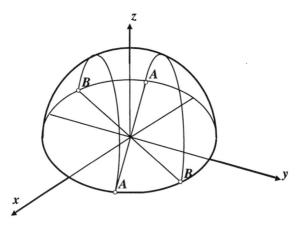

Fig. 8.11. The real projective plane $\mathbb{P}^2(\mathbb{R})$ is the surface obtained from the upper hemisphere, with diametrically opposite points on the equator identified. A highway on the real projective plane will have some curious properties, as explained in the text.

corresponding to lines in the xy-plane, correspond to two points, diametrically opposite on the equator. Thus we have to identify diametrically opposite equatorial points, and in doing so we obtain a representation of the surface $\mathbb{P}^2(\mathbb{R})$.

We shall return to the procedure for identifying points later, this technique requires some firm and stringent foundations which we have not yet developed. At this point one may view the identification as an informal explanation.

The surface $\mathbb{P}^2(\mathbb{R})$ has some very interesting properties. Thus for instance, it is a surface with only one side! This is seen as follows. In Figure 8.11 we may imagine that a highway is constructed, as indicated by the strip drawn over the upper hemisphere shown there. We may now drive off on this highway, approaching the equator. We then actually approach the points at infinity, judged from our initial position. As we move along, however, the perception of where infinity is located changes, so as to always be off in the distance. The line at infinity would be like the rainbow, always moving away as we attempt to chase after it. We will eventually return to the starting point. We will then be driving on the other side of the road, in a somewhat disturbing way: The car will be upside down, *under the pavement*. This is best understood by cutting out the highway from $\mathbb{P}^2(\mathbb{R})$, and examining it more closely. Indeed, we get a strip from the upper hemisphere, fairly approximated by a rectangular strip of paper. Then the diagonally opposite points at the short sides will have to be identified, as they were diametrically opposite equatorial points on the hemisphere. Thus we twist and glue the strip, obtaining the *Möbius strip* shown in Figure 8.12.

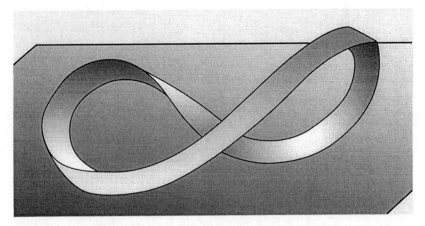

Fig. 8.12. The highway on the real projective plane is nothing but the *Möbius strip*.

It follows as well, from the considerations above, that the points in the usual Euclidian plane $\mathbb{R}^2$ may be represented by the points of the northern hemisphere, where the equator is not included. Thus the points at infinity, which are added to the Euclidean plane to give the real projective plane, are precisely the points obtained by identifying diametrically opposite points on the equator. This is the projective line of points at infinity. Now any projective line may be identified with a circle, as we have seen. Thus, whenever we identify diametrically opposite points on a circle, we get another circle as result.

When we identify the Euclidian plane $\mathbb{R}^2$ with the upper hemisphere, then the lines are represented as the circles of intersection between the northern hemisphere of the sphere and planes through the origin $(0,0,0)$. That is to say, the lines correspond to the northern pieces of *great circles* on the sphere. We thus have a model for Euclidian geometry in which the points are the points on the northern hemisphere, and the lines are the northern parts of great circles. It is not difficult to verify that Euclid's postulates are satisfied by arguing directly with these definitions. However, we of course know this already by the way this last model was constructed from $\mathbb{R}^2$, where the axioms do hold.

We may take this one step further, and *project* this model onto the xy-plane, parallel with the z-axis. This yields a third model for the Euclidian plane, in which the points are the points in the interior of a fixed circle, and the lines are half ellipses, with the longest axis along a diameter of the fixed circle. The points on the boundary of the fixed circle is not included. Figure 8.13 illustrates the situation.

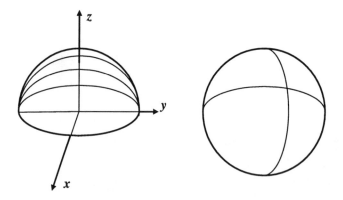

Fig. 8.13. Two models for the Euclidian plane.

A variation of this model is obtained by letting the lines be circle arcs, with a diameter of the fixed circle as a cord. It is shown in Figure 8.14.

Finally, there is another way of representing the points of the Euclidian plane: It is to use *all* the points of the entire sphere, *except* for the *North Pole*. Taking out the North Pole leaves us with a *punctured sphere*.

The correspondence between the points in the Euclidian plane $\mathbb{R}^2$ and the points on the punctured sphere is given by *projection with center at the North Pole, N:* A point A on the punctured sphere is mapped to the point $pr_N(A) = B$ obtained as the point of intersection between the xy-plane and the line passing through the points N and A. See Figure 8.15.

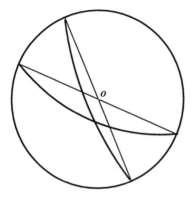

Fig. 8.14. The model for the Euclidian plane in which the lines are circle arcs.

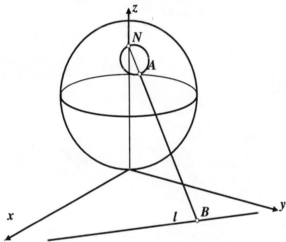

Fig. 8.15. The punctured sphere projected onto the plane.

The xy-plane is tangent to the sphere at the origin $(0,0,0)$, the South Pole. The line ℓ will correspond to a small circle as indicated, namely to the circle of intersection between the sphere and the plane determined by the line ℓ and the North Pole, the center of projection.

Thus the entire sphere, including the North Pole, may be viewed as another kind of *compactification* of the Euclidian plane $\mathbb{R}^2$ than the real projective plane $\mathbb{P}^2(\mathbb{R})$ which we have already seen. Here we only add *one single point* at infinity. We say that the sphere is the *one point compactification* of $\mathbb{R}^2$.

Actually this is not out of line at all with the framework of projective geometry. In fact the sphere is nothing but *the projective line over the complex numbers,* which we denote by $\mathbb{P}^1(\mathbb{C})$.

9 Models for Non-Euclidian Geometry

9.1 Three Types of Geometry

The absolutely most fundamental model in geometry is of course the real Euclidian plane $\mathbb{R}^2$. But there are many others, even alternative ones for Euclidian geometry. Indeed, in the last chapter we saw how we get several models for the Euclidian plane. One of them was given by representing the points as the points in the interior of a fixed circle, and letting the lines be circle arcs with the cord along a diameter of the fixed circle.

We now show how we may realize the two other possibilities, in which the Fifth Postulate of Euclid does not hold. The models we are going to describe do closely resemble the above-mentioned models for Euclidian geometry, but the subtle changes from it make a lot of difference.

The three possibilities were named by Felix Klein, in 1871, as *hyperbolic* geometry, *parabolic* geometry and as *elliptic geometry*. Here parabolic geometry stands for the usual Euclidian plane, whereas the other two will be described in the following sections. The reason for these names comes from Klein's program of classifying geometry according to the groups of transformations leaving its geometric properties unchanged – *invariant*. A further discussion along these lines does, however, fall outside the scope of this book.

9.2 Hyperbolic Geometry

The *hyperbolic plane* is characterized by satisfying all the axioms and postulates of the Euclidean plane, except for the Fifth Postulate, which is replaced by the assertion that

Postulate 9.1 (Hyperbolic plane) *Given a line and a point outside it. Then there are at least two lines through the point which do not meet the line.*

Hyperbolic geometry is a special case of neutral geometry. We obtain a model for the hyperbolic plane by letting the set of points $\mathcal{P}$ be the set of points inside a fixed circle, and the set of lines $\mathcal{L}$ be circular arcs inside the fixed circle, which are all perpendicular to the fixed circle. The situation is shown in Figure 9.1.

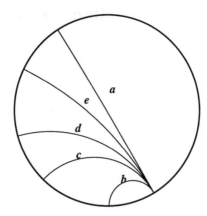

Fig. 9.1. The points in this model for the hyperbolic plane are the points inside a fixed circle, and the lines are the circular arcs which are perpendicular to the fixed circle. We show five parallel lines a, b, c, d and e, meeting at "a point at infinity", loosely speaking since this point at the boundary is not part of the model: The lines are parallel, so they do not meet.

A point P is incident with such a line if it lies on the circular arc, so we keep the usual concept of incidence. This will be so for all the models discussed in this chapter.

In Figure 9.2 we consider a line a and a point P which does not lie on a – which is not incident with a. We see the two lines b and c through P which do not meet a, two lines through P parallel with a. The two lines shown are extremal cases as far as parallels to a through P is concerned: Between them we find an infinite number of parallels through P to a, one of them shown and denoted by d. b and c are referred to as *bounding parallels*.

This model is due to *Henri Poincaré*, and it shows some of its beauty by accurately representing the *angle between lines*. Indeed, a distance – or *metric* – may be defined in the model, but it is quite different from the usual distance between points in the plane. On the other hand, the angle which may be defined between lines in hyperbolic geometry, is the one measured between the circular arcs representing them in this model. We express this by saying that Poincaré's model is *conformal*. From this, we see that the angular sum of a hyperbolic triangle is less than two right angles. This is illustrated in Figure 9.3. We return to the concept of *distance* in this model in Section 9.4

If we keep the center of the circle fixed and let the radius tend to infinity, then this model will approach the usual Euclidean plane. Thus the Euclidian plane is a limiting case of the hyperbolic plane. On the other hand, if we let the radius tend to infinity but keep a point on the circumference fixed, then

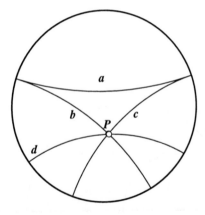

Fig. 9.2. The two bounding parallels to a through P.

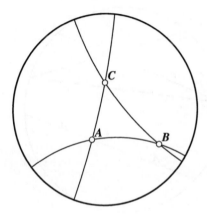

Fig. 9.3. A hyperbolic triangle has angular sum less than two times a right angle.

we get *the upper half-plane model* for the hyperbolic plane, shown in Figure 9.4.

Another model for the hyperbolic plane has been devised by Klein. He also uses the points in the interior of a fixed circle, but defines the "lines" as being all possible *cords* to this fixed circle. This model is illustrated in Figure 9.5.

Klein's model does not have the appealing property of Poincaré's, in that the angle between lines is faithfully represented. In fact, the angle between the cords are the usual ones from the Euclidian plane, thus using them we would get the angular sum of a triangle equal to two right angles, while as asserted above the angular sum of a hyperbolic triangle is less than this. But there are two significant features of Klein's model. First of all, it is manifestly a model

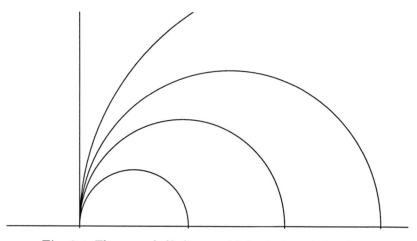

Fig. 9.4. The upper half plane model for the hyperbolic plane.

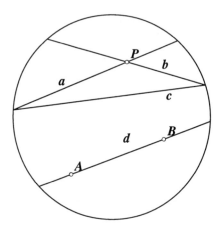

Fig. 9.5. Klein's model for the hyperbolic plane, with two points A and B which determine a unique line d, as well as a point P outside the line c, with the bounding parallels a and b as they are represented in this model.

for hyperbolic geometry developed entirely within Euclidian geometry. Thus consistency of Euclidian geometry *implies the consistency of hyperbolic geometry*. This was an important observation at the time when Klein developed the model. It proves that hyperbolic geometry is at least as consistent as the Euclidian counterpart. The latter being universally accepted, this dispels any objections to hyperbolic geometry from a mathematical point of view.

The second nice feature of this model, is that we may give an elegant description of the *distance* between two points. It uses a fundamental and very interesting concept from *projective geometry*. The concept we use is that of the *cross ratio of four collinear points*. We consider two points P and

Q in Klein's model, and mark points S and T *outside* the model, namely the points of intersection of the line through P and Q in the ordinary Euclidian plane with the fixed circle.

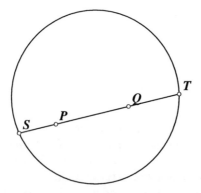

Fig. 9.6. We compute the distance between the points P and Q in Klein's model for the hyperbolic plane as the logarithm of the cross ratio of the four points indicated, as described in the text.

In general the *cross ratio* of four such points, on the same line, is defined as the fraction

$$[T : S; P : Q] = \frac{\frac{TP}{PS}}{\frac{TQ}{QS}}$$

Here, for instance, TP is the distance from T to P, normally with a sign, so we endow the line with an orientation, a positive direction. If the cross ratio is -1, then the ratio is called a *harmonic ratio*, if it is 1 then we speak of an *antiharmonic ratio*.

In the present situation we shall only consider absolute values, so the cross ratio is always non-negative. Klein now defined the *distance* between P and Q as the absolute value of the logarithm of the cross ratio:

$$d(P,Q) = \left| \log \left(\frac{TP}{PS} \Big/ \frac{TQ}{QS} \right) \right|$$

If one of the points P or Q approaches the rim of the fixed circle while the other stays put, then $d(P,Q) \longrightarrow \infty$.

The general concept of distance will be discussed in Section 9.4.

Finally, we recall from Section 4.1 the following alternative version of the Parallel postulate, due to Gauss:

There exists a triangle, the contents of which is greater than any given area.

Thus in a hyperbolic plane there is an upper bound to the areas of triangles.

9.3 Elliptic Geometry

The version of the Fifth Postulate which defines *elliptic geometry* is the following:

Postulate 9.2 (Elliptic plane) *Given a line ℓ and a point P outside it. Then all lines through P meet ℓ.*

We may not introduce this axiom into the system of Hilbert, minus the parallel postulate: Elliptic geometry is not a special case of neutral geometry. But by a suitable modification of some of Hilbert's other axioms, we get a firm axiomatic base for elliptic geometry as well.

We have already constructed a model for this geometry, namely $\mathbb{P}^2(\mathbb{R})$, the real projective plane. As we have seen before, we may identify $\mathbb{P}^2(\mathbb{R})$ with the points inside a fixed circle, with the points on the circumference included this time, but with diametrically opposite points identified. The set of *lines* in this model is the set of half ellipses, end points identified, where the longest axis coincides with a diameter in the fixed circle. See Figure 9.7. This model for elliptic geometry is attributed to Klein.

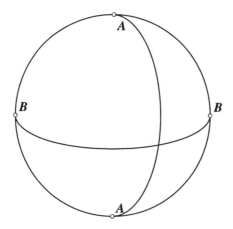

Fig. 9.7. Model for the elliptic plane. Diametrically opposite points on the circumference are identified. The lines are half ellipses, end points identified, where the longest axis coincides with a diameter in the fixed circle.

As was the case for Poincaré's model for hyperbolic geometry, Klein's model for elliptic geometry is conformal. Using this, we see that in the elliptic plane the angular sum of any triangle is greater than two right angles.

We do not go into the details of distance and angular measure in the hyperbolic and the elliptic plane. Instead we refer to M. Greenbergs interesting book [10]. But by building on Hilbert's axioms with the appropriate versions of the Parallel Postulate[1], we get geometries with all features from the Euclidean case, except of course Euclid's Fifth Postulate. In particular there is distance and angular measure. In the Euclidean and hyperbolic cases there is an absolute measure of angles, namely the *radian*. Analogously, in elliptic geometry there is an absolute measure of *length*, in other words of distance. All lines in the elliptic plane have the same finite length. Of course right angles exist in all three versions of geometry.

9.4 Euclidian and Non-Euclidian Geometry in Space

We may construct models for Euclidian, hyperbolic and elliptic *space* by letting the points be all points in the interior of a fixed sphere in the Euclidian and hyperbolic case, and in the elliptic case with the addition of the points at the surface with diametrically opposite points identified. The lines are defined as for the corresponding planes.

The concept of a metric is quite general, and it is one of the most fruitful abstractions undertaken in the development of modern mathematics. So we shall take a few moments to explain it.

We start out with some set S. We make no assumptions about the set S whatsoever, the abstract concept of a metric may be defined in complete generality. If it is to be of any use, however, we need to restrict our attention somewhat, but this is a subject which we shall not pursue here.

A metric defined in S is a *real valued function*, $d(P,Q) \in \mathbb{R}$ of two variables $P, Q \in S$, such that the following is true:

Distance is always non-negative: For all $P, Q \in S$,

$$d(P,Q) \geq 0.$$

If two points are different, then the distance between them is never zero. In an equivalent formulation:

$$d(P,Q) = 0 \iff P = Q$$

The distance from me to you is of course the same as the distance from you to me:

$$d(P,Q) = d(Q,P)$$

The final *axiom for a metric* is usually referred to as the *triangular condition* – the distance traveled from point P to point R is never shortened by passing by some third point Q:

---

[1] And, as already noted, some additional modifications in the elliptic case.

$$d(P, R) \leq d(P, Q) + d(Q, R)$$

In the Euclidian plane $\mathbb{R}^2$ the metric is the usual distance between two points. Of course one readily sees that the three axioms above are satisfied in this case. More formally, we define the metric by writing the function explicitly in terms of the coordinates of the points involved. Denoting the metric in the Euclidian plane by ρ_2, we have

$$\rho_2(P, Q) = \sqrt{(x_1 - x_2)^2 + (y_1 - y_2)^2},$$

where $P = (x_1, y_1)$ and $Q = (x_2, y_2)$. It is an easy exercise in applying the Pythagorean theorem to verify that this yields the usual distance, namely the length of the line segment joining the two points.

Quite analogously, the metric in the Euclidian 3-space $\mathbb{R}^3$ is the usual distance, the length of the line segment joining the two points. This metric is given by the formula

$$\rho_3(P, Q) = \sqrt{(x_1 - x_2)^2 + (y_1 - y_2)^2 + (z_1 - z_2)^2},$$

where now

$$P = (x_1, y_1, z_1) \text{ and } Q = (x_2, y_2, z_2)$$

Again, this is the usual distance. The proof is not quite as simple as in the planar case, here we need to appeal to Pythagoras *twice*.

In general a metric space is defined as a set S, in which there is given a metric as explained above. The set $\mathbb{R}^n$, which consists of all n-tuples $(x_1, x_2, \ldots, x_n)$ where $x_1, \ldots, x_n$ are all real numbers, can also be made into a metric space by defining a metric by

$$\rho_n((x_1, \ldots, x_n), (y_1, \ldots, y_n)) = \sqrt{(x_1 - y_1)^2 + \ldots + (x_n - y_n)^2}$$

It is easy to see that the first two axioms for a metric are satisfied. The third is a little bit more tricky to verify, and is left as a challenge to the reader.

The *abstract notion of a metric space*, which we have explained above, admits many *"spaces"* which we intuitively would reject as both strange and unusual. Thus for instance, we may take *any set* S, say the set of all US senators, and make it into a *metric space* by defining the distance σ as follows:

$$\sigma(P, Q) = \begin{cases} 0 \text{ if } P = Q \\ 1 \text{ if } P \neq Q \end{cases}$$

The metric thus defined clearly satisfies the three axioms. We refer to this metric as the *discrete metric* on the set S. This metric space clearly provides no real information beyond the description of the set S. To pursue this digression one step further, one might try to introduce politics in this, and instead define the distance between two senators from the same party to be 0, and to be 1 for senators from different parties. But this does not yield a metric space. Rather, we obtain what is known as a *pseudo-metric* space.

In the model for Euclidian space, the points are the points in the interior of a fixed sphere. We may introduce a metric in this set, given by a somewhat complicated formula which will not be given here. But with this metric the interior of the fixed sphere becomes a metric space, and if we let the radius of the fixed sphere tend to infinity, then the metric we defined for the interior of the sphere will approach the usual distance in Euclidian 3-space.

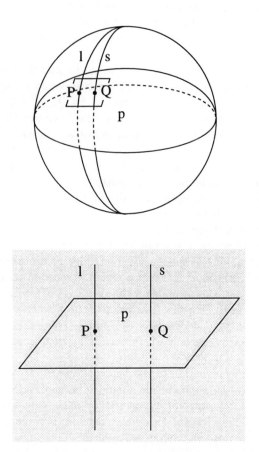

Fig. 9.8. The Euclidian universe inside a fixed sphere. The half ellipsoid with two of its axes along two diameters of the sphere, is a "plane" in this model and is denoted by p. Two "lines" l and s perpendicular to the plane p are shown, intersecting the plane p in the two points P and Q indicated. They also illustrate the Fifth Postulate of Euclid: l and s are parallel, being both perpendicular to the same p, and the line s is the unique parallel to l which passes through the point Q. Below we have cut out a small piece of this space, locally it looks like the normal Euclidian space.

If we imagine that beings in this universe inside the fixed sphere starts at the center, with a spaceship 100 meters long, and travels along towards the boundary at a fixed speed, then an observer outside the sphere would find that 1. The spaceship would shrink in size as the boundary was approached, and 2. The speed would appear to decrease towards zero as the boundary was approached. To the beings inside the universe, however, the size of the spaceship as well as its speed would appear unchanged. Thus they would have no way of knowing that they actually lived in a *bounded universe*. In fact, from their point of view their universe would appear unbounded. Only for the outside observer would the boundedness of this universe be manifest. In this model the lines are halves of ellipses, with the longest axis along a diameter of the space, analogous to what we have for the planar model. The *planes* will be halves of ellipsoids, as shown in the illustration given in Figure 9.8.

There is a close connection between the metric and the curves designated as *lines*. Indeed, the line-segment joining the two points P and Q is precisely the path yielding the shortest route from P to Q. As *light* will travel the shortest route from one point to another, the lines are therefore nothing but all the possible *light rays* we may have in the universe we are studying. Another name for a such curve is *a geodesic curve* or just *a geodesic*.

The explicit formula for the metric is complicated in this and most other serious cases, and is therefore usually omitted. We may do so because of the following important phenomenon: Suppose that our set S is a subset of $\mathbb{R}^3$ (or $\mathbb{R}^n$, for that matter). It then turns out that to specify a metric is equivalent to prescribing how a measuring rod of unit length will shrink or expand as it is moved around within the space. In the present case this means, moved towards the surface of the sphere, the boundary of the universe. Having a formula for this "shrinkage" makes it possible to reconstruct the metric, and indeed as it turns out, this formula for the shrinkage of a small measuring rod is a more convenient tool for the description of the geometry of a metric space than the distance-formula itself. The reader should note that this explanation is of an informal nature, thus mathematically deficient in many respects. We will, however, return to it in the following section.

We now turn our attention to the *hyperbolic universe*. Again, the points of this universe are the points inside a fixed sphere, of radius R, say.

In this set we may introduce a metric by specifying that the size of a measuring rod of unit length, located at a point $P = (x, y, z)$ of distance $r = \sqrt{x^2 + y^2 + z^2}$ from the center, is equal to

$$\delta_P = 1 - \frac{r^2}{R^2}$$

We see that as long as the measuring rod is located at the origin, deemed as the center of the universe by the outside observer, then our "universe-

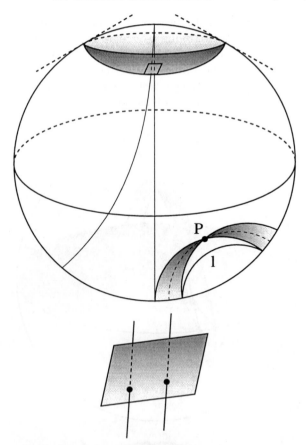

Fig. 9.9. The hyperbolic universe inside a fixed sphere. Lines are circular arcs, perpendicular to the surface of the sphere. Two lines perpendicular to the shaded hyperbolic plane are shown, one of them happens to be a diameter to the sphere, a special case of a hyperbolic line. As in the Euclidian case above we have shown a small piece of the space around where the two lines pass through the plane. In this universe there are infinitely many parallels to a given line ℓ through a point P outside it, the situation is shown in the lower right corner of our universe.

dwellers" inside the sphere and the outside observer will see it as being of the same length, namely 1.

But as the universe-dwellers take the rod and move away from the center, a disagreement developers: While it retains the length of 1 as far as the universe-dweller is concerned, the outside observer will see it as shrinking. Correspondingly, even if the insider has kept up a decision to move at a constant speed, the outsider will perceive the speed as diminishing when the distance from the center increases.

It can be shown that the shortest route between two points P and Q in this universe is attained by following a circular arc which passes through P and Q, and in addition is perpendicular to the surface of the sphere.

This is the "lines" we have already described for Klein's model for the hyperbolic plane. So these are the geodesic curves, the "straight lines", in the hyperbolic universe. The situation is illustrated in Figure 9.9.

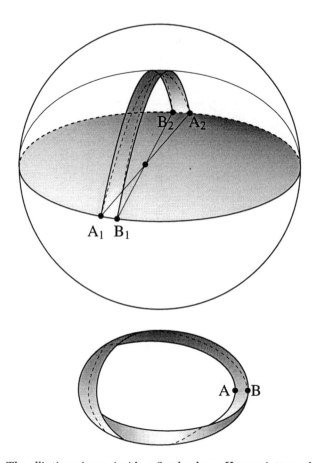

Fig. 9.10. The elliptic universe inside a fixed sphere. Here points on the surface of the sphere are in the space, diametrically opposite ones being identified. Thus the lines are closed curves, of a finite length. Here A_1 is identified with A_2, B_1 with B_2, and we get the indicated "highway". Its edges are not straight lines. But we can see two straight lines in the illustration, namely half ellipses joining A_1 to A_2 and B_1 to B_2. The situation is shown below the sphere.

We similarly obtain a model for the *elliptic universe*. Here we include the points on the surface of the sphere, but with diametrically opposite points identified. The "lines" described in the Euclidian case are kept in this model for the elliptic universe, except that we add to the line the one point obtained by identifying the two points where the half ellipse meets the surface of the sphere.

Again, this elliptic 3-space is closely related to the elliptic plane, which has been explained earlier.

We may introduce a metric in this space as well, and the "lines" of the model then become the geodesics.

9.5 Riemannian Geometry

Among the concepts and properties which Euclid did not state, but implicitly took for granted, we find the notion of *distance* or as we say today, the metric discussed in the preceeding section. Tied to this concept is the notion of *transformations* preserving the metric, as well as transformations preserving *shape* or *shape and size* of geometric objects: Any figure may be moved without altering its shape, form or size. Making this precise, turning these ideas into mathematics, has been on the agenda of a number of renowned mathematicians.

Another towering mathematician, *Georg Riemann*, presented his probationary lecture at the University of Göttingen in 1854. Here he outlines a new course in the understanding of space and geometry.

Riemann's very fundamental idea has already been informally hinted at in the previous section, in the form of the *shrinking measuring rod*. His idea was to express the *distance between two infinitesimally close points*. We first illustrate this idea by working with the space $\mathbb{R}^3$. Here the usual distance between two points

$$P = (a_1, b_1, c_1) \text{ and } Q = (a_2, b_2, c_2)$$

is give by

$$\rho_3(P, Q) = \sqrt{(a_2 - a_1)^2 + (b_2 - b_1)^2 + (c_2 - c_1)^2}$$

as we have seen in the previous section. For our purpose now it is more practical to write this definition as follows:

$$\rho_3(P, Q)^2 = (a_2 - a_1)^2 + (b_2 - b_1)^2 + (c_2 - c_1)^2$$

If we let the two points approach each other, to become very close, this relation may be written as

$$(\Delta s)^2 = (\Delta x_1)^2 + (\Delta x_2)^2 + (\Delta x_3)^2$$

As we pass to the infinitesimal, we get the expression

$$ds^2 = dx_1^2 + dx_2^2 + dx_3^2$$

This relation is very useful when we wish to calculate the length of a curve-segment. Suppose a curve-segment C is given on parametric form as

$$x_1 = \varphi(t), x_2 = \chi(t), x_3 = \psi(t), t \in [t_1, t_2]$$

Then the length of an infinitesimal piece of the curve is given by

$$ds = \left(\sqrt{\frac{d\varphi}{dt}(t)^2 + \frac{d\chi}{dt}(t)^2 + \frac{d\psi}{dt}(t)^2} \right) dt$$

and hence the length of the curve-segment is given by the integral formula, well known from calculus to some of the readers:

$$L = \int_{t_1}^{t_2} \left(\sqrt{\frac{d\varphi}{dt}(t)^2 + \frac{d\chi}{dt}(t)^2 + \frac{d\psi}{dt}(t)^2} \right) dt$$

Frequently a formula as above for ds is referred to as a metric. A general Riemannian metric on $\mathbb{R}^3$ is of the form

$$ds^2 = g_{1,1}dx_1^2 + g_{1,2}dx_1dx_2 + g_{1,3}dx_1dx_3+$$

$$g_{2,1}dx_2dx_1 + g_{2,2}dx_2^2 + g_{2,3}dx_2dx_3+$$

$$g_{3,1}dx_3dx_1 + g_{2,3}dx_2dx_3 + g_{3,3}dx_3^2,$$

where the $g_{i,j}$ are functions of x_1, x_2, x_3 in general. As $dx_idx_j = dx_jdx_i$, we may adjust the $g_{i,j}$'s such that $g_{i,j} = g_{j,i}$ for all i and j.

For the reader with some knowledge of linear algebra we note the convenient organization of these functions in a *symmetric* 3×3 *matrix*:

$$\mathbf{g} = \begin{bmatrix} g_{1,1} & g_{1,2} & g_{1,3} \\ g_{2,1} & g_{2,2} & g_{2,3} \\ g_{3,1} & g_{3,2} & g_{3,3} \end{bmatrix}$$

Here $g_{i,j} = g_{j,i}$. The metric may expressed as

$$ds^2 = [dx_1, dx_2, dx_3]\mathbf{g} \begin{bmatrix} dx_1 \\ dx_2 \\ dx_3 \end{bmatrix}$$

In Riemannian geometry it was assumed that the determinant of **g** be non zero, and that $ds^2 > 0$. This last condition is, however, abolished when these ideas are used in Einstein's General Relativity, which we come to later.

In physics and applied mathematics the matrix **g** is usually referred to as the *(covariant) metric tensor*. We shall not use this language here, the concept of covariant and contravariant tensors is with some justification perceived as rather murky and obscure when it is first encountered. However, solid knowledge of linear algebra makes it crystal clear and perfectly obvious.

The generalization of these ideas to $\mathbb{R}^n$ is rather immediate, we write

$$ds^2 = \sum_{i=1}^{i=n} \sum_{j=1}^{j=n} g_{i,j} dx_i dx_j,$$

where as before one assumes that $g_{i,j} = g_{j,i}$ for all i and j.

Frequently the space under consideration is not simply some subset of $\mathbb{R}^n$, but rather is one which may be *patched together* of small pieces of Euclidian n-space $\mathbb{R}^n$. Thus for instance, a spheric surface can be viewed as being pieced together of small patches from $\mathbb{R}^2$, as may a toric surface, or more generally a "torus with n holes in it". Such a surface is called *a compact Riemann surface of genus n*. This is illustrated in Figure 9.11.

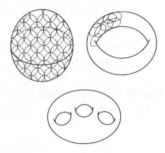

Fig. 9.11. Some surfaces as pieced together by patches from $\mathbb{R}^2$: A sphere, a torus and a compact Riemann surface of genus 3.

Each patch from Euclidian n-space comes with the coordinates $x_1, \ldots, x_n$, which are now only valid within each patch. When two patches overlap, it is necessary to have a set of *transition functions* between them. That is to say, in the intersection between the patch U with coordinate functions $x_1, \ldots, x_n$ and V with $y_1, \ldots, y_n$, we have

$$x_1 = \tau_1(y_1, \dots, y_n)$$
$$x_2 = \tau_2(y_1, \dots, y_n)$$
$$\cdots$$
$$x_i = \tau_i(y_1, \dots, y_n)$$
$$\cdots$$
$$x_n = \tau_n(y_1, \dots, y_n)$$

The mathematics involved in getting all the details of this right is too complicated to explain here, but the metric of this patched together-space is given on each patch as above, the **g** only being valid within its patch. Then the transition to an overlapping patch is expressed in terms of the transition functions.

In terms of the metric tensor **g** we may express the so-called *Riemannian curvature* of the space. We shall not be specific in this exposition, only mention that while Euclidian space has curvature zero, elliptic space will have a positive curvature while hyperbolic space has negative curvature.

A remarkable application of these ideas may be found in *Einstein's theory of general relativity*. As we have seen, the *straight lines* must be the *geodesic curves* in the space, that is to say the path traveled by *light*. Thus light is intimately tied to the geometry of space, and one remarkable consequence is the mathematical necessity of regarding the speed of light as a *universal constant*, determined by the geometry of space itself. This had indeed been observed for some time prior to Einstein's theory, and had been regarded as a puzzling paradox.

Einstein also realized that the concept of *universality of time* had to be abandoned. Instead, time is incorporated into the geometry of space, as a fourth dimension.

The four-dimensional space-time universe is conceived as a subspace of $\mathbb{C}^4$, four-dimensional complex space. The four real dimensions in it are arranged with the three spatial axes along the first three real axes in the three first complex dimensions, and the fourth, the time-axis, along the complex axis of the fourth. The basic transformations from one coordinate system to another are rotations in this four-dimensional complex space.

We will not pursue these ideas here, as they require more technical mathematics than the scope of the present book warrants.

10 Making Things Precise

10.1 Relations and Their Uses

In Section 8.3 we saw how a model for the projective plane may be constructed by taking the northern hemisphere of a spherical surface, including the equator, and then *identifying* diametrically opposite points on the equator.

For a construction of this type to make sense mathematically, we need a stringent base for being able to identify points in this way. We simply *can not* identify points according to any odd rule we may think of. We need the following important concept:

Definition 6 (Equivalence relation). *Let $\mathcal{M}$ be a set, where there is given a relation $m \sim n$, which may be satisfied between two elements m and n from $\mathcal{M}$. We say that $\sim$ is an* equivalence relation *if the following conditions are satisfied:*

$$(R) \quad \textit{For all } m \in \mathcal{M} \textit{ we have that } m \sim m$$
$$(S) \qquad\quad m \sim n \Longrightarrow n \sim m$$
$$(T) \qquad m \sim n \textit{ and } n \sim r \Longrightarrow m \sim r$$

In this definition (R) is referred to as *reflexivity*, (S) as *symmetry* and (T) is called *the condition of transitivity*.

These three important properties express the essence of the identification process: The relation $\sim$ will subdivide the set $\mathcal{M}$ into a collection of disjoint *classes of mutually equivalent* elements. We have the following

Proposition 2. *Let $\sim$ be an equivalence relation in $\mathcal{M}$. We put*

$$[m] = \{r \in \mathcal{M} \mid r \sim m\},$$

in other words, the set $[m]$ consists of all elements r in $\mathcal{M}$ such that the relation $r \sim m$ holds true. For two arbitrary elements m and $n \in \mathcal{M}$ we then have that

$$[m] = [n] \textit{ or } [m] \cap [n] = \emptyset,$$

that is to say, two sets in the subdivision which are distinct are disjoint. Finally, we have that

$$[m] = [n] \text{ if and only if } m \sim n$$

Definition 7. $[m]$ *is referred to as the* equivalence class *of the element* m.

Proof. We first prove the last biimplication.

Proof for $\Longrightarrow$: Assume that $[m] = [n]$. Since $n \in [n]$ because of (S), we get $n \in [m]$, thus $n \sim m$ by the definition of $[m]$ stated in the proposition. But by (S) this gives $m \sim n$.

Proof for $\Longleftarrow$: Assume that $m \sim n$. We shall prove that $[m] = [n]$. For this we have to prove the two inclusions $[m] \subseteq [n]$ and $[n] \subseteq [m]$. From our assumption that $m \sim n$ we know by (S) that we also have $n \sim m$, so it suffices to prove the first inclusion, the other then following simply by interchanging the roles of m and n. So we prove that

$$m \sim n \Longrightarrow [m] \subseteq [n]$$

So let $r \in [m]$, that is to say, assume that $r \sim m$. Out assumption that $m \sim n$ then gives $r \sim n$, on account of (T). Thus we have $r \in [n]$, as claimed.

We next prove the first part of the proposition, namely that $[m]$ and $[n]$ either coincide or else have no common element. Assume that $[m]$ and $[n]$ are different subsets, but nevertheless have a common element, say r:

$$r \in [m] \cap [n]$$

We shall prove that this leads to a contradiction: By assumption we first of all have that $r \sim m$ and $r \sim n$. But by (S) we then get that $m \sim r$, which together with $r \sim n$ gives $m \sim n$ because of (T). But above we have already shown the biimplication in the assertion of the proposition, so this gives

$$[m] = [n]$$

which is a contradiction. $\square$

10.2 Identification of Points

Whenever we have a set with an equivalence relation $\sim$, we may subdivide the set $\mathcal{M}$ in a *set of equivalence classes*. This new set of equivalence classes is denoted by $\mathcal{M}/\sim$:

$$(\mathcal{M}/\sim) = \{[m]| \; m \in \mathcal{M}\}$$

The set $\mathcal{M}/\sim$ is thus obtained by *identifying equivalent points* in $\mathcal{M}$. We have an important surjective mapping

$$\mathcal{M} \longrightarrow \mathcal{M}/\sim$$

$$m \mapsto [m]$$

which *carries out the identification of equivalent points in $\mathcal{M}$*.

We note that the classes need not have the same number of elements, and indeed, some of the classes could be infinite, others finite and some could even consist of a single element. We consider a few examples.

1. The relation of being equal, $=$, is an equivalence relation, the three conditions of course being obvious in this case.
2. The relations of ordering $\geq$ and $\leq$ defined for real numbers are not equivalence relations. They do satisfy (R) and (T), but not (S).
3. The so-called strong relations of ordering, $<$ and $>$ only obey (T), and they therefore are also not equivalence relations.
4. For any set Ω we can define the set of all subsets, denoted $\mathcal{P}^{\Omega}$. In this set we have the various inclusion-relations $\subseteq, \supseteq, \subset$ and $\supset$. They are not equivalence relations either, but behave as the relations of ordering.
5. In the set $\mathcal{P}^{\Omega}$ we put $P \sim Q$ if P has the same number of elements as Q, which may be ∞. This is an equivalence relation on $\mathcal{P}^{\Omega}$.
6. We consider the set of all smooth surfaces in $\mathbb{R}^3$. If S and T are two such surfaces, we put $S \sim T$ if S can be deformed into T smoothly without breaking anything. This is an informal description of an important equivalence relation known as *topological equivalence*.
7. We say that two geometric figures in the pane $\mathbb{R}^2$ are *congruent* if one may be placed on the other by a translation followed by a rotation in the plane. This is an equivalence relation.
8. We say that two geometric figures in the plane $\mathbb{R}^2$ are *similar* if one may be placed on the other by a translation followed by a rotation in the plane and a shrinkage or an enlargement. This also is an equivalence relation.

We now draw the conclusion which was announced as part of our motivation for introducing the machinery of equivalence relations. Namely, we show how we may construct the set of points which went into the model for the projective plane, by identifying the appropriate points on the northern hemisphere, equator included:

Proposition 3. *Let $\mathcal{M}$ be the set of points on the northern hemisphere, including the equator. We define the relation $\sim$ by setting $P \sim Q$ if and only if*

$P = Q$ or P and Q are diametrically opposite points on the equator.

This is an equivalence relation.

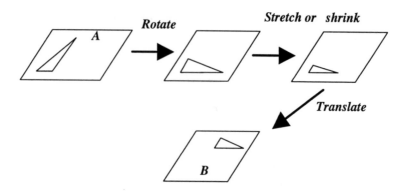

Fig. 10.1. The equivalence relation of similarity applied to two triangles: The triangle A is similar to B.

Proof. Except for the case when some point is on the equator, this is nothing but the relation $=$, and the three conditions are clear outside the equator. Moreover, whether P is on the equator or not, clearly $P \sim P$, so (R) holds.

If P is on the equator and $P \sim Q$, then Q must be on the equator, and it is either equal to P or diametrically opposite to P. In either case it follows that $Q \sim P$, thus (S) is proven.

Finally we show (T), assume $P \sim Q$ and $Q \sim R$. If P is not on the equator, then $P = Q$, and so $P \sim R$. So assume that P is on the equator. If $P = Q$, we are finished as before. If not, then P and Q are diametrically opposite points. Then either $R = Q$ or $R = P$, in both cases it follows that $P \sim R$. □

10.3 Our Number System

One of the most amazing applications of the concept of equivalence relations, is the systematic and mathematically rigorous construction of our system of numbers, starting from the natural numbers

$$\mathbb{N} = \{1, 2, 3, \dots\},$$

which we take as intuitively given. Actually, they may be constructed too, from the Zermelo-Fraenkel-Skolem axioms for the Set-theory. But we take the natural numbers as given intuitively, as has been the case from time immemorial.

We proceed one step at the time: First the set of natural numbers $\mathbb{N}$ is extended to the *set of integers* $\mathbb{Z}$. Thereafter $\mathbb{Z}$ is extended to the *rational numbers* $\mathbb{Q}$, which again is extended to the *real numbers* $\mathbb{R}$. Finally we briefly introduce the complex numbers $\mathbb{C}$.

Passing from the natural numbers, the set of which is denoted by $\mathbb{N}$, to the integers $\mathbb{Z}$ was historically accomplished in two steps: First, the discovery of the integer *zero* was more important and more difficult than most of us realize today. Secondly, the concept of *negative numbers* was historically very difficult, and was achieved in comparatively recent times. We shall, however, not retrace this historical path. Instead, we follow another line, and will introduce the integers as the result of *an attempt to solve an impossible problem.* Ludicrous as it may seem, there are few more fruitful endeavors throughout the development leading to present day mathematics than this: Attempting to solve an impossible problem.

Let $\mathcal{M} = \mathbb{N} \times \mathbb{N}$, and write

$$(a, b) \sim (c, d) \text{ whenever } a + d = b + c.$$

It is a fairly easy exercise to verify that this is an equivalence relation on the set $\mathcal{M}$. Now define

$$\mathbb{Z} = \mathcal{M}/\sim .$$

We introduce operations called *addition* and *multiplication* in the set $\mathcal{M}$ by putting

$$(a, b) + (c, d) = (a + c, b + d) \text{ and } (a, b) \cdot (c, d) = (ac + bd, ad + bc)$$

These, apparently strange, definitions are motivated by the idea that the pair (a, b) of natural numbers should correspond to the *integer* (not yet defined) $a - b$. We now see the motivation for the definition of the equivalence relation $\sim$ given above: $(a, b) \sim (c, d)$ if $a - b = c - d$, or phrased in terms of natural numbers: $a + d = b + c$. The relation $\sim$ defined above has a very important property which goes beyond that of merely being an equivalence relation: It is a *congruence relation* for the two operations addition and multiplication defined above: That is to say, the following two implications hold:

$$(a, b) \sim (c, d) \implies (a, b) + (e, f) \sim (c, d) + (e, f) \text{ for all } (e, f) \in \mathcal{M},$$

$$(a, b) \sim (c, d) \implies (a, b) \cdot (e, f) \sim (c, d) \cdot (e, f) \text{ for all } (e, f) \in \mathcal{M}$$

These two properties are easily verified, as is the important consequence that

$$[(a, b) + (c, d)] = [(a', b') + (c', d')] \text{ and } [(a, b) \cdot (c, d)] = [(a', b') \cdot (c', d')]$$

whenever

$$(a, b) \sim (a', b') \text{ and } (c, d) \sim (c', d')$$

Thus addition and multiplication as defined in $\mathcal{M}$ induce addition and multiplication in $\mathbb{Z}$. We finally note that $[(a, b)] = [a + n, b + n]$ for all natural numbers n, and in particular that adding $[n, n]$ to any $[(a, b)]$ produces no change: Thus we denote this element (the same for all choices of n) by 0. Moreover, the element $[(a + n, n)]$ is independent of n, thus this element, this *integer*, is identified with the *natural number a*. We have obtained an injective mapping, or as we say an *embedding*

$$\varphi : \mathbb{N} \hookrightarrow \mathbb{Z},$$

which satisfies the important conditions

$$\varphi(a + b) = \varphi(a) + \varphi(b) \text{ and } \varphi(ab) = \varphi(a) \cdot \varphi(b)$$

In language from abstract algebra we say that φ is an injective *homomorphism of the semi-ring* $\mathbb{N}$ *into the ring* $\mathbb{Z}$. We shall not pursue these algebraic concepts much further here, except to note that if a, b are natural numbers, then the equation

$$a + X = b,$$

which does not always have a solution in $\mathbb{N}$, does indeed always have a unique solution in $\mathbb{Z}$, namely the integer $[(b, a)]$. Identifying, as we always do, a with $\varphi(a)$, we then have that $[(b, a)] = b - a$.

We next pass to *the Rational Numbers*. This time we put

$$\mathcal{M} = \{(a, b) \mid a, b \in \mathbb{Z}, b \neq 0\}$$

as we now have the integers $\mathbb{Z}$ at our disposal. Here we define

$$(a, b) \sim (c, d) \text{ whenever } ad = bc$$

which again is easily seen to be an equivalence relation, enabling us to define the quotient-set as above:

$$\mathbb{Q} = \mathcal{M} / \sim$$

Again, analogously to the above, we define the operations addition and multiplication in $\mathcal{M}$ by

$$(a, b) + (c, d) = (ad + bc, bd) \text{ and } (a, b) \cdot (c, d) = (ac, bd),$$

and again, $\sim$ is a congruence relation for the two operations addition and multiplication, i.e. the two implications

$$(a, b) \sim (c, d) \implies (a, b) + (e, f) \sim (c, d) + (e, f) \text{ for all } (e, f) \in \mathcal{M},$$

$$(a, b) \sim (c, d) \implies (a, b) \cdot (e, f) \sim (c, d) \cdot (e, f) \text{ for all } (e, f) \in \mathcal{M}$$

hold. Thus the addition and multiplication defined in $\mathcal{M}$ induce addition and multiplication in $\mathbb{Q}$. We define an injective mapping

$$\varphi : \mathbb{Z} \hookrightarrow \mathbb{Q}$$

by putting $\varphi(t) = [(t, 1)]$, and verify right away that $\varphi(s + t) = \varphi(s) + \varphi(t)$ and $\varphi(st) = \varphi(s) \cdot \varphi(t)$. By means of φ we may identify $\mathbb{Z}$ with at subset of $\mathbb{Q}$, in such a way that addition and multiplication in $\mathbb{Q}$ reduces to the one we already know for $\mathbb{Z}$ on this subset. Finally, the equation

$$aX = b,$$

which is not always solvable in $\mathbb{Z}$ when b and $a \neq 0$ are integers, is now solvable in $\mathbb{Q}$, for $a \neq 0, b \in \mathbb{Q}$.

To treat the *real numbers* we let $\mathcal{F}$ be the set of all sequences of rational numbers,

$$\mathcal{F} = \{\{x_n\}, n = 1, 2, \ldots, x_n \in \mathbb{Q}\}$$

We consider the subset $\mathcal{M} \subset \mathcal{F}$ of all the so-called *Cauchy sequences*: $\mathcal{M}$ is the set of all sequences $x = \{x_n, n = 1, 2, \ldots\}$ which satisfy the condition

$$\forall \epsilon > 0 \exists n_0 \in \mathbb{N} \text{ such that } m, n \geq n_0 \implies |x_m - x_n| < \epsilon$$

or in plain language, more readable but less precise: The absolute difference between any two members of the sequence can be made arbitrarily small, once the indices are chosen sufficiently big. An example of such a sequence would be

$$x_1 = 3, x_2 = 3.1, x_3 = 3.14, x_4 = 3.141, x_5 = 3.1415, \ldots,$$

x_n being π to the first n digits. The informal idea is that for any Cauchy sequence x all the x_n are in $\mathbb{Q}$, they are rational numbers, while the sequence *converges to* a real number. Moreover, all real numbers are obtainable in this manner. One of several approaches to *defining* the real numbers is to use this idea. We introduce an equivalence relation in the set $\mathcal{M}$ by

$$x \sim y \Leftrightarrow (\forall \epsilon > 0 \exists n_0 \in \mathbb{N} \text{ such that } n \geq n_0 \implies |x_n - y_n| < \epsilon)$$

Informally, the absolute difference $|x_n - y_n|$ is arbitrarily small for all sufficiently big values of n. It is easy to see that this is indeed an equivalence relation.

We now proceed by defining addition and multiplication in the set $\mathcal{M}$ in the obvious manner, namely

$$x + y = z \text{ where } z_n = x_n + y_n \text{ and } x \cdot y = z \text{ where } z_n = x_n y_n$$

It actually requires a proof that the operations so defined yield new Cauchy sequences. It also must be proven that $\sim$ is a congruence relation for these operations. We omit these verifications, not too difficult but of some mild complexity. In the end we define $\mathbb{R} = \mathcal{M}/\sim$. $\mathbb{R}$ is an extension of $\mathbb{Q}$ by the injective mapping

$$\varphi : \mathbb{Q} \hookrightarrow \mathbb{R} \text{ given by } \varphi(r) = [x] \text{ where } x_n = r \text{ for all n.}$$

The extension of $\mathbb{Q}$ to $\mathbb{R}$ makes it possible to solve a general problem, insoluble in general within $\mathbb{Q}$:

Problem. Given a Cauchy sequence $x = \{x_n, n = 1, 2, \dots\}$. Find $\lim_{n \to \infty} x_n$.

We finally come to *the complex numbers*. We define the set of complex numbers by $\mathbb{C} = \mathbb{R}^2 = \mathbb{R} \times \mathbb{R}$, and define addition and multiplication in $\mathbb{C}$ by

$$(a, b) + (c, d) = (a + c, b + d), (a, b) \cdot (c, d) = (ac - bd, ad + bc)$$

It is a straightforward while uninspiring exercise to check that these operations satisfy the usual properties of addition and multiplication. Moreover, $\mathbb{C}$ becomes an extension of $\mathbb{R}$ by the injective mapping

$$\varphi : \mathbb{R} \hookrightarrow \mathbb{C} \text{ given by } \varphi(a) = (a, 0)$$

We identify $\mathbb{R}$ with this subset of $\mathbb{C}$. The whole point in this construction is, amazingly, contained in the following little computation:

$$(0, 1) \cdot (0, 1) = (0 \cdot 0 - 1 \cdot 1, 0 \cdot 1 + 1 \cdot 0) = (-1, 0)$$

Writing $(0, 1) = i$, we have

$$i^2 = -1,$$

having identified $\mathbb{R}$ with a subset of $\mathbb{C}$ as described above. We also note that

$$(a, b) = a + ib,$$

where $i = \sqrt{-1}$, which is the usual form of a complex number. The extension of $\mathbb{R}$ to $\mathbb{C}$ makes it possible to solve the following problem, insoluble in $\mathbb{R}$:

Problem. Solve the equation $X^2 + 1 = 0$.

In fact, the set of complex numbers represents a much stronger enlargement of $\mathbb{R}$. We have the following beautifully theorem, known as *the Fundamental Theorem of Algebra:*

Theorem 15. *A polynomial*

$$X^n + a_1 X^{n-1} + \cdots + a_{n-1}X + a_n = P(X)$$

where the coefficients $a_1, \ldots, a_n$ are complex numbers, may be factored into a product of (possibly repeated) factors of the type $X - x$, where $x \in \mathbb{C}$. In particular, all equations of the type

$$X^n + a_1 X^{n-1} + \cdots + a_{n-1}X + a_n = 0$$

where the coefficients $a_1, \ldots, a_n$ are complex numbers, have solutions in $\mathbb{C}$.

11 Projective Space

Projective space is not merely a dry and theoretical invention. On the contrary, it is a living reality. It would be impossible to create perspective and depth in a painting without an understanding of projective space. When we are painting or making pictures of physical objects in space, like buildings, roads or other objects, we need to project the three-dimensional space with all it contains onto a two-dimensional canvas, piece of paper, or photographic film.

Fig. 11.1. A side street at Berkeley, in the mid 80's. We notice the parallel lines in space projected onto a bunch of lines intersecting in a point located about the middle of the left edge of the picture.

Then two parallel lines in nature will be represented in the picture as two lines *intersecting* at a point located at *the line of perspectivity*. We might say that the Euclidian 3-space is projected onto the *projective plane*.

An example is shown in Figure 11.1, a photo from a side street at Berkeley in the mod 80's. In this otherwise totally uninteresting picture where nothing happens, there is a conspicuous collection of *parallel lines*: The power lines, the features on the facades of the buildings on the right hand side of the street, then the lines giving the outlines of the cars parked along the sidewalks, the street itself, the sidewalks. All these lines are projected, by the process of taking the photo, onto the photographic film. And the projections intersect at a point around the middle of the left hand edge of the picture. That is the projection of a point at infinity of the 3-dimensional space. So even if we cannot see this "point at infinity" of the usual 3-space, we capture its projection onto the film!

11.1 Coordinates in the Projective Plane

We shall now explain how it is possible to introduce *coordinates* for the projective plane, in a similar manner to what we have for the Euclidian plane $\mathbb{R}^2$.

Recall that the points in $\mathbb{P}^2(\mathbb{R})$ are the *lines* through $(0,0,0)$ in $\mathbb{R}^3$. Such a line α is uniquely determined by a *vector* $(a, b, c) \neq (0, 0, 0)$. This vector gives the direction of the line, and as the line passes through the origin, it is given as follows:

$$\alpha = \{(x, y, z) \mid x = at, y = bt, z = ct, \text{ where } t \in \mathbb{R}\}.$$

We refer to this as *the line α on parametric form*. Thus to every value of the parameter t there corresponds a uniquely determined point $P(t) \in \alpha$, and conversely, to every $P \in \alpha$ there corresponds a uniquely determined value of $t, t = t_P$, such that

$$P = P(t_P)$$

We may also describe some other space curves in the same manner. Then the curve C is given by

$$C = \{(x, y, z) \mid x = f(t), y = g(t), z = g(t), \text{ where } t \in \mathbb{R}\}$$

which defines the curve C in $\mathbb{R}^3$. For example, a line which *does not* pass through the origin, will have the following parametric form:

$$\alpha = \{(x, y, z) \mid x = x_0 + at, y = y_0 + bt, z = z_0 + ct, t \in \mathbb{R}\}$$

where (x_0, y_0, z_0) is a point on the line which we may chose as we like. Thus clearly one and same line may be given on different parametric forms. Choosing another point than (x_0, y_0, z_0) gives another parametric form, and the vector defining the direction of the line may be replaced by any non-zero multiple.

We return to the lines through the origin. As mentioned above, if we change the vector (a, b, c) to some (a', b', c') which is proportional to the original one, then the new parametric form yields the same line. Thus it is only *the ratio* between the numbers a, b and c which is important. We let $(a : b : c)$ denote *the set of all such vectors* proportional to (a, b, c):

$$(a : b : c) = \{(a', b', c')| \ a' = ra, b' = rb, c' = rc, \text{ where } r \neq 0\}$$

Proceeding formally, we define a relation in the set

$$\mathcal{M} = \mathbb{R}^3 - (0, 0, 0):$$

by putting

$$(a, b, c) \sim (a', b', c') \text{ whenever } \exists r \neq 0 \text{ such that } a' = ra, \ b' = rb, c' = rc$$

Thus for instance

$$(1 : 2 : 3) = (2 : 4 : 6) = (\pi : 2\pi : 3\pi)$$

This relation is, as is immediately checked, an equivalence relation, and $(a : b : c)$ is the equivalence class defined by the element (a, b, c). These equivalence classes are in bijective correspondence with the lines through the origin, i.e., with the set of points of $\mathbb{P}^2(\mathbb{R})$.

Further, it is clear that if $a \neq 0$, then we may assume that $a = 1$: We just choose $r = \frac{1}{a}$ so that $(a, b, c) = r(1, \frac{b}{a}, \frac{c}{a})$.

The same applies to the other coordinates as well, so if a, b and c are $\neq 0$, then

$$(a : b : c) = (1 : \frac{b}{a} : \frac{c}{a}) = (\frac{a}{b} : 1 : \frac{c}{b}) = (\frac{a}{c} : \frac{b}{c} : 1)$$

We now define the *projective coordinates* of a point $\alpha \in \mathbb{P}^2(\mathbb{R})$:

Definition 8. *If the point $\alpha \in \mathbb{P}^2(\mathbb{R})$ is given as the line in $\mathbb{R}^3$ defined on parametric form as*

$$\alpha = \{(x, y, z) \ | x = at, y = bt, z = ct, t \in \mathbb{R}\}$$

then we denote this point by $(a : b : c)$. The ratio $a : b : c$ is referred to as the projective coordinates of the point α, this name is also applied to the tuple (a, b, c), which is only determined up to a constant multiple.

The relation between the projective and the usual (*"affine"*) coordinates of a point in $\mathbb{R}^2$ is given by the following

Proposition 4. *Let as before $\mathbb{R}^2$ be identified with the points in $\mathbb{P}^2(\mathbb{R})$ which correspond to lines which are not contained in the xy-plane. Under the identification performed in Proposition 1 we get that*

$$(x : y : 1) = (x, y)$$

Proof. We may choose $(a, b, c) = (x, y, 1)$ in the parametric form of the line which corresponds to the point. $(x, y, 1)$ is then the point of intersection between this line and the plane given by $z = 1$. The claim now follows from Proposition 1.

11.2 Projective n-Space

In this section we shall define the *n-dimensional* projective space $\mathbb{P}^n(\mathbb{R})$. We then consider the set

$$\mathcal{M} = \mathbb{R}^{n+1} - \{(0, \dots, 0)\}$$

and as for the case $n = 2$ define a relation by

$$(a_1, a_2, \dots, a_{n+1}) \sim (b_1, b_2, \dots, b_{n+1})$$

whenever $\exists r \in \mathbb{R}$ such that

$$a_i = r b_i \quad \text{for all } i = 1, 2, \dots, n+1.$$

We easily verify that this is an equivalence relation.

Definition 9. *The set $\mathcal{M}/\sim$ is denoted by $\mathbb{P}^n(\mathbb{R})$ and referred to as the projective n-space over the reals $\mathbb{R}$*

As in the case $n = 2$ we use the following notation for the equivalence classes:

$$[(a_1, a_2, \dots, a_{n+1})] = (a_1 : a_2 : \dots : a_{n+1}).$$

Whenever $a_{n+1} \neq 0$, we may assume that $a_{n+1} = 1$, and with this assumption the other coordinates are uniquely determined. Thus we may identify the sets

$$\{(a_1 : a_2 : \dots : a_{n+1}) | \, a_{n+1} \neq 0\}$$

and

$$\mathbb{R}^n$$

The set

$$\{(a_1 : a_2 : \ldots : a_{n+1}) | \ a_{n+1} = 0\}$$

is referred to as *the points at infinity* in $\mathbb{P}^n(\mathbb{R})$. This subset may in turn be identified with $\mathbb{P}^{n-1}(\mathbb{R})$ in the obvious manner by ignoring the last coordinate, which is zero. We obtain the following description of $\mathbb{P}^n(\mathbb{R})$:

$$\mathbb{P}^n(\mathbb{R}) = \mathbb{R}^n \cup \mathbb{P}^{n-1}(\mathbb{R})$$

and we say that $\mathbb{P}^n(\mathbb{R})$ *is obtained by adjoining to* $\mathbb{R}^n$ *a space* $\mathbb{P}^{n-1}(\mathbb{R})$ *of points at infinity.*

By dividing up $\mathbb{P}^{n-1}(\mathbb{R})$ similarly, and repeating the process all the way down to $\mathbb{P}^0(\mathbb{R})$, we get

$$\mathbb{P}^n(\mathbb{R}) = \mathbb{R}^n \cup \mathbb{R}^{n-1} \cup \ldots \cup \mathbb{R} \cup \mathbb{P}^0(\mathbb{R}).$$

But $\mathbb{P}^0(\mathbb{R})$ consist of only one point: In fact, if a and b are real non-zero numbers, then

$$(a) \sim (b)$$

since

$$b = \frac{b}{a} a.$$

Thus $\mathbb{P}^0(\mathbb{R}) = \{pt\}$. This yields

$$\mathbb{P}^n(\mathbb{R}) = \mathbb{R}^n \cup \mathbb{R}^{n-1} \cup \ldots \cup \mathbb{R} \cup \{pt\}$$

where union is disjoint, and $\{pt\}$ denotes a set which only consist of the single point pt. This union is some times referred to as the cell decomposition of $\mathbb{P}^n(\mathbb{R})$.

11.3 Affine and Projective Coordinate Systems

In the previous paragraph we saw that a point in $\mathbb{P}^n(\mathbb{R})$ is given by $n+1$ projective coordinates: $P = (a_1 : \ldots : a_n : a_{n+1})$. Before we proceed, we shall switch to another notation for the coordinates: Instead of writing $P = (a_1 : a_2 : \ldots : a_{n+1})$ we put $P = (a_0 : a_1 : \ldots : a_n)$, where we have relabeled the last coordinate and put it up front: $a_0 = a_{n+1}$. This is customary in the most recent literature.

In this notation $\mathbb{R}^n$ is the subset of $\mathbb{P}^n(\mathbb{R})$ consisting of the points $(a_1, \ldots, a_n) = (1 : a_1 : \ldots : a_n)$.

For the remainder of the paragraph we only treat the case $n = 2$, thus the affine and the projective *plane*. However, everything we show may easily be proven in the general case as well.

Let $L = b_0 X_0 + b_1 X_1 + b_2 X_2$. An expression of this type is referred to as a *linear form* in X_0, X_1 and X_2. We then put

$$V_+(L) = \{(a_0 : a_1 : a_2) \in \mathbb{P}^2(\mathbb{R}) \mid b_0 a_0 + b_1 a_1 + b_2 a_2 = 0\}$$

and

$$D_+(L) = \{(a_0 : a_1 : a_2) \in \mathbb{P}^2(\mathbb{R}) \mid b_0 a_0 + b_1 a_1 + b_2 a_2 \neq 0\}$$

Some books omit the $+$ in this terminology. However, we will keep it and reserve V and D without the $+$ for the sets

$$V(L) = \{(a_0, a_1, a_2) \in \mathbb{R}^3 \mid b_0 a_0 + b_1 a_1 + b_2 a_2 = 0\}$$

and

$$D(L) = \{(a_0, a_1, a_2) \in \mathbb{R}^3 \mid b_0 a_0 + b_1 a_1 + b_2 a_2 \neq 0\}$$

By switching to a new projective coordinate system, $D_+(L)$ may be identified in a natural way with $\mathbb{R}^2$, while $V_+(L)$ is identified with the projective line $\mathbb{P}^1(\mathbb{R}) = \mathbb{R} \cup \{pt\}$. We shall now explain this in some detail.

An *affine* coordinate system in $\mathbb{R}^2$ is a skew coordinate system. That is to say, the axes are not necessarily orthogonal to one another, and the scales may be different on the x-axis and the y-axis. The difference between a Cartesian coordinate system and an affine one is shown in Figure 11.2. The affine coordinate system do not necessarily have the x- and y-axes oriented counter clockwise, as is the case with the Cartesian system.

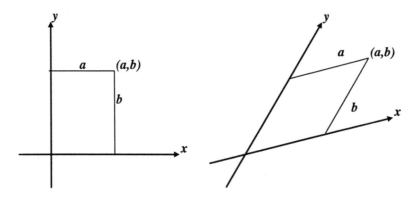

Fig. 11.2. A Cartesian coordinate system to the left, and a more general affine one to the right.

The relation between an old affine coordinate system with coordinates denoted by (x, y), and a new one $(\overline{x}, \overline{y})$ is given by

$$\bar{x} = \alpha + ax + by$$

$$\bar{y} = \beta + cx + dy$$

where a, b, c, d, α and β are real numbers, such that

$$\begin{vmatrix} a & b \\ c & d \end{vmatrix} = ad - bc \neq 0$$

The $\bar{x}$-axis is given, in the old coordinate system, by the equation

$$\beta + cx + dy = 0$$

while the $\bar{y}$-axis is given by

$$\alpha + ax + by = 0.$$

The coordinates of the old origin in the new system is (α, β).

The condition $ad - bc = 1$ will ensure that if the old system is a Cartesian one, then so is the new.

It is clear that a curve in $\mathbb{R}^2$ given by an equation of degree m in the coordinates x, y will be given by an equation of the same degree m in the coordinates $\bar{x}, \bar{y}$. Indeed, we get

$$x = \frac{d(\bar{x} - \alpha) - b(\bar{y} - \beta)}{ad - bc}, \ y = \frac{-c(\bar{x} - \alpha) + a(\bar{y} - \beta)}{ad - bc}$$

Rather than to view this as moving from one coordinate system to another, we may regard it as describing a mapping from the plane to itself, a so called affine transformation:

$$\mathbb{R}^2 \longrightarrow \mathbb{R}^2$$

$$(x, y) \longmapsto (\bar{x}, \bar{y})$$

where

$$\bar{x} = \alpha + ax + by$$

$$\bar{y} = \beta + cx + dy$$

and a, b, c, d, α and β are real numbers such that

$$\begin{vmatrix} a & b \\ c & d \end{vmatrix} = ad - bc \neq 0$$

Some of the geometric properties of curves or other figures are preserved by the affine transformations: That is to say, such properties as incidence, collinearity, passing through a fixed point or being a straight line, are preserved. Among properties *not* preserved we mention *the property of being a circle*, while being an ellipse *is preserved*. The property for two lines to form a right angle is not preserved, while the property for a line to *bisect* the angle formed by two given lines in equal parts, is preserved. The property for a point to divide a line segment in equal parts is an affine property, while the property of dividing the segment in a part of preassigned length is not. Finally, the property for a line to be *tangent* to a given curve is preserved by an affine transformation. The properties preserved by the affine transformations are referred to as *affine* properties.

The identity transformation is affine, as it is given by $\alpha = \beta = 0$ and the identity matric where $a = d = 1$, $b = c = 0$. The composition of two affine transformations is again an affine transformation, being given by adding the vectors (α, β) and taking the matrix-product of the two matrices, and finally the inverse of an affine transformation is therefore given by $-(\alpha, \beta)$ and the inverse matrix. Thus it is also affine. We express this by saying that *the affine transformations form a transformation-group*.

One may well take another, but equivalent, point of view and say that the affine properties are the ones which are preserved under affine changes of coordinate systems.

This represents a great advantage when we wish to find simple proofs for theorems in affine geometry. For example, suppose we wish to prove the following result:

The medians of a triangle intersect in one point. *Let there be given a triangle ABC. Draw the medians, i.e., the line joining A to the mid point of BC, the line joining B to the mid point of AC and the line joining C to the mid point of AB. Then these three lines meet in a single point.*

A proof is provided by considering the *center of gravity* of the triangle: Since the triangle will ballance on a "knifes edge" along the medians, the center of gravity lies on the three median lines, hence they intersect in that point.

Simple as this proof is, it does presuppose quite a bit of calculus. A purely algebraic proof runs as follows: We may choose an affine coordinate system such that $A = (-1, 0)$, $B = (1, 0)$ and $C = (0, \sqrt{3})$. This is always possible by considerations which will be explained below. Since the property of the triangle we wish to prove may be checked by algebra only, and is independent of the coordinate system, we have reduced the question to proving the property for a triangle where the vertices have these coordinates in a *Cartesian coordinate system.* But that is an *equilateral* triangle with sides equal to 2. For an equilateral triangle the claim is obvious by symmetry. The situation is illustrated in Figure 11.3.

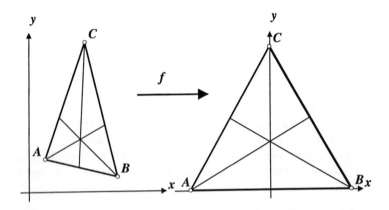

Fig. 11.3. Simplifying the geometry by a change of coordinate system.

For the projective plane we may proceed in the same way. We introduce a new coordinate system by

$$\overline{x}_0 = \alpha_{0,0} x_0 + \alpha_{0,1} x_1 + \alpha_{0,2} x_2$$

$$\overline{x}_1 = \alpha_{1,0} x_0 + \alpha_{1,1} x_1 + \alpha_{1,2} x_2$$

$$\overline{x}_2 = \alpha_{2,0} x_0 + \alpha_{2,1} x_1 + \alpha_{2,2} x_2$$

where we assume that the determinant is $\neq 0$:

$$A = \begin{vmatrix} \alpha_{0,0} & \alpha_{0,1} & \alpha_{0,2} \\ \alpha_{1,0} & \alpha_{1,1} & \alpha_{1,2} \\ \alpha_{2,0} & \alpha_{2,1} & \alpha_{2,2} \end{vmatrix} \neq 0.$$

As in the affine case, we may view this as describing a transformation

$$\mathbb{P}^2(\mathbb{R}) \longrightarrow \mathbb{P}^2(\mathbb{R})$$

$$(a_0 : a_1 : a_2) \mapsto (\overline{x}_0 : \overline{x}_1 : \overline{x}_2)$$

given as above. This is referred to as a *projective transformation*, and as in the affine case these transformations form a *transformation-group*: The identity mapping is a projective transformation, the composition of two projective transformations is again a projective transformation, and the inverse of a projective transformation is again a projective transformation.

All projective properties are preserved by projective transformations. We may also phrase this as saying that the projective properties are independent of choice of projective coordinate system.

However the points at infinity are not preserved. Being at infinity is not a projective property: This is definitely dependent on the coordinate system. In the original coordinate system the points at infinity are the points in the subset $V_+(x_0)$, as we have identified the affine plane $\mathbb{R}^2$ with $D_+(x_0)$. In the new coordinate system the points at infinity will be the points in the subset $V_+(\overline{x}_0) = V_+(\alpha_{0,0}x_0 + \alpha_{0,1}x_1 + \alpha_{0,2}x_1)$, and the subset $D_+(\overline{x}_0) = D_+(\alpha_{0,0}x_0 + \alpha_{0,1}x_1 + \alpha_{0,2}x_2)$ is now identified with $\mathbb{R}^2$.

The simplifying new coordinate system which we introduced in affine space $\mathbb{R}^2$ to prove the property of the medians, comes from the following proposition:

Proposition 5. *Given four points P_1, P_2, P_3 and $P_4 \in \mathbb{P}^2(\mathbb{R})$, such that no three of them are collinear, i.e., such that the four points constitute an arc of four. Then there exists a projective coordinate system in $\mathbb{P}^2(\mathbb{R})$ such that*

$$P_1 = (1:0:0), P_2 = (0:1:0), P_3 = (0:0:1), P_4 = (1:1:1)$$

Before we proceed to the proof, we state a more general result, this times in terms of *projective transformations*:

Corollary 2. *Given four points P_1, P_2, P_3 and $P_4 \in \mathbb{P}^2(\mathbb{R})$ as in the proposition, as well as another set of four points P_1', P_2', P_3' and $P_4' \in \mathbb{P}^2(\mathbb{R})$ with the same property. Then there exists a projective transformation G of $\mathbb{P}^2(\mathbb{R})$ onto itself, mapping P_i to P_i' for $i = 1,2,3$, and 4.*

The proposition implies the corollary, since it implies the existence of a projective transformation F mapping P_1 to $(1:0:0)$, P_2 to $(0:1:0)$, etc., and a projective transformation F' mapping P_1' to $(1:0:0)$, P_2' to $(0:1:0)$, etc. Take G equal to the composition $(F')^{-1} \circ F$.

Proof (of the proposition). We give the proof below. But this proof may well be skipped, at least at a first reading. It is of a purely algebraic nature, and belongs to the field of linear algebra more than to projective geometry. But the result itself is very useful, and saves us from a lot of tedious reasoning. By going through this, we do away with these technical matters once and for all.

We may write the transition from one coordinate system to another as a matrix multiplication as follows:

$$\left\{ \begin{matrix} \alpha_{0,0} & \alpha_{0,1} & \alpha_{0,2} \\ \alpha_{1,0} & \alpha_{1,1} & \alpha_{1,2} \\ \alpha_{2,0} & \alpha_{2,1} & \alpha_{2,2} \end{matrix} \right\} \cdot \left\{ \begin{matrix} x_0 \\ x_1 \\ x_2 \end{matrix} \right\} = \left\{ \begin{matrix} \overline{x}_0 \\ \overline{x}_1 \\ \overline{x}_2 \end{matrix} \right\}$$

We now write

$$P_1 = (a_{1,0} : a_{1,1} : a_{1,2})$$

$$P_2 = (a_{2,0} : a_{2,1} : a_{2,2})$$

$$P_3 = (a_{3,0} : a_{3,1} : a_{3,2})$$

$$P_4 = (a_{4,0} : a_{4,1} : a_{4,2})$$

The aim is to find a matrix $\{\alpha_{i,j}\}$, with determinant $\neq 0$, such that the following four conditions are satisfied for suitable choices of r, s, t, u, all $\neq 0$:

$$\left\{ \begin{matrix} \alpha_{0,0} & \alpha_{0,1} & \alpha_{0,2} \\ \alpha_{1,0} & \alpha_{1,1} & \alpha_{1,2} \\ \alpha_{2,0} & \alpha_{2,1} & \alpha_{2,2} \end{matrix} \right\} \cdot \left\{ \begin{matrix} a_{1,0} \\ a_{1,1} \\ a_{1,2} \end{matrix} \right\} = \left\{ \begin{matrix} r \\ 0 \\ 0 \end{matrix} \right\}$$

$$\left\{ \begin{matrix} \alpha_{0,0} & \alpha_{0,1} & \alpha_{0,2} \\ \alpha_{1,0} & \alpha_{1,1} & \alpha_{1,2} \\ \alpha_{2,0} & \alpha_{2,1} & \alpha_{2,2} \end{matrix} \right\} \cdot \left\{ \begin{matrix} a_{2,0} \\ a_{2,1} \\ a_{2,2} \end{matrix} \right\} = \left\{ \begin{matrix} 0 \\ s \\ 0 \end{matrix} \right\}$$

$$\left\{ \begin{matrix} \alpha_{0,0} & \alpha_{0,1} & \alpha_{0,2} \\ \alpha_{1,0} & \alpha_{1,1} & \alpha_{1,2} \\ \alpha_{2,0} & \alpha_{2,1} & \alpha_{2,2} \end{matrix} \right\} \cdot \left\{ \begin{matrix} a_{3,0} \\ a_{3,1} \\ a_{3,2} \end{matrix} \right\} = \left\{ \begin{matrix} 0 \\ 0 \\ t \end{matrix} \right\}$$

$$\left\{ \begin{matrix} \alpha_{0,0} & \alpha_{0,1} & \alpha_{0,2} \\ \alpha_{1,0} & \alpha_{1,1} & \alpha_{1,2} \\ \alpha_{2,0} & \alpha_{2,1} & \alpha_{2,2} \end{matrix} \right\} \cdot \left\{ \begin{matrix} a_{4,0} \\ a_{4,1} \\ a_{4,2} \end{matrix} \right\} = \left\{ \begin{matrix} u \\ u \\ u \end{matrix} \right\}$$

Replacing the matrix and the parameters r, s, t and u by their respective inverses, we find that it suffices to find a matrix $\{\beta_{i,j}\}$ such that

$$\left\{ \begin{matrix} \beta_{0,0} & \beta_{0,1} & \beta_{0,2} \\ \beta_{1,0} & \beta_{1,1} & \beta_{1,2} \\ \beta_{2,0} & \beta_{2,1} & \beta_{2,2} \end{matrix} \right\} \cdot \left\{ \begin{matrix} 1 \\ 0 \\ 0 \end{matrix} \right\} = \left\{ \begin{matrix} r a_{1,0} \\ r a_{1,1} \\ r a_{1,2} \end{matrix} \right\}$$

$$\left\{\begin{array}{ccc} \beta_{0,0} & \beta_{0,1} & \beta_{0,2} \\ \beta_{1,0} & \beta_{1,1} & \beta_{1,2} \\ \beta_{2,0} & \beta_{2,1} & \beta_{2,2} \end{array}\right\} \cdot \left\{\begin{array}{c} 0 \\ 1 \\ 0 \end{array}\right\} = \left\{\begin{array}{c} sa_{2,0} \\ sa_{2,1} \\ sa_{2,2} \end{array}\right\}$$

$$\left\{\begin{array}{ccc} \beta_{0,0} & \beta_{0,1} & \beta_{0,2} \\ \beta_{1,0} & \beta_{1,1} & \beta_{1,2} \\ \beta_{2,0} & \beta_{2,1} & \beta_{2,2} \end{array}\right\} \cdot \left\{\begin{array}{c} 0 \\ 0 \\ 1 \end{array}\right\} = \left\{\begin{array}{c} ta_{3,0} \\ ta_{3,1} \\ ta_{3,2} \end{array}\right\}$$

$$\left\{\begin{array}{ccc} \beta_{0,0} & \beta_{0,1} & \beta_{0,2} \\ \beta_{1,0} & \beta_{1,1} & \beta_{1,2} \\ \beta_{2,0} & \beta_{2,1} & \beta_{2,2} \end{array}\right\} \cdot \left\{\begin{array}{c} 1 \\ 1 \\ 1 \end{array}\right\} = \left\{\begin{array}{c} ua_{4,0} \\ ua_{4,1} \\ ua_{4,2} \end{array}\right\}$$

Clearly the first three of these conditions will be satisfied, for all r, s and t, if we put

$$\left\{\begin{array}{ccc} \beta_{0,0} & \beta_{0,1} & \beta_{0,2} \\ \beta_{1,0} & \beta_{1,1} & \beta_{1,2} \\ \beta_{2,0} & \beta_{2,1} & \beta_{2,2} \end{array}\right\} = \left\{\begin{array}{ccc} ra_{1,0} & sa_{2,0} & ta_{3,0} \\ ra_{1,1} & sa_{2,1} & ta_{3,1} \\ ra_{1,2} & sa_{2,2} & ta_{3,2} \end{array}\right\}$$

Indeed, multiplication of this matrix by the column vector

$$\left\{\begin{array}{c} 1 \\ 0 \\ 0 \end{array}\right\}$$

to the right yields the first column of the matrix. In the same way the two other column vectors with the 1 in the middle and at the bottom, respectively, yield the second and third column of the matrix. To get a matrix which satisfies the final fourth condition, it therefore will suffice to determine r, s and t such that

$$\left\{\begin{array}{ccc} ra_{1,0} & sa_{2,0} & ta_{3,0} \\ ra_{1,1} & sa_{2,1} & ta_{3,1} \\ ra_{1,2} & sa_{2,2} & ta_{3,2} \end{array}\right\} \cdot \left\{\begin{array}{c} 1 \\ 1 \\ 1 \end{array}\right\} = \left\{\begin{array}{c} a_{4,0} \\ a_{4,1} \\ a_{4,2} \end{array}\right\}$$

This amounts to an equation for r, s and t which may be written as

$$\left\{\begin{array}{ccc} a_{1,0} & a_{2,0} & a_{3,0} \\ a_{1,1} & a_{2,1} & a_{3,1} \\ a_{1,2} & a_{2,2} & a_{3,2} \end{array}\right\} \cdot \left\{\begin{array}{c} r \\ s \\ t \end{array}\right\} = \left\{\begin{array}{c} a_{4,0} \\ a_{4,1} \\ a_{4,2} \end{array}\right\}$$

So we have a system of equations with three equations and three unknowns, with determinant $\neq 0$, since the points P_1, P_2 and P_3 are not collinear. But in Section 5.6 we showed Cramer's Theorem which yields that such a system has a unique solution, (r, s, t). This completes the proof. □

12 Geometry in the Affine and the Projective Plane

In this chapter we shall, among other things, prove the classical theorems of *Desargues, Pappus and Pascal*. These theorems are valid in the projective plane $\mathbb{P}^2(\mathbb{R})$, and we shall give simple algebraic proofs, which fully take advantage of the strength inherent in *Analytic Planar Geometry*.

12.1 The Theorem of Desargues

The most important result in Desargues' book, which we told about in section 5.4, is a typical example of what we call an *incidence theorem*. It is frequently referred to as *Desargues' Perspective Theorem* and says the following:

Theorem 16 (Desargues). *Let two triangles ABC and $A'B'C'$ be given in $\mathbb{P}^2(\mathbb{R})$, such that $A \neq A'$, $B \neq B'$ and $C \neq C'$. Then if the lines through corresponding vertices pass through the same point, the intersections of (the prolongation of) corresponding sides will intersect in points lying on the same line.*

Proof. We introduce notation as in Figure 12.1.

We now preform a very common trick, which immediately reduces the proof to a very simple and elementary fact which is quite well known from high school math. We remember that the question of whether three points are collinear or not certainly is independent of the chosen projective coordinate system in $\mathbb{P}^2(\mathbb{R})$. Now choose a coordinate system in $\mathbb{P}^2(\mathbb{R})$ such that the points P and Q lie *on the line at infinity*. We then have to prove that the point R also lies at the line at infinity. This will prove the claim, since when they all lie at infinity, they are collinear, all being on the same line, namely the one at infinity.

This will signify that the lines AB and $A'B'$ are parallel, and that the same holds for AC and $A'C'$, both intersecting in points at infinity. We have to prove that BC and $B'C'$ also are parallel. But this is a well known fact from elementary planar geometry. The situation is illustrated in Figure 12.2.

Here two of the pairs of triangles with top O are similar. But then so is the third pair, and the claim follows. $\square$

This theorem has the following

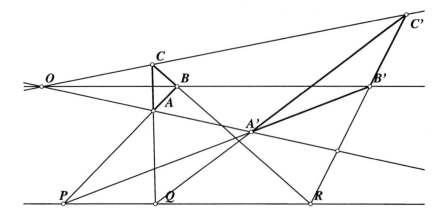

Fig. 12.1. Desargues' Theorem.

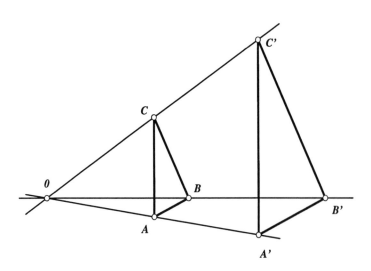

Fig. 12.2. Desargues' Theorem, after a good choice of coordinate system in $\mathbb{P}^2(\mathbb{R})$.

Corollary 3. *Let two triangles ABC and $A'B'C'$ be given in $\mathbb{P}^2(\mathbb{R})$, and assume that $A \neq A'$, $B \neq B'$ and $C \neq C'$. If the intersections of (the prolongations of) corresponding sides lie on the same line, then the lines through corresponding vertices all pass through the same point.*

Proof. This is an immediate consequence of the *Principle of Duality* for $\mathbb{P}^2(\mathbb{R})$, which we shall prove in the next section. Here we find that the dual result is actually a *converse* to the assertion of the theorem. The figure in the proof above is actually *self dual*. We leave the verification of this beautiful fact to the reader.   □

12.2 Duality for $\mathbb{P}^2(\mathbb{R})$

For $\mathbb{P}^2(\mathbb{R})$ we have an *extended principle of duality*. Recall that the usual principle of duality in axiomatic projective geometry says the following: Any statement about points, lines and incidence, which may be deduced from the system of axioms, is transformed into another statement which also may be deduced from the axioms if the words *point* and *line* are interchanged. We refer to the latter statement as the *dual* of the former. But here we are not dealing with the axiomatic system, but rather with a special *model* for it. The collection of statements covered by the original principle of duality only consists of those which may be deduced from the axiomatic system given in Chapter 8, Section 8.1. And the model $\mathbb{P}^2(\mathbb{R})$ has many more true statements than that, in fact the Theorem of Desargues, which we have just proved in $\mathbb{P}^2(\mathbb{R})$ may not be deduced from that system of axioms. Indeed, this statement is usually taken as one of the additional axioms needed to finally arrive at the full axiomatic description of $\mathbb{P}^2(\mathbb{R})$, in the spirit of Hilbert.

The study of so-called *non-Desargian planes* is an exotic interest pursued by some mathematicians. We shall not indulge in it here, but who knows: Some day this may turn out to be useful.

For $\mathbb{P}^2(\mathbb{R})$ we do, however, have the following *stronger* principle of duality:

Theorem 17 (Duality for $\mathbb{P}^2(\mathbb{R})$). *If P denotes a true statement about* $\mathbb{P}^2(\mathbb{R})$ *dealing with points, lines and incidence, then the dual statement* $P^\vee$ *is also a true statement.*

Proof. The statement P may be translated into a collection of relations between the projective coordinates of the points involved and the coefficients of the equations of the lines involved. The relations will all be of the form

$$A_0\alpha_0 + A_1\alpha_1 + A_2\alpha_2 = 0,$$

where $A = (A_0 : A_1 : A_2)$ is a point and α_0, α_1 and α_2 are the coefficients in the equation for a line in $\mathbb{P}^2(\mathbb{R})$, so the line is given by

$$\alpha_0 X_0 + \alpha_1 X_1 + \alpha_2 X_2 = 0.$$

If we denote this line by ℓ, then the statement $A \; I \; \ell$, or $A \in \ell$, is equivalent to the relation above. For every point in $\mathbb{P}^2(\mathbb{R})$ we now let correspond a *line* in $\mathbb{P}^2(\mathbb{R})$ given by the equation whose coefficients are the coordinates of the point, and to every line we let correspond the point whose projective coordinates are the coefficients of the equation giving the line.

We then have the following beautiful and simple situation:

The collection of algebraic relations between the coordinates of points in $\mathbb{P}^2(\mathbb{R})$ and coefficients of lines in $\mathbb{P}^2(\mathbb{R})$ which expresses the truth of the statement P is the same as the collection of relations which expresses the truth of $P^\vee$

This completes the proof. Note that this proof remains valid even if the collections of lines and/or points are infinite. □

12.3 Naive Definition and First Examples of Affine Plane Curves

For most purposes the following definition of an affine algebraic curve in $\mathbb{R}^2$ will suffice: It is a subset of $\mathbb{R}^2$ given as the set of points (x, y) which satisfies an equation

$$f(x, y) = 0,$$

where $f(X, Y)$ is a polynomial with real coefficients in the variables, or as we should rather say, *in the transcendentals X and Y*.

This definition suffices for most purposes, but it becomes insufficient when the need arises to consider *curves occuring with a certain multiplicity*. Thus for instance, the x-axis is given by the equation $y = 0$. But in some considerations it is necessary, at least it is *practical* to be able to consider the x-axis with *multiplicity 2*. This geometrical object would then have the equation $y^2 = 0$.

We will return to this below, for now we content ourselves with the naive definition given above.

12.4 Straight Lines

In general straight lines are given by linear equations in x and y,

$$Ax + By + C = 0,$$

where A and B are not both zero. The curve given by this equation will intersect the x-axis in the point $(-\frac{C}{A}, 0)$ provided that $A \neq 0$. The curve given by $By = C$ is a line parallel with the x-axis, as in this case $B \neq 0$. Similarly, if $B \neq 0$ the line intersects the y-axis in the point $(0, -\frac{C}{B})$.

Give two distinct points in $\mathbb{R}^2$, (x_1, y_1) and (x_2, y_2). Then there is a unique line ℓ passing through them, and in High School we are usually taught some more or less cumbersome ways of finding the equation of this line. Knowing some *linear algebra* greatly facilitates this task, we now explain how.

We seek A, B and C such that ℓ is given by the equation

$$Ax + By + C = 0.$$

We then must have

$$Ax_1 + By_1 + C = 0$$

and

$$Ax_2 + By_2 + C = 0.$$

Here the numbers A, B and C are not all zero. Now we may regard the three relations as equations in the *unknowns* A, B, and C, with coefficients x, y and 1 for the first equation, $x_1, y_1, 1$ for the second and $x_2, y_2, 1$ for the third. As $A = B = c = 0$ also is a solution, the solution to the system is not unique, thus by Cramer's Theorem from Section 5.6 we find that *the determinant of the system* is zero. Thus the equation of the line passing through the points (x_1, y_1) and (x_2, y_2) may be written as a determinant as follows:

$$\begin{vmatrix} x & y & 1 \\ x_1 & y_1 & 1 \\ x_2 & y_2 & 1 \end{vmatrix} = 0.$$

12.5 Conic Sections in the Affine Plane $\mathbb{R}^2$

A *Conic Section* is, as we know, a curve in the plane $\mathbb{R}^2$ of degree 2. The general form of the equation of such a curve is usually written as

$$q(x, y) = Ax^2 + 2Bxy + Cy^2 + 2Dx + 2Ey + F = 0.$$

One may wonder why some of the coefficients are written as 2 times some other constant. The answer to this is that writing the equation in this form is a great convenience later, in that many expressions we need will take a simpler form.

All such curves may be obtained by cutting a cone with a plane. This will be shown in Section 14.5.

We classify the conic sections as the ellipses, the parabolas and the hyperbolas. In addition to these, we have some *degenerate cases*, in that an ellipse may shrink to a point, a hyperbola may degenerate to two lines and a parabola collapse to one or degenerate into two parallel lines.

We remind the reader that in the three non-degenerate cases, the equation can be brought on one of the three so-called *canonical forms* shown in figures 12.4, 12.5 and 12.6

In fact, we consider changes of coordinate system in $\mathbb{R}^2$ from x, y to x', y' of the form

$$x' = (x - a) \cos(v) + (y - b) \sin(v)$$

$$y' = -(x - a) \sin(v) + (y - b) \cos(v).$$

This corresponds to a new coordinate system with origin in (a, b) and rotated an angle v, see Figure 12.3.

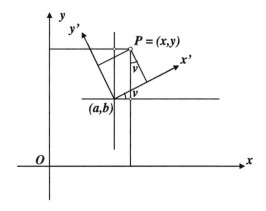

Fig. 12.3. A new coordinate system.

Such a shift may also be regarded as a transformation of the plane onto itself, known as a *rigid motion*. Throughout this chapter, by the term *a new coordinate system in* $\mathbb{R}^2$ we mean a new coordinate system of this type.

We divide the non-degenerate conic sections into three classes, namely *ellipses, parabolas* and *hyperbolas*. After a change of coordinate system, any ellipse may be described by an equation of the type

$$(\frac{x}{a})^2 + (\frac{y}{b})^2 = 1$$

A hyperbola is a conic section which may be transformed into the form

$$(\frac{x}{a})^2 - (\frac{y}{b})^2 = \pm 1,$$

after a change of coordinate system (and here the $\pm$ is really redundant.)

Finally a parabola is a conic section which may be brought on the form

$$px = y^2$$

In Figure 12.4 we have indicated the *half-axes* of the ellipse. Their lengths are, respectively, a and b, and if they are equal then evidently we have a circle of radius $a = b$. We have not indicated the *focal points* of the ellipse,

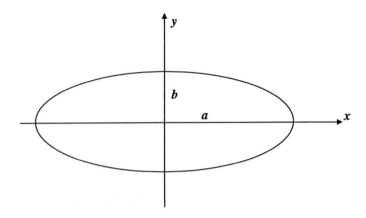

Fig. 12.4. The ellipse given by $\left(\frac{x}{a}\right)^2 + \left(\frac{y}{b}\right)^2 = 1$.

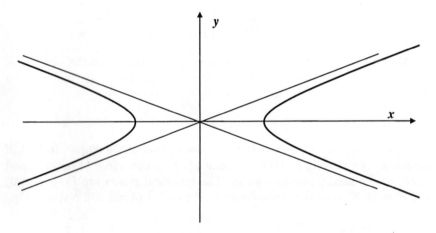

Fig. 12.5. The hyperbola given by $\left(\frac{x}{a}\right)^2 - \left(\frac{y}{b}\right)^2 = 1$. The asymptotes are also shown, those are the lines given by $\left(\frac{x}{a}\right)^2 - \left(\frac{y}{b}\right)^2 = 0$.

but their locations are at the longest axis of the ellipse, at a distance c from the center, i.e., the point of intersection of the two axes, where

$$a^2 = b^2 + c^2$$

The focal points have two interesting properties. First of all, we have the

Proposition 6. *The ellipse is the locus of all points such that the sum of their distances from F_1 and F_2 is constant, namely $2a$ where a is the length of the longest half axis.*

Proof. Assume first that the ellipse is given by the equation

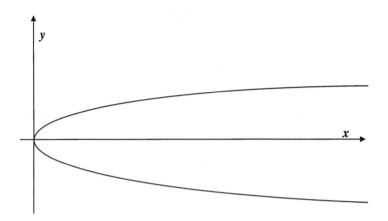

Fig. 12.6. The parabola given by $px = y^2$.

$$\left(\frac{x}{a}\right)^2 + \left(\frac{y}{b}\right)^2 = 1$$

Denote a point on the ellipse by $P = (x, y)$. We show that the sum of the distances to the focal points is $2a$. We remark that there is an angle φ such that

$$x = a\cos(\varphi), y = b\sin(\varphi),$$

and in this way we obtain a parametric description of the ellipse. Indeed, if the point is on the ellipse, then the point $(\frac{x}{a}, \frac{y}{b})$ is on a circle of radius 1 and center at the origin, and conversely. The two focal points are $F_1 = (-c, 0)$ and $F_2 = (c, 0)$, thus the distances from P to these points are, respectively

$$PF_1 = \sqrt{(x+c)^2 + y^2} = \sqrt{(a\cos(\varphi) + c)^2 + b^2\sin^2(\varphi)}$$

and

$$PF_2 = \sqrt{(x-c)^2 + y^2} = \sqrt{(a\cos(\varphi) - c)^2 + b^2\sin^2(\varphi)}$$

which after a short computation yields

$$PF_1 = a + c\cos(\varphi) \text{ and } PF_2 = a - c\cos(\varphi)$$

Hence the claim follows.

Next, let F_1 and F_2 be two points, and let the distance between them be $2c$. Let $r > 2c$, and put $a = \frac{r}{2}$. Choose a coordinate system with origin at the mid point between F_1 and F_2, and x-axis along F_1F_2 directed towards F_2. Let $P = (x, y)$, then the condition that $F_1P + PF_2 = r$ is expressed as

$$\sqrt{(c+x)^2 + y^2} + \sqrt{(c-x)^2 + y^2} = 2a$$

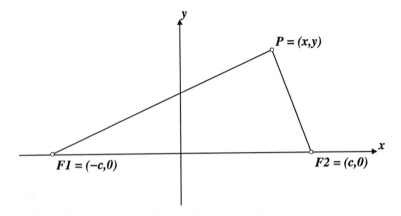

Fig. 12.7. The points F_1 and F_2 and the point $P = (x, y)$.

see Figure 12.7.

Frequently one finds that the argument from this point on is rendered somewhat clumsily: One first squares the left hand side, then isolates the remaining root-expression, and then squares a second time. After messy computations the expression then simplifies to yield the equation for the ellipse.

It is better to perform a standard trick, with which the ancients were well acquainted: We simply multiply both sides of the equality above by *the difference* of the two root expressions:

$$(\sqrt{(c+x)^2 + y^2} + \sqrt{(c-x)^2 + y^2})(\sqrt{(c+x)^2 + y^2} - \sqrt{(c-x)^2 + y^2})$$

$$= 2a(\sqrt{(c+x)^2 + y^2} - \sqrt{(c-x)^2 + y^2})$$

which yields

$$((c+x)^2 + y^2) - ((c-x)^2 + y^2) = 2a(\sqrt{(c+x)^2 + y^2} - \sqrt{(c-x)^2 + y^2})$$

or

$$4cx = 2a(\sqrt{(c+x)^2 + y^2} - \sqrt{(c-x)^2 + y^2})$$

thus

$$\sqrt{(c+x)^2 + y^2} - \sqrt{(c-x)^2 + y^2} = 2\frac{c}{a}x$$

Adding this to the original equality yields

$$\sqrt{(c+x)^2 + y^2} = a + \frac{c}{a}x$$

which when squared yields

$$(c+x)^2 + y^2 = (a + \frac{c}{a}x)^2$$

and after a short computation this becomes

$$\frac{x^2}{a^2} + \frac{y^2}{a^2 - c^2} = 1$$

so letting $b = \sqrt{a^2 - c^2}$ we get the equation of the ellipse on its usual form.
□

Remark Note that the parametric form of the ellipse given in the proof above is not the one obtained through introducing polar coordinates for the ellipse, but rather the one deduced from polar coordinates for the circle. Polar coordinates for the ellipse yields more complicated expressions. In fact, the angle φ is the so-called *eccentric angle of the point* $P = (x, y)$.

It also should be pointed out that strictly speaking the first part of the proof, involving the parametric form, is redundant: Indeed, the last part of the argument can be made to work both ways. We leave this analysis to the reader.

Using Proposition 6 we may draw an ellipse with given focal points and given half axis a by means of two needles, a piece of string and a pencil as shown in Figure 5.2 in Chapter 5.

By means of Proposition 6 it is also easy to explain the reason for calling the two points F_1 and F_2 the *focal points*. The reader may contemplate the illustration in Figure 12.8.

Ellipses form an important class of curves. As we know, the planets move around the Sun in orbits which are approximate ellipses, with the Sun at one of the focal points. Many comets have other types of orbits, parabolas or hyperbolas. Those constitute the two other classes of non-degenerate conic sections.

With a and b as before we define the *eccentricity* e by

$$e^2 = 1 - (\frac{b}{a})^2$$

Thus if the ellipse is a circle then $e = 0$.

Parabolas may, by a change of coordinates in $\mathbb{R}^2$, be given by an equation of the form

$$px = y^2$$

This class of conic sections can be obtained by deforming an ellipse in the following way: One of the focal points is kept fixed, the other is moved towards infinity. The remaining focal point will then be the focusing point for a ray

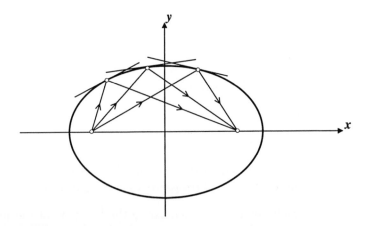

Fig. 12.8. Assume that the ellipse is a mirror. Then an observer at one focal point sees the other in any direction. Similarly, a source of light at one of the points will emit light rays which are reflected in rays which are focused at the other one. This is the second interesting property of the focal points of an ellipse.

of incoming signals, light or radio waves, which is parallel with the axis. We will see, by a closer analysis which we omit here, that under this stretching-procedure the eccentricity e will approach 1, as the largest half-axis tends to infinity faster than the smaller half axis. Thus parabolas are of eccentricity 1. Parabolas are given by a similar geometric property to the one which determines the ellipses, by a *focal point* F and a line L referred to as the *directrix* of the parabola. We have the following:

Proposition 7. *The parabola is the locus of all points such that their distances from a fixed point F and a fixed line L are equal.*

Proof. We choose a coordinate system so that the line L has equation $x + c = 0$ and $F = (c, 0)$. Then the condition becomes

$$\sqrt{(x-c)^2 + y^2} = x + c$$

which is equivalent to

$$y^2 = 4cx$$

☐

Hyperbolas form a class of conic sections which after a change of coordinate system in $\mathbb{R}^2$ are given by

$$(\frac{x}{a})^2 - (\frac{y}{b})^2 = 1$$

The eccentricity e is given by

$$e^2 = 1 + (\frac{b}{a})^2$$

In particular $e > 1$.

Moreover, there are two focal points as indicated in Figure 12.9, where the distance c from the origin is given by

$$c^2 = a^2 + b^2,$$

For hyperbolas we have the

Proposition 8. *The hyperbola is the locus of all points such that the difference of their distances from two fixed points F_1 and F_2 is constant.*

The proof is omitted. It is very similar to the last part of the proof of Proposition 6, using the last part of the remark after the proof.

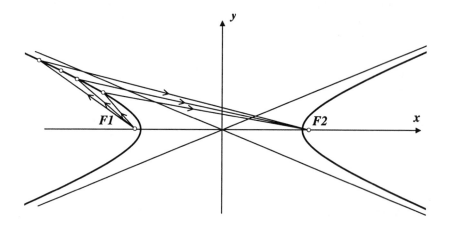

Fig. 12.9. The hyperbola have similar reflection properties to that of the ellipse.

Finally, the reflection property for the focal points of the hyperbola is illustrated in Figure 12.9.

The equation

$$(\frac{x}{a})^2 - (\frac{y}{b})^2 = 0,$$

yields two lines, the so-called asymptotes of the hyperbola whose equation it resembles. As x and y become very large, the two lines will be indistinguishable from the hyperbola itself: The curve approaches these lines as the point moves towards infinity. In precise terms, the two asymptotes are the tangents of the curve at its two points at infinity. We return to this phenomenon later in this chapter.

We next consider the following problem, which is a continuation of what we did regarding the equation of a line passing through two given, distinct points.

Consider 5 distinct points in $\mathbb{R}^2$, $P_i = (x_i, y_i)$, $i = 1, 2, \ldots, 5$. If the points are in *"sufficiently general position"*, then there is a unique, non-degenerate conic section, in other words a non-degenerate curve of degree 2, passing through them. Here we may immediately make precise the requirement *"in sufficiently general position."* It is the condition that no three of them be collinear. If three are collinear, but not four, then there is a unique *degenerate* conic section passing through them, namely the union of two lines, and if four are collinear but not all five then there are four degenerate conics through them, if all five are collinear then there are infinitely many degenerate conics passing through them, all having the line containing them as a component.

To find the equation of the unique conic section passing through the sufficiently general points, we may proceed in an analogous manner to what we did for lines. Indeed, the equation is

$$\begin{vmatrix} x^2 & xy & y^2 & x & y & 1 \\ x_1^2 & x_1y_1 & y_1^2 & x_1 & y_1 & 1 \\ x_2^2 & x_2y_2 & y_2^2 & x_2 & y_2 & 1 \\ x_3^2 & x_3y_3 & y_3^2 & x_3 & y_3 & 1 \\ x_4^2 & x_4y_4 & y_4^2 & x_4 & y_4 & 1 \\ x_5^2 & x_5y_5 & y_5^2 & x_5 & y_5 & 1 \end{vmatrix} = 0.$$

If uniqueness fails, the coefficients of this equation are all zero. If the conic section is degenerate, the equation factors as a product of two linear polynomials.

12.6 Constructing Points on Conic Sections by Compass and Straightedge

Constructions involving *conic sections* are *asymptotic Euclidian constructions*. In other words, using compass and straightedge in the legal way, we may construct as many points on any conic section as we wish, and thus a construction involving general conic sections may be completed to any prescribed degree of accuracy using compass and straightedge.

Proposition 6 gives a method for constructing as many points as we wish on an ellipse by the *Euclidian tools*. The construction is given in Figure 12.10. The longest axis is PQ.

A point A on the ellipse satisfies $F_1A + AF_2 = r = PQ$. We proceed as follows: We subdivide the line segment F_1F_2 into n pieces, not necessarily of equal length. Call a point in the subdivision S. We then draw a circle with center F_1 and radius PS, and another circle with center F_2 and radius SQ. The two points of intersection of these two circles then lie on the ellipse with largest axis r and focal points F_1 and F_2.

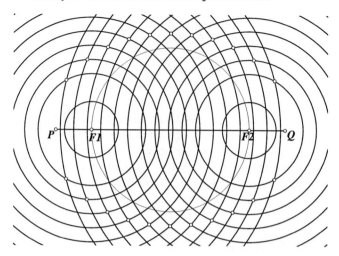

Fig. 12.10. The construction of points on an ellipse with given focal points F_1 and F_2, and given largest axis $r = 2a = PQ$.

Similarly we may construct points on a parabola as shown in Figure 12.11.

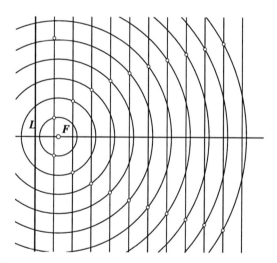

Fig. 12.11. The construction of points on a parabola with a given focal point F and directrix L.

A point on the parabola is the intersection between a circle with center at F and radius ρ, and lines parallel with L, at a distance equal to ρ from L. Varying ρ we thus get as many points as we wish on the parabola.

Finally we show how to construct points on a hyperbola. We start out by marking the two focal points F_1 and F_2. About F_1 and F_2 we draw families of

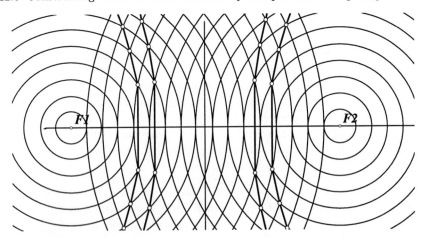

Fig. 12.12. The construction of points on a hyperbola with given focal points F_1 and F_2 and given difference between the distances from a point on the hyperbola to the two focal points.

circles of radius Δ, 2Δ, 3Δ, ..., $n\Delta$ for some small fixed distance Δ. We then obtain points on a family of hyperbolas, corresponding to fixed differences in distance to F_1 and F_2. This is shown in Figure 12.12.

This principle is utilized in the navigational system GPS, the letters stand for *Global Positioning System*. The system works in three dimensions, but the same principle is used in the older, and now essentially outdated, LORAN-system which is two-dimensional. Then we have several radio beacons, and a ship with a receiver and appropriate equipment is able to compute the difference between the distances to any two transmitters it is receiving. Thus the navigator may pinpoint the ship's position to several such hyperbolas, three signals will leave four possible locations, and in normal circumstances three of them may be ruled out from other information. GPS works with satellites located at F_1 and F_2, and we now get the position as being on a *surface* from receiving two satellites. The surface is obtained by rotating the hyperbola about the line F_1F_2, such a surface is a special case of what we call a *hyperboloid surface*.

With three satellites we get the possible positions on certain curves, to get points we need a minimum of four satellites. Then in principle there are eight possibilities. An added difficulty is that we are working with approximations. For these and other reasons we need to receive a larger number of satellites to get a good position.

12.7 Further Properties of Conic Sections

In the seventeenth century a conic section was given a definition as the locus of points with a certain fundamental relation to a fixed point and a fixed line. Due to Johan de Witt and John Wallis this definition may in some sense be understood as a precursor to the full algebraization of the subject. It is, however, used less frequently today. It has the advantage of being very geometric in nature, and to elucidate the difference between the three classes of conic sections, while at the same time providing a unified treatment. Here we give it in the form of a proposition:

Proposition 9. *Given a fixed point F and a line ℓ. Let $0 \leq e$ be a real number. Let $\mathcal{C}$ denote the set of point in $\mathbb{R}^2$ such that the ratio of the distance from F to the distance from ℓ is constant and equal to e. If $e = 0$[1] then the curve is a circle, if $0 < e < 1$ then it is an ellipse, for $e = 1$ it is a parabola and for $e > 1$ it is a hyperbola. All conic sections can be obtained in this way.*

Proof. We shall only sketch the proof here, but return to it in Section 12.8. For now, the case of $e = 0$ will give us some trouble, and right now we shall only indicate how this is dealt with. So assume that $e > 0$. We may choose a coordinate system such that F is the origin and such that ℓ is parallel with the y-axis, and intersects the x-axis at a distance p from the origin, where p may be positive, zero or negative. Then the condition is

$$\frac{\sqrt{x^2 + y^2}}{x + p} = e$$

This yields after a short computation

$$(1 - e^2)x^2 - 2pe^2 x + y^2 = e^2 p^2,$$

and from this everything follows *except* for the case $e = 0$. But this case is troubling, since it would appear that the assertion is false as stated in this case. As indeed it is!

The point is that one must understand the assertion somewhat differently from what one straightforwardly would assume. Namely, in this case the line ℓ has to be chosen at infinity. But a finite point has an infinite distance to a point at infinity. One way to resolve this is to let e tend to 0 as p tends to infinity. We will not pursue this further, however. □

We return to the concept of *degeneration* for conic sections.[2]

---

[1] $e = 0$ is strictly speaking not covered by the proposition as it is stated. See remark at the end of the proof. But this is the form in which this result is often stated.

[2] The proof of the following theorem may be omitted on the first reading of this book. We give a self-contained proof, somewhat long but avoiding the machinery

Theorem 18. *The equation*

$$q(x, y) = Ax^2 + 2Bxy + Cy^2 + 2Dx + 2Ey + F = 0$$

yields a non-degenerate conic section if and only if the following determinantal criterion is satisfied:

$$\begin{vmatrix} A & B & D \\ B & C & E \\ D & E & F \end{vmatrix} \neq 0.$$

Proof. We need the notion of a *non-singular* point of a plane curve, we return to a refined treatment of this important concept in Section 13.4:

Definition 10. *Let Z be a plane curve given by the equation*

$$f(x, y) = 0.$$

Let (x_0, y_0) be a point on the curve such that the two partial derivatives do not both vanish,

$$\left(\frac{\partial f}{\partial x}(x_0, y_0), \frac{\partial f}{\partial y}(x_0, y_0) \right) \neq (0, 0).$$

Such a point is called a non-singular point on the curve. At all non-singular points we define the tangent line[3] by the equation

$$\frac{\partial f}{\partial x}(x_0, y_0)(x - x_0) + \frac{\partial f}{\partial y}(x_0, y_0)(y - y_0) = 0.$$

A point which is not non-singular is called a singular point.

Remark. In general the property for a point on a curve to be singular or non-singular is independent of the choice of affine coordinate system, which we shall prove as Proposition 17. A simpler case, which is immediate to verify, is that when checking singularity or non-singularity for a point P on some curve, we may assume $P = (0,0)$ without loss of generality, modifying the equation appropriately, of course.

Now let $\mathcal{C}$ be the conic section given by the equation above, and let (x_0, y_0) be a point on it. Then

$$\frac{\partial q}{\partial x}(x_0, y_0) = 2Ax_0 + 2By_0 + 2D, \quad \frac{\partial q}{\partial y}(x_0, y_0) = 2Bx_0 + 2Cy_0 + 2E.$$

of linear algebra, namely the theorem asserting that *any real, symmetric matrix may be diagonalized by an orthogonal non-singular matrix.* The reader familiar with this material, and with the concept of projective closure to be introduced later, will find the proof to be a straightforward exercise using that theory.

[3] This concept will be explained in more detail in Section 13.5.

Thus we observe that the point $(x_0, y_0) \in \mathcal{C}$ is a non-singular point if and only if it does not lie on both of the lines given by

$$Ax + By + D = 0, Bx + Cy + E = 0.$$

Since the equation of $\mathcal{C}$ may be written as

$$Ax^2 + 2Bxy + Cy^2 + 2Dx + 2Ey + F =$$
$$x(Ax + By + D) + y(Bx + Cy + E) + Dx + Ey + F,$$

these considerations provide the proof of the

Lemma 1. *The point $(x_0, y_0) \in \mathbb{R}^2$ is a singular point of the curve given by the equation*

$$q(x, y) = Ax^2 + 2Bxy + Cy^2 + 2Dx + 2Ey + F = 0$$

if and only if

$$Ax_0 + By_0 + D = 0$$
$$Bx_0 + Cy_0 + E = 0$$
$$Dx_0 + Ey_0 + F = 0$$

We next prove the following proposition:

Proposition 10. *A non-degenerate conic section in $\mathbb{R}^2$ does not have any singular points. Moreover, if $\mathcal{C}$ is a degenerate conic section in $\mathbb{R}^2$ without singular points, then $\mathcal{C}$ consists of two distinct parallel lines.*

Proof. Assume that (x_0, y_0) were a singular point. We may, without loss of generality, assume that $P = (0, 0)$, so the equation becomes

$$Ax^2 + 2Bxy + Cy^2 + 2Dx + 2Ey = 0.$$

The origin being a singular point, however, we have $D = E = 0$, thus the equation is

$$Ax^2 + 2Bxy + Cy^2 = 0.$$

The zero locus of this equation in $\mathbb{R}^2$ may consist of the origin alone, which is a degenerate conic section. Alternatively, the equation may be written as

$$(a_1 x + b_1 y)(a_2 x + b_2 y) = 0$$

which either represents a double line, or two lines through the origin, both cases being degenerate conics.

Next, assume that $\mathcal{C}$ is degenerate but without singular points. We may assume that $(0,0) \in \mathcal{C}$. Thus the case of $\mathcal{C}$ being a single point, i.e. the origin, is excluded as the equation would then be $Ax^2 + 2Bxy + Cy^2 = 0$, so $(0,0)$ would be singular. Hence $\mathcal{C}$ consists of two (possibly coinciding) lines. Then the equation factors as

$$Ax^2 + 2Bxy + Cy^2 + 2Dx + 2Ey = (ax + by)(cx + dy + e).$$

If the two lines given by $ax + by = 0$ and $cx + dy + e = 0$ coincide, then we may write the equation as $(ax + by)^2 = 0$, all of whose points are singular. If the lines are distinct but intersect, the point of intersection will have to be a singular point as is easily checked. Thus the lines are parallel, and the claim is proven. $\square$

We may now prove Theorem 18. Assume first that

$$\begin{vmatrix} A & B & D \\ B & C & E \\ D & E & F \end{vmatrix} \neq 0.$$

Then by Lemma 1 $\mathcal{C}$ has no singular points: Indeed, by Cramer's Theorem 4 in Section 5.6 the only solution to the system of equations would be $(x_0, y_0, z_0) = (0, 0, 0)$. Thus by Proposition 10 it is non-degenerate, unless it consists of two parallel lines, i.e.

$$Ax^2 + 2Bxy + Cy^2 + 2Dx + 2Ey = (ax + by)(ax + by + e)$$

where $e \neq 0$. Thus $A = a^2, B = ab, C = b^2, D = \frac{1}{2}ae, E = \frac{1}{2}be$ and $F = 0$. But with these values the determinant is zero.

Conversely, assume that $\mathcal{C}$ is non-degenerate. Evidently we may assume $(0,0) \in \mathcal{C}$ without loss of generality, so $F = 0$. If

$$\begin{vmatrix} A & B & D \\ B & C & E \\ D & E & F \end{vmatrix} = 0$$

then by Cramer's Theorem 4 in Section 5.6, there are real numbers x_0, y_0 and z_0 not all zero such that

$$\begin{aligned} Ax_0 + By_0 + Dz_0 &= 0 \\ Bx_0 + Cy_0 + Ez_0 &= 0 \\ Dx_0 + Ey_0 + Fz_0 &= 0 \end{aligned}$$

If $z_0 \neq 0$, we may assume $z_0 = 1$, dividing through the equations. Then (x_0, y_0) is a singular point on $\mathcal{C}$, by Lemma 1. This contradicts $\mathcal{C}$ being non-degenerate, by Proposition 10.

If $z_0 = 0$ in all triples (x_0, y_0, z_0) satisfying the system of equations above, then we choose one solution $(x_0, y_0, 0) \neq (0, 0, 0)$, and note that $(tx_0, ty_0, 0)$ also is a solution for all $t \in \mathbb{R}$. Since we have

$$
\begin{aligned}
Atx_0 + Bty_0 &= 0 \\
Btx_0 + Cty_0 &= 0 \\
Dtx_0 + Ety_0 &= 0
\end{aligned}
$$

we find that $(tx_0, ty_0) \in \mathcal{C}$ for all $t \in \mathbb{R}$ since

$$
tx_0(Atx_0 + Bty_0) + ty_0(Btx_0 + Cty_0) + 2(Dtx_0 + Ety_0)
$$

$$
= q(tx_0, ty_0) = 0.
$$

But then $\mathcal{C}$ contains the line joining $(0, 0)$ and (x_0, y_0), contradicting the assumption of it being non-degenerate. $\square$

We conclude this section by discussing tangents to non-degenerate conic sections in $\mathbb{R}^2$, as well as the related concepts of *pole and polar line*.

So let $P = (x_0, y_0)$ be a point on the non-singular conic section given by the equation

$$
q(x, y) = Ax^2 + 2Bxy + Cy^2 + 2Dx + 2Ey + F = 0.
$$

Then the tangent at P is given by

$$
(Ax_0 + By_0 + D)(x - x_0) + (Bx_0 + Cy_0 + E)(y - y_0) = 0,
$$

which after a short calculation, using that $q(x_0, y_0) = 0$, takes the following beautiful form:

$$
Ax_0x + B(y_0x + x_0y) + Cy_0y + D(x + x_0) + E(y + y_0) + F = 0.
$$

This equation is also important when the point is not on C. Indeed, we have the following

Proposition 11. *Let $P = (x_0, y_0)$ be a point and let $\mathcal{C}$ be the conic section given by the equation*

$$
q(x, y) = Ax^2 + 2Bxy + Cy^2 + 2Dx + 2Ey + F = 0.
$$

There are two tangent lines to $\mathcal{C}$ passing through P, coinciding if P is on C. Let the two points of tangency be Q_1 and Q_2. Then the line p passing through Q_1 and Q_2 is given by the equation

$$
Ax_0x + B(y_0x + x_0y) + Cy_0y + D(x + x_0) + E(y + y_0) + F = 0.
$$

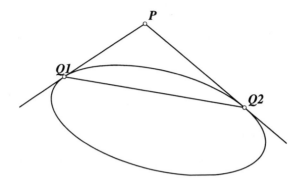

Fig. 12.13. The line joining the two points of tangency.

Proof. The situation is shown in Figure 12.13.

Let $Q_1 = (x_1, y_1)$ and $Q_2 = (x_2, y_2)$, then the two tangents in question will have equations

$$Ax_1x + B(y_1x + x_1y) + Cy_1y + D(x + x_1) + E(y + y_1) + F = 0,$$

$$Ax_2x + B(y_2x + x_2y) + Cy_2y + D(x + x_2) + E(y + y_2) + F = 0.$$

These lines pass through $P = (x_0, y_0)$, thus

$$Ax_1x_0 + B(y_1x_0 + x_1y_0) + Cy_1y_0 + D(x_0 + x_1) + E(y_0 + y_1) + F = 0,$$

$$Ax_2x_0 + B(y_2x_0 + x_2y_0) + Cy_2y_0 + D(x_0 + x_2) + E(y_0 + y_2) + F = 0.$$

But this demonstrates that the line whose equation is given in the assertion of the proposition, does indeed pass through the two points Q_1 and Q_2. Hence the claim follows. □

Definition 11. *The point P and the line p in Proposition 11 are called the pole and the polar line corresponding to each other.*

Remark If we take a point inside an ellipse, then there will be no real points of tangency, even though we get a well defined polar line using the equation. But if we compute the *complex* points of tangency, we find that corresponding coordinates are complex conjugates, and we get a real line joining them.

Moreover, if we choose the center of the unit circle, say, then the "line" given by the formula is just $0 = 1$, which has no points on it. The explanation for this is that the polar of the center is the *line at infinity*. Thus we see here both the need for computing complex points as well as for considering points at infinity.

The latter is the subject of the next section.

12.8 Conic Sections in the Projective Plane

We shall now consider the *projective closure* of the conic sections in $\mathbb{R}^2$. We substitute

$$x = \frac{X_1}{X_0}, y = \frac{X_2}{X_0}$$

into the equation for a conic section $\mathcal{C}$,

$$q(x,y) = Ax^2 + 2Bxy + Cy^2 + 2Dx + 2Ey + F = 0,$$

which after clearing denominators, yields the following equation

$$Q(X_0, X_1, X_2) =$$

$$AX_1^2 + 2BX_1X_2 + CX_2^2 + 2DX_0X_1 + 2EX_0X_2 + FX_0^2 = 0.$$

This is a *homogeneous equation*. Recall that a polynomial $F(X_0, X_1, X_2)$ is said to be homogeneous if *all monomials which occur in it are of the same degree*. If $F(X_0, X_1, X_2)$ is a homogeneous polynomial of degree 2, then as is immediately verified

$$F(ta_0, ta_1, ta_2) = t^2 F(a_0, a_1, a_2)$$

for all real numbers t, a_0, a_1, a_2. Hence we get the equation for a curve in $\mathbb{P}^2(\mathbb{R})$, which when intersected with $D_+(X_0) = \mathbb{R}^2$ gives back the original curve. We call this curve in $\mathbb{P}^2(\mathbb{R})$ the *projective closure* of $\mathcal{C}$.

We first wish to determine its points at infinity. Those are the points $C \cap V_+(X_0)$. The point $(u : v : 0)$ is in $\mathcal{C}$ if

$$Au^2 + 2Buv + Cv^2 = 0,$$

and we immediately get the

Proposition 12. *1. $\mathcal{C}$ has no real points at infinity if $B^2 - AC < 0$.*
2. $\mathcal{C}$ has one real point at infinity if $B^2 - AC = 0$.
3. $\mathcal{C}$ has two points at infinity if $B^2 - AC > 0$.
   Thus 1. corresponds to a possibly degenerate ellipse, 2. to a possibly degenerate parabola and 3. to a possibly degenerate hyperbola.

In general, whenever

$$F(X_0, X_1, X_2) = 0$$

is a homogeneous equation, it gives a curve in $\mathbb{P}^2(\mathbb{R})$. Let $P = (a_0, a_1, a_2)$ be a point on it. In Section 14.4 we show that the equation

$$\frac{\partial F}{\partial X_0}(a_0, a_1, a_2)X_0 + \frac{\partial F}{\partial X_1}(a_0, a_1, a_2)X_1 + \frac{\partial F}{\partial X_2}(a_0, a_1, a_2)X_2 = 0$$

yields the tangent line at P.

Definition 12. *If the partial derivatives involved in the equation for the tangent line all vanish at some point on the curve, then the point is said to be a singular point. If they do not all vanish, the point is called non-singular.*

The equation for the tangent to the conic section in $\mathbb{P}^2(\mathbb{R})$ given by the equation $Q(X_0, X_1, X_2) = 0$ at the point $P = (a_0, a_1, a_2)$ is

$$(Ax_1 + Bx_2 + Dx_0)X_1 + (Bx_1 + Cx_2 + Ex_0)X_2$$
$$+(Dx_1 + Ex_2 + Fx_0)X_0 = 0$$

or written on a more appealing form

$$Ax_1X_1 + B(x_1X_2 + x_2X_1) + Cx_2X_2 + D(x_0X_1 + x_1X_0)$$
$$+E(x_0X_2 + x_2X_0) + Fx_0X_0 = 0$$

This is similar to what we found in the affine case.

If the point P is singular, then its projective coordinates constitute a non-trivial solution of the homogeneous system of equations which we encountered in the previous section:

$$Au + Bv + Dw = 0$$
$$Bu + Cv + Ew = 0$$
$$Du + Ev + Fw = 0$$

In $\mathbb{P}^2(\mathbb{R})$ we have the best way of understanding the relation between *pole and polar*. With the conic section given by

$$AX_1^2 + 2BX_1X_2 + CX_2^2 + 2DX_0X_1 + 2EX_0X_2 + FX_0^2 = 0$$

and a point $(a_0 : a_1 : a_2)$, the polar is given by the equation

$$Ax_1X_1 + B(x_1X_2 + x_2X_1) + Cx_2X_2 + D(x_0X_1 + x_1X_0)$$
$$+E(x_0X_2 + x_2X_0) + Fx_0X_0 = 0.$$

Now we may compute the polar line of a circle with respect to the center. Let the circle be given by

$$x^2 + y^2 = R^2$$

The projective closure is

$$X_1^2 + X_2^2 - R^2 X_0^2 = 0,$$

and the point is $P = (1 : 0 : 0)$. The polar line is given by

$$-R^2 X_0 = 0.$$

Thus we get the line at infinity, as claimed earlier.

We are now also in the position to clarify the remaining aspects of the definition from 9 We chose a coordinate system such that F is the origin and such that ℓ is parallel with the y-axis, and intersects the x-axis at a distance p from the origin, where p may be positive, zero or negative. Then the condition defining the curve is

$$\frac{\sqrt{x^2 + y^2}}{x + p} = e,$$

which yields the equation

$$(1 - e^2)x^2 - 2pe^2 x + y^2 = e^2 p^2.$$

The projective closure of this curve in $\mathbb{P}^2(\mathbb{R})$ is given by the equation

$$(1 - e^2)X_1^2 - 2pe^2 X_0 X_1 + X_2^2 - e^2 p^2 X_0^2 = 0,$$

and the origin becomes $(1 : 0 : 0)$. The polar of our conic section with respect to the point $(a_0 : a_1 : a_2)$ is given by

$$(1 - e^2)x_1 X_1 - pe^2 x_0 X_1 - pe^2 x_1 X_0 + x_2 X_2 - e^2 p^2 x_0 X_0 = 0.$$

With $x_0 = 1, x_1 = x_2 = 0$, this becomes the line

$$pe^2 X_1 + e^2 p^2 X_0 = 0,$$

or

$$X_1 + pX_0 = 0,$$

and in $\mathbb{R}^2$ this is the line $x = -p$, thus the directrix. If furthermore we assume $e = 0$, then the second equation gives a line, whereas the first one does not. As indicated in Section 12.7, we let e tend to 0 and p tend to infinity to capture this situation. Thus $\frac{1}{p} \to 0$ and the line becomes $X_0 = 0$. This line is the line at infinity.

We sum up our findings:

Proposition 13. *The polar of a focal point of a conic section is a directrix.*

We note, however, that the concepts of eccentricity, focal point and directrix depends on the representation of the curve in $\mathbb{R}^2$. Thus for instance, the circle given by $x^2 + y^2 = 1$ has the projective closure in $\mathbb{P}^2(\mathbb{R})$

$$X_1^2 + X_2^2 - X_0^2 = 0,$$

which when restricted to $D_+(X_1)$ yields a hyperbola.

12.9 The Theorems of Pappus and Pascal

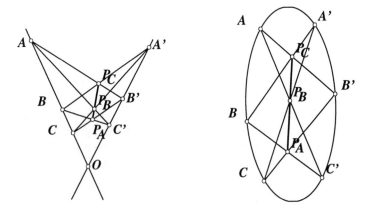

Fig. 12.14. Illustration to Pappus' and Pascal's Theorems, with notation.

Let $\mathcal{C}$ be a conic section in $\mathbb{P}^2(\mathbb{R})$, and let A, B, C as well as A', B', C' be points on $\mathcal{C}$. We assume $\mathcal{C}$ to be non-degenerate, or degenerate as two separate lines ℓ_1 and ℓ_2, called the *components* of the degenerate conic section. In the former case we may choose the 6 points freely on $\mathcal{C}$, but in the latter case the points may be chosen according to the following restriction: No point may be chosen at the intersection of the two components, and the first 3 must lie on ℓ_1 and the last 3 on ℓ_2. See Figure 12.14, where the notation to be used in the sequel is introduced.

The points P_A, P_B and P_C are the points of intersection between, respectively, the lines CB' and $C'B$, the lines AC' and $A'C$, and between AB' and $A'B$. The subscript marking the point is the letter missing in the designation of the lines.

We have already encountered *Pappus of Alexandria* in Section 4.12. The following result is referred to as *Pappus' Theorem* in the degenerate case and as *Pascal's Theorem* (named after Blaise Pascal) in the non-degenerate case:

Theorem 19. *The points P_A, P_B and P_C are collinear.*

Proof. Assume first that the conic section is non-degenerate. By Proposition 5 we may normalize the coordinate system by selecting four points in $\mathbb{P}^2(\mathbb{R})$, no three of them being collinear, and then introduce a new projective coordinate system in $\mathbb{P}^2(\mathbb{R})$ in which these four points become expressed as $(0:0:1)$, $(1:0:0)$, $(1:1:1)$, and $(0:1:0)$. As straight lines and collinearity are of course independent of the chosen coordinate system, we may choose the four points so as to obtain a very simple equation for the conic section, and then hopefully be able to prove the result easily. This is the strategy.

We choose as our points, in this order, C', A, B', and a point on the tangent line to the conic section at the point C'. In the projective coordinate system which this gives rise to, the tangent to $\mathcal{C}$ at C' is the line $V_+(X_0)$, the line at infinity, when we restrict to $D_+(X_0) = \mathbb{R}^2$. C' is then the point at infinity in the direction of the y-axis. The point A is the origin. $B' = (1,1)$, and this is all our information on the points.

We can now draw several conclusions. First of all, we have a non-degenerate conic section in $\mathbb{R}^2$ which has exactly one point at infinity, thus it is a parabola. Moreover, that point at infinity is the infinite point on the x-axis, and the curve passes through the origin, thus the parabola is of the type $y = px^2$. It also passes through the point $(1,1)$, hence $p = 1$. Thus the conic section $\mathcal{C}$ restricted to $\mathbb{R}^2$ is given by the equation

$$y = x^2.$$

The situation which we have arrived at is illustrated in Figure 12.15.

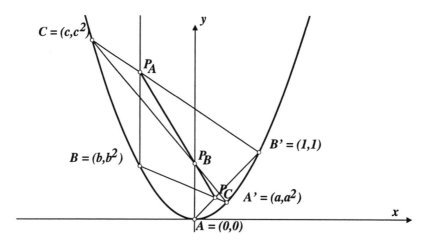

Fig. 12.15. The normalized situation.

We now put $C = (c, c^2), B = (b, b^2)$ and $A' = (a, a^2)$. Here $c, b < 0$, of course, while $a > 0$.

It now is an elementary exercise to check that the points P_A, P_B, P_C are as follows:

$$P_A = (b, b + bc - c), P_B = (0, -ac), P_C = \left(\frac{ab}{a+b-1}, \frac{ab}{a+b-1}\right).$$

Indeed, for P_A clearly the first coordinate is b. The line CB' has the equation

$$y = (c+1)x - c,$$

thus the claimed second coordinate follows. P_B lies on the y-axis, and the equation for the line $A'C$ is

$$y = (a + c)x - ac,$$

and the second coordinate follows as claimed. Finally, P_C is the intersection between the two lines

$$y = x \text{ and } y = (a + b)x - ab$$

We thus get

$$x(a + b - 1) = ab,$$

and we get P_C as claimed. To prove that these three points are collinear, we compute the determinant

$$\begin{vmatrix} 1 & b & b + bc - c \\ 1 & 0 & -ac \\ 1 & \frac{ab}{a+b-1} & \frac{ab}{a+b-1} \end{vmatrix} = 0,$$

and the proof is complete in the non-degenerate case.

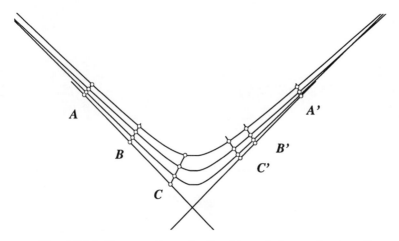

Fig. 12.16. Degenerating a family of hyperbolas to the given lines.

We next turn to the degenerate case, which we deduce from the non-degenerate case by a classical method known as *degeneration*. We fit a family of hyperbolas having the two lines as asymptotes into the picture, as shown in Figure 12.16. The algebraic details on how this is done should be clear by now: We may assume that the lines intersect at the origin, and that their equations may be written as

$$(\frac{y}{b})^2 - (\frac{x}{a})^2 = 0.$$

We then consider the family of hyperbolas

$$(\frac{y}{b})^2 - (\frac{x}{a})^2 = \epsilon,$$

where ϵ is a positive number which we let tend to zero. Clearly, then, the family of hyperbolas will tend to the limiting position of the two lines, their common asymptotes. Now we may chose 6 points on each hyperbola, denote them by $A(\epsilon), \ldots, C'(\epsilon)$, in such a way that $A(\epsilon)$ tends to A as $\epsilon \longrightarrow 0$, etc. Since we have collinearity for the three points in question for each value of ϵ, the same must hold in the limit. This completes the proof. □

13 Algebraic Curves of Higher Degrees in the Affine Plane $\mathbb{R}^2$

13.1 The Cubic Curves in $\mathbb{R}^2$

The simplest curve of *higher degree*, by which we mean degree higher than 2, is the curve known as the *cubical parabola*. The parabola we have encountered before has the equation $y = x^2$, after a suitable change of coordinate system in $\mathbb{R}^2$. It has the graph shown in Figure 13.1.

Similarly, any curve which may be brought on the from

$$y = x^3,$$

if necessary after such a change of variables, is referred to as a *cubical parabola*. It has the graph shown in Figure 13.2.

The next step in complexity is a curve which may be brought on the form

$$y^2 = x^3$$

It is called a *semi cubical parabola*. It has the graph displayed in Figure 13.3.

Planar curves of degree 3 already constitute a much richer and interesting group of geometrical objects than the ones of degree 2. And no one may claim that conic sections is not an interesting group of geometrical objects!

Two important concepts are those of *degenerate curves* and the related process of *degeneration of a family of curves*.

We say that a curve is degenerate if it decomposes into the union of two curves of lower degrees. For a cubic curve this would signify the curve being the union of a conic section and a line, or of three lines (some possibly coinciding). An example of such a curve is shown in Figure 13.4.

The simplest example of a degenerate cubic curve would be the y-axis with multiplicity 3. Its equation is $x^3 = 0$. We have not yet made the notion of curves with multiplicity precise, this comes in Section 13.3. But we may already at this point consider a family of semi cubical parabolas, *degenerating to* the triple y-axis. Namely, consider the curves depending on the parameter t, as $t \to 0$:

$$ty^2 = x^3.$$

We show some members of this family in Figure 13.5. The values of t in the plots are $t = 10, 4, 1, 0.1$.

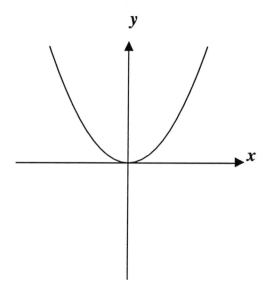

Fig. 13.1. The parabola, given by the equation $y = x^2$.

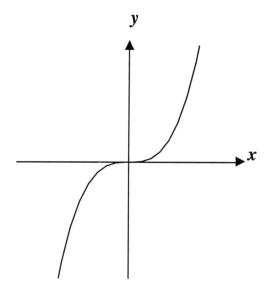

Fig. 13.2. The cubical parabola, given by the equation $y = x^3$.

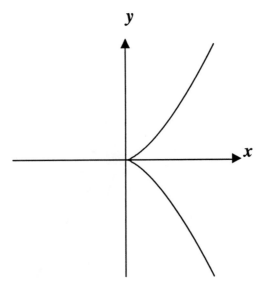

Fig. 13.3. The semi cubical parabola, given by the equation $y^2 = x^3$.

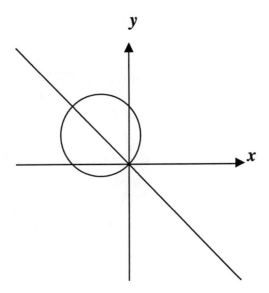

Fig. 13.4. A degenerate cubic, given by the equation $x^3 + x^2 y + x y^2 + y^3 + x^2 - y^2 = 0$.

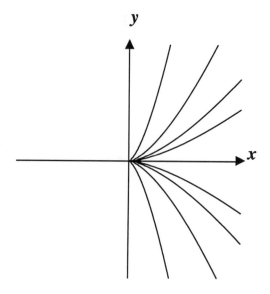

Fig. 13.5. Degeneration of $ty^2 = x^3$ to the triple y-axis.

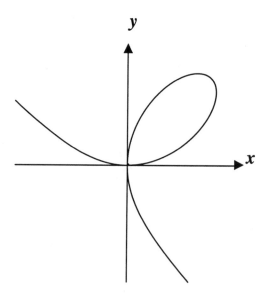

Fig. 13.6. The Folium of Descartes, given by the equation $x^3 + y^3 = 3xy$.

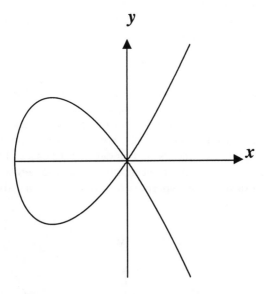

Fig. 13.7. The usual nodal cubic given by $y^2 - x^3 - x^2 = 0$.

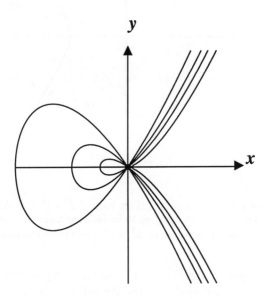

Fig. 13.8. The family degenerating the usual nodal cubic given by $y^2 - x^3 - x^2 = 0$ to a semi cubical parabola.

We note also that when $t \to \infty$, then the limit is the x-axis with multiplicity 2, since this degeneration is equivalent to letting u tend to 0 for the family given by the equation

$$y^2 = ux^3$$

We have used the term degeneration loosely, without a formal definition. The idea we intend to convey by this, is to have one curve, say the semi cubical parabola $y^2 = x^3$, be a member of a family of curves depending on a parameter, all but a finite number of which are of the same type. Then the exceptional members are understood as degenerate cases. This is, of course, the way we may view two intersecting lines as a degenerate hyperbola, or a double line as a degenerate hyperbola or a degenerate parabola, and so on.

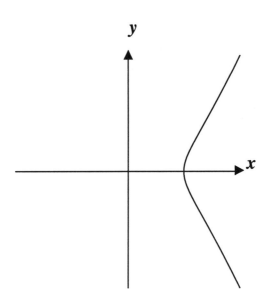

Fig. 13.9. An unusual "nodal cubic" given by $y^2 - x^3 + x^2 = 0$. Actually, the origin is on the curve, but that point appears to be isolated from the main part of it. But there are complex points, invisible in $\mathbb{R}^2$, which establish the connection.

Two more types of non-degenerate curves of degree three exist, up to a *projective change of coordinate system*. We will explain this *projective equivalence* for curves in $\mathbb{P}^2(\mathbb{R})$ (and in $\mathbb{R}^2$) later, in Section 14.5. The simple *affine equivalence* for two curves means that one may be obtained from the other by a suitable *affine transformation*. This kind of equivalence is more complicated than the projective equivalence, there are more equivalence classes of

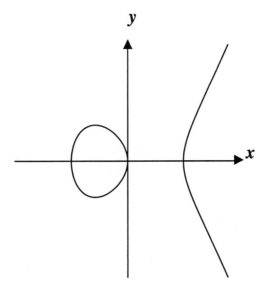

Fig. 13.10. The elliptic cubic given by $y^2 - x^3 + x = 0$.

affine cubic curves under this affine equivalence. But from our point of view, the projective equivalence is more interesting than the affine one.

The first of the remaining classes of cubic curves is represented by the *Folium of Descartes*. It was given as an example by Descartes in an argument over how to properly define tangent lines to curves. The Folium is shown in Figure 13.6.

We also give another curve, belonging to the same class as the Folium under projective equivalence, but to a separate class under the affine equivalence. It looks somewhat similar to the semi cubical parabola. In fact, the latter may be obtained by deforming the former.

This is the simplest and most used example of a *nodal cubic curve* in $\mathbb{R}^2$. It is shown in Figure 13.7. The deformation referred to is obtained from the family

$$y^2 - x^3 - tx^2 = 0,$$

and in Figure 13.8 we see some of the corresponding plots, for $t = 0, 0.5, 2$.

We also plot the curve given by

$$y^2 - x^3 + x^2 = 0$$

in Figure 13.9.

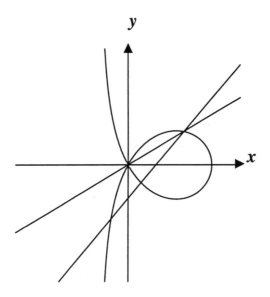

Fig. 13.11. The Trisectrix of Maclaurin given by $x^3 + xy^2 + y^2 - 3x^2 = 0$. Two lines are drawn through a point on it, located in the first quadrant. One line passes through the origin, the other through the point $(1, 0)$. Then if v is the angle contained by the x-axis and the former, $u = 3v$ will be the angle contained by the x-axis and the latter line.

The third class of cubic curves in $\mathbb{R}^2$ under the projective equivalence, is the class of *elliptic cubic curves*. But by all means: They are not ellipses! Ellipses are conics, elliptical curves are called so for an involved reason, ultimately coming from the computation of the length of a curve segment of an ellipse. Figure 13.10 shows an elliptic cubic curve in $\mathbb{R}^2$.

Elliptic curves constitute an important class of curves. But an extensive investigation of this theme falls outside the scope of the present book. We only mention two points. First of all, elliptic curves are tied closely to the *torus surfaces*. But this link comes from the inclusion of *complex points* on a curve. This will be explained in Section 13.3. The second point is that we need to investigate the behavior of the curves at infinity, also to be explained in Section 13.3. In fact, adding the points at infinity, the points of an elliptic cubic curve as the one we give here, form a very interesting *Abelian group!* But this theory also falls outside the scope of the present book.

A variation of the Folium of Descartes is provided by the *Trisectrix of Maclaurin*. As the name indicates, this is a curve which may be used to trisect an angle. We show the situation in Figure 13.11.

Another curve looks like a *clover leaf*. It has equation

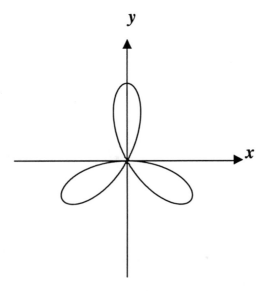

Fig. 13.12. The Clover Leaf Curve has equation $(x^2 + y^2)^2 + 3x^2y - y^3 = 0$.

$$(x^2 + y^2)^2 + 3x^2y - y^3 = 0$$

and is shown in Figure 13.12.

13.2 Curves of Degree Higher Than Three

Our first curve of degree higher than three was the Clover Leaf Curve. Another such interesting curve is the famous *Airplane Wing Curve*. Amazingly it looks very similar to a section through the wing of an airplane, and by modifying the coefficients it can be made to model various types of wings. The curve has the equation

$$x^4 + x^2y^2 - 2x^2y - xy^2 + y^2 = 0.$$

It is useful in the study of the aerodynamic behavior of different airplane wing designs, and is shown in Figure 13.13.

We finally show, in Figure 13.14, a curve of degree four with two points where it crosses itself, that is to say with two *singular points*. It is given by the equation

$$2x^4 - 3x^2y + y^4 - 2y^3 + y^2 = 0.$$

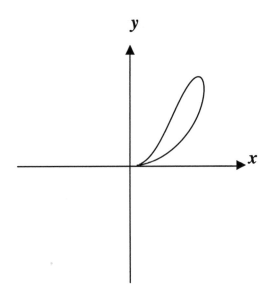

Fig. 13.13. The Airplane Wing Curve, with equation $x^4 + x^2y^2 - 2x^2y - xy^2 + y^2 = 0$.

13.3 Affine Algebraic Curves

We now have a sufficient base of *examples* to appreciate some more general theory. We already mentioned the need to incorporate complex points in connection with elliptic cubic curves above. We now take a closer look.

Let the curve C be given by the equation $f(x, y) = 0$. We then study the set of pairs (u, v) of *complex numbers* such that $f(u, v) = 0$. So we consider the zero locus of $f(x, y) = 0$ in $\mathbb{C}^2$. We may denote this set by $C(\mathbb{C})$, and the curve considered as a subset of $\mathbb{R}^2$ we may denote by $C(\mathbb{R})$. If we identify $\mathbb{C}^2$ with $\mathbb{R}^4$, this locus is identified with a *surface* defined by two equations. Namely, writing

$$u = x_1 + ix_2, v = x_3 + ix_4$$

and

$$f(u, v) = f_1(x_1, x_2, x_3, x_4) + if_2(x_1, x_2, x_3, x_4)$$

then f_1 and f_2 are polynomials with real coefficients in four variables, and the set of all complex points on the curve is given as

$$C(\mathbb{C}) = \left\{ (a_1, a_2, a_3, a_4) \in \mathbb{R}^4 \,\middle|\, \begin{array}{l} f_1(a_1, a_2, a_3, a_4) = 0 \\ f_2(a_1, a_2, a_3, a_4) = 0 \end{array} \right\}$$

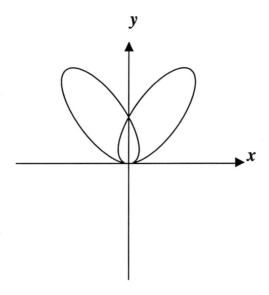

Fig. 13.14. The curve with equation $2x^4 - 3x^2y + y^4 - 2y^3 + y^2 = 0$.

This is a surface in four-space, in $\mathbb{R}^4$, defined by two polynomials. In many situations we really need to include all complex points of a curve, although we usually still confine ourselves to sketch the real points only. And even if the complex points form a surface in $\mathbb{R}^4$, it is important to keep in mind that we really are studying a *curve* in the plane, and not a surface in four space. Indeed, of we switch to regard our object under study as a surface in $\mathbb{R}^4$, then *it will also have complex points*, thus yielding a *fourfold in* $\mathbb{R}^8$, and so on. Thus we have to remember that we are studying complex points on a curve in the plane, rather than the real points of a surface in four space.

The further important extension is to include the *points at infinity* of a curve. This is a somewhat more technical matter, which we come to in Chapter 14, where we study *projective curves*. But first we give more details on the affine case.

We are given a curve in the plane $\mathbb{R}^2$ as the set of zeroes of the equation

$$f(x, y) = 0$$

where $f(x, y)$ is a polynomial in the variables x and y:

$$f(x, y) = a_{0,0} + xa_{1,0} + ya_{0,1} + x^2 a_{2,0} + xya_{1,1} + y^2 a_{0,2} + \dots$$

$$\dots + x^d a_{d,0} + x^{d-1} y a_{d-1,1} + \dots + y^d a_{0,d}$$

Some, but not all, of the coefficients may be zero. The largest integer d such that not all $a_{d-i,i}$ are zero is the *degree* of the polynomial, and this is by definition the *degree of the curve*. But here we have a problem, best elucidated by an example.

The equation

$$y = 0$$

defines the x-axis. But so does the equation

$$y^2 = 0$$

at least as a *point-set*. But algebraically we need to distinguish between these two cases. The former equation defines the x-axis as a line, whereas the latter defines a *double line* along the x-axis: Informally speaking, it defines twice the x-axis.

The situation becomes even more difficult when we consider complicated polynomials. Thus for example we may consider the curve defined by the equation

$$(y^2 - x^3 - x^2)(y^2 - x^2) = 0.$$

When we are given the equation on this partly factored form, it is not difficult to see what we get: It is the nodal cubic curve displayed in Figure 13.7 together with the two lines defined by $y = \pm x$. But suppose that we are given the following equation, on expanded form

$$3\,y^2x^4 - 3\,y^4x^2 + y^6 - x^7 + 2\,x^5y^2 - x^3y^4 - x^6 = 0$$

then it is not so easy to understand the situation. Using some PC-program to plot this curve, we should get the same picture as above. But this result is quite deceptive. Indeed, if we *factor* the left hand side of the equation, say again by some PC-program, we find that the equation becomes

$$(x^3 + x^2 - y^2)(x + y)^2(x - y)^2 = 0$$

which certainly defines the same point set, but reveals that this time the two lines occurring should *be counted with multiplicity 2*.

Recall that an *irreducible polynomial* in x and y is a polynomial $p(x, y)$ which may not be factored as a product of two polynomials, both non-constants. Thus for instance $p(x, y) = x^3 + x^2 - y^2$ is irreducible, as is $r(x, y) = x + y$ and $s(x, y) = x - y$. A special case of an important theorem is the following:

Theorem 20 (Unique factorization of polynomials). *Any polynomial in x and y with real (respectively complex,) coefficients, may be factored as a product of powers of irreducible polynomials with real (respectively complex,) coefficients. These irreducible polynomials are unique except for possibly being proportional by constant factors.*

We make the following definition:

Definition 13 (The factorization in irreducible polynomials). *The irreducible factorization of $f(x, y)$ is defined as an expression*

$$f(x, y) = p_1(x, y)^{n_1} \cdots p_r(x, y)^{n_r}$$

where n_i are positive integers and all $p_i(x, y)$ are irreducible and no two are proportional by constant factor.

This factorization is unique up to constant factors, by the theorem.

Remark 1. Theorem 20 also holds for a polynomial in any number of variables, 1 up to any N. Definition 13 is also unchanged in the general case.

A polynomial may be irreducible as a polynomial with real coefficients, but *reducible* when considered as a polynomial with *complex* coefficients. This is the case for the polynomial

$$g(x, y) = x^2 + y^2,$$

which may not be factored as a polynomial with real coefficients, while

$$x^2 + y^2 = (x + iy)(x - iy)$$

The curve given by this polynomial has another interesting feature: As a curve in $\mathbb{R}^2$ it consists only of the origin, while it consists of two (complex) *lines* in $\mathbb{C}^2$, with equations $y = \pm ix$. They have only one real point on them, namely their point of intersection which is the origin. We would consider this as a degenerate case, say as a member of a family of circles, where the radius has shrunk to zero.

We shall mainly be concerned with *real curves* in this book, but as we see the picture may be quite different when we pass to the complex case. We are now ready to make the following definition:

Definition 14 (Real Affine Curve). *A real affine plane curve C is the set of points $(a, b) \in \mathbb{R}^2$ which are zeroes of a polynomial $f(x, y)$ with real coefficients. The irreducible polynomials $p_i(x, y)$ occuring in the irreducible factorization of $f(x, y)$ referred to in Definition 13 define subsets C_i of C called the irreducible components of C. The exponent n_i of $p_i(x, y)$ in the factorization of $f(x, y)$ is called the multiplicity of the irreducible component.*
In other words, C_i occurs with multiplicity n_i in C.

Remark This definition suffices for our purpose in this book, but it should not be concealed that it does represent a simplification. Indeed, according to the definition the "real affine curve" defined by $x^2 + y^2 = 0$ is the same as the one defined by $x^2 + 2y^2 = 0$. For a variety of reasons this is undesirable.

One solution is to simply *define* a curve in $\mathbb{R}^2$ as being an equivalence class of polynomials, two polynomials being regarded as equivalent if one is a non-zero constant multiple of the other. This is mathematically sound, but only applies to a special geometric situation, where one geometric object, here the curve, is contained in another geometric object of one dimension higher, here the plane, and is defined by one "equation". In any case this point of view belongs to a somewhat more advanced treatment of algebraic geometry than the one offered in the present book.

Another solution is to include all the complex points on a real curve. This would be a logical next step, having digested the present treatment of curves.

Finally we may consider more general "points", which takes us way beyond the scope of this book and into Grothendiecks theory of Schemes, [11].

By a change of variables, which corresponds to a change of coordinate system,

$$\overline{x} = \alpha_{1,1}x + \alpha_{1,2}y$$

$$\overline{y} = \alpha_{2,1}x + \alpha_{2,2}y$$

the curve given by $f(x,y) = 0$ is expressed by the equation $\overline{f}(\overline{x},\overline{y}) = 0$, where $\overline{f}(\overline{x},\overline{y})$ is obtained by substituting the expressions obtained by solving for x and y,

$$x = \beta_{1,1}\overline{x} + \beta_{1,2}\overline{y}$$

$$y = \beta_{2,1}\overline{x} + \beta_{2,2}\overline{y}$$

into $f(x,y)$.

There are curves in the affine plane $\mathbb{R}^2$ which are not affine algebraic, but nevertheless form an important subject in geometry. We have encountered some of them: The Archimedian spiral and the quadratrix of Hippias. They both were invented to solve some of the Classical Problems. They are not defined by a polynomial equation. Some other simple examples are the curves defined by $y = \sin(x)$ or by $y = e^x$. This class of curves is called *the Transcendental Curves*. In this book we will confine the general theory to treating the algebraic curves, that is to say the ones defined by a polynomial equation.

13.4 Singularities and Multiplicities

We now return to some general concepts introduced in Section 12.7, where we needed it to understand the degeneracy of conic sections. Consider an algebraic affine curve K with equation

$$f(x, y) = 0.$$

Furthermore, let (a, b) be a point on the curve, i.e., $f(a, b) = 0$. We note that the following definition relies heavily on the *equation* of the curve, not just the curve as a subset of $\mathbb{R}^2$:

Definition 15. (a, b) *is said to be a smooth, or a non-singular, point on K if*

$$\left(\frac{\partial f}{\partial x}(a, b), \ \frac{\partial f}{\partial y}(a, b) \right) \neq (0, 0).$$

Otherwise (a, b) is called a singular point on K. A curve all of whose points are non-singular is referred to as a non-singular curve.

The vector $\left(\frac{\partial f}{\partial x}, \frac{\partial f}{\partial y} \right)$ is referred to as the Jacobian vector (for short, the Jacobian) of the polynomial $f(x, y)$. Thus by definition a singular point is a point on the curve at which the Jacobian evaluates to the zero vector.

In Section 12.7 we saw that a non-degenerate conic section is a non-singular curve. We look at the situation in more detail by the examples below.

Examples 1) We first look at some simple *conic sections*, and start out with a circle of radius $R > 0$, which has the equation

$$x^2 + y^2 = R^2$$

Here $f(x, y) = x^2 + y^2 - R^2$, and

$$\left(\frac{\partial f}{\partial x}, \frac{\partial f}{\partial y} \right) = (2x, 2y)$$

Evidently no point outside the origin can be a singular point of the circle, and as $R > 0$, every point on the circle is therefore smooth. We note that the same proof shows that an ellipse on standard form,

$$\left(\frac{x}{a} \right)^2 + \left(\frac{y}{b} \right)^2 = 1$$

is smooth everywhere as well.

A (non-degenerate) hyperbola on standard form, which is given as

$$\left(\frac{x}{a} \right)^2 - \left(\frac{y}{b} \right)^2 = 1$$

similarly has Jacobian

$$\left(\frac{2}{a^2}x, -\frac{2}{b^2}y \right)$$

which also does not vanish outside the origin, showing that a hyperbola is smooth.

A degenerate hyperbola is one which has collapsed to the asymptotes, hence a curve with equation

$$\left(\frac{x}{a}\right)^2 - \left(\frac{y}{b}\right)^2 = 0.$$

This curve has the same Jacobian as in the non-degenerate case, but now the origin actually lies on the curve, which therefore has the origin as its only singular point. Of course this degenerate hyperbola consists of two irreducible components which are lines intersecting at the origin, and that point is singular.

Our final conic section is the parabola with equation

$$ay - x^2 = 0$$

where $a \neq 0$. The Jacobian is $(-2x, a)$, so the only possibility of getting the zero vector at a point would be to have $x = 0$ and $a = 0$. For $a \neq 0$ we therefore have no singular points. If $a = 0$, then the equation yields the y-axis with multiplicity 2, and we see that then *all points on the curve are singular*.

2) We next turn to the *nodal cubic curve* with equation

$$y^2 - x^3 - x^2 = 0$$

which is plotted in Figure 13.7. The Jacobian is

$$(-3x^2 - 2x, 2y)$$

and thus (x, y) is a singular point if and only if the two additional equations below are satisfied:

$$-3x^2 - 2x = 0$$

$$2y = 0.$$

Thus $(x, y) = (0, 0)$ or $(x, y) = (-\frac{2}{3}, 0)$, and only the former lies on the curve, so the only singular point is $(0, 0)$.

3) If $f(x, y)$ is any polynomial, then all points on the curve given by $f(x, y)^n = 0$ for n an integer greater than 1, will have all its points singular. This follows at once, since the Jacobian is

$$\left(nf(x, y)^{n-1}\frac{\partial f}{\partial x}, nf(x, y)^{n-1}\frac{\partial f}{\partial y}\right)$$

At this point it is highly recommended that the reader examines the curves plotted in Section 13.1, and determines their singular points.

13.5 Tangency

Now let (a, b) be a smooth point on the curve K. Then we find the equation for the tangent line at that point as follows. We first consider *the parametric form* for any line through (a, b) with direction given by the vector (u, v):

$$L = \left\{ (x, y) \, \middle| \, \begin{array}{l} x = a + ut \\ y = b + vt \end{array} \text{ where } t \in \mathbb{R} \right\}$$

This line will have the point (a, b) in common with K. We wish to determine other points of intersection. To do so we substitute the expressions for x and y in the parametric form for L into the equation for K, and get

$$f(a + ut, b + vt) = 0.$$

Expanding the left hand side in a Taylor series we obtain

$$f(a, b) + t(u\tfrac{\partial f}{\partial x}(a, b) + v\tfrac{\partial f}{\partial y}(a, b)) +$$
$$t^2(u^2\tfrac{\partial^2 f}{\partial x^2}(a, b) + uv\tfrac{\partial^2 f}{\partial x \partial y}(a, b) + v^2\tfrac{\partial^2 f}{\partial y^2}(a, b)) + \ldots = 0$$

which since $f(a, b) = 0$ gives

$$t(u\tfrac{\partial f}{\partial x}(a, b) + v\tfrac{\partial f}{\partial y}(a, b)) +$$
$$t^2(u^2\tfrac{\partial^2 f}{\partial x^2}(a, b) + uv\tfrac{\partial^2 f}{\partial x \partial y}(a, b) + v^2\tfrac{\partial^2 f}{\partial y^2}(a, b)) + \ldots = 0$$

The points of intersection between the curve and the line are found by solving this equation for t. Of course we have $t = 0$ as one solution, and we see that this solution will occur with multiplicity 1 if and only if

$$u\frac{\partial f}{\partial x}(a, b) + v\frac{\partial f}{\partial y}(a, b) \neq 0.$$

Such values of u, v exist if and only if (a, b) is a smooth point on the curve. In that case there is exactly one line which *does not intersect the curve with multiplicity 1*, namely the line corresponding to u and v such that

$$u\frac{\partial f}{\partial x}(a, b) + v\frac{\partial f}{\partial y}(a, b) = 0.$$

By substituting

$$ut = x - a$$

$$vt = y - b$$

in this equation, we get *the equation for the tangent* to the curve at the point (a, b)

$$(x - a)\frac{\partial f}{\partial x}(a, b) + (y - b)\frac{\partial f}{\partial x}(a, b) = 0.$$

We next turns to the question of what happens at *a singular point*. So let $P = (a, b)$ be a singular point on the curve K. Since the situation is more complicated than in the case when P is smooth, we introduce new variables by

$$\overline{x} = x - a, \overline{y} = y - b$$

In other words, we shift the variables so that the new origin falls in P, $P = (0, 0)$. We then find a new polynomial g such that

$$f(x, y) = g(\overline{x}, \overline{y})$$

by substituting $x = \overline{x} + a$ and $y = \overline{y} + b$ into $f(x, y)$. The curve is also given by the equation

$$g(\overline{x}, \overline{y}) = 0.$$

Since the origin is a point on the curve given by $g(\overline{x}, \overline{y}) = 0$, it is clear that the polynomial $g(\overline{x}, \overline{y})$ has no constant term. We now collect the terms of $g(\overline{x}, \overline{y})$ which are of lowest total degree, and denote the sum of those terms by $h(\overline{x}, \overline{y})$.

Thus for example, if

$$g(\overline{x}, \overline{y}) = 2\overline{x}\overline{y}^2 - 5\overline{x}^2\overline{y} + 10\overline{x}^9\overline{y}^2 + 15\overline{x}^2\overline{y}^{12},$$

then

$$h(\overline{x}, \overline{y}) = 2\overline{x}\overline{y}^2 - 5\overline{x}^2\overline{y}.$$

This piece of a polynomial consisting of all terms of lowest total degree is called the *initial part* of the polynomial. Of course the sum of the degrees of the two variables $\overline{x}$ and $\overline{y}$ is the same for all terms occuring in $h(\overline{x}, \overline{y})$. If the point $P = (a, b)$ is smooth, then the Taylor expansion around the point (a, b) immediately shows that the polynomial $h(\overline{x}, \overline{y})$ is nothing but

$$\frac{\partial g}{\partial x}(0, 0)\overline{x} + \frac{\partial g}{\partial y}(0, 0)\overline{y} =$$

$$\frac{\partial f}{\partial x}(a, b)(x - a) + \frac{\partial f}{\partial y}(a, b)(y - b)$$

Thus the concept introduced below generalizes the tangent at a smooth point, to a concept which applies to *singular points as well*.

With notations as above the polynomial $h(\bar{x}, \bar{y})$ defines a curve which is a finite union of lines through the point $(0,0)$. In terms of x and y, the equation

$$h(x - a, y - b) = 0$$

defines a finite union of lines through $P = (a, b)$, some of them occuring with multiplicity > 1. Indeed, we have

$$h(\bar{x}, \bar{y}) = a_0\bar{x}^m + a_1\bar{x}^{m-1}\bar{y} + \cdots + a_i\bar{x}^{m-i}\bar{y}^i + \cdots + a_m\bar{y}^m$$

where not all a_i vanish. If (α_0, β_0) satisfies $h(\alpha_0, \beta_0) = 0$, then we also have $h(s\alpha_0, s\beta_0) = 0$ for all real numbers s, as one immediately verifies since all the monomials of h are of the same total degree m.

These lines are called the *lines of tangency* at the point $P = (a, b)$. If P happens to be smooth, then there is only one line, occuring with multiplicity 1.

Definition 16. *The curve given by* $h(x - a, y - b) = 0$ *is referred to as the* (affine) tangent cone *of K at P.*

Any line through $P = (a, b)$ may, as we have seen, be written on parametric form as

$$x - a = ut, y - b = vt$$

and its intersections with the curve is determined by the equation

$$f(a + ut, b + vt) = g(ut, vt) = 0.$$

The multiplicity of the root $t = 0$ in this equation is referred to as *the multiplicity of intersection* between the curve and the line at the point $P = (a, b)$.

A moments reflection will convince the reader of an important fact: *All lines through* $P = (a, b)$ *which do not coincide with one of the lines of tangency, intersect the curve with multiplicity equal to the number m.* This number m is of course only dependent upon the polynomial $f(x, y)$ and the point $P = (a, b)$.

In fact, we may assume that $P = (0, 0)$. An arbitrary line through $(0, 0)$ has the parametric form

$$L = \left\{ (x, y) \;\middle|\; \begin{matrix} x = ut \\ y = vt \end{matrix} \text{ where } t \in \mathbb{R} \right\}$$

To find all points of intersection between this line and the curve K, we substitute the expressions for x and y into $f(x, y)$ and get

$$f(a + ut, b + vt) = 0.$$

This gives

$$h(ut, vt) + R(ut, vt) = 0.$$

where $R(x, y)$ denotes $f(x, y) - h(x, y)$. Thus the points of intersection are given by the roots of the equation

$$t^m(h(u, v) + t\varphi(t)) = 0.$$

One of the roots is $t = 0$, and this solution will occur with multiplicity $\geq m$, where equality holds if and only if

$$h(u, v) \neq 0.$$

thus if and only if L is not one of the lines of tangency.
    We conclude with the

Definition 17 (Multiplicity of a point on a curve). *The number m referred to above is called the multiplicity of the point P at K.*

    We thus have the observation

Proposition 14. *A point on an affine algebraic curve is smooth if and only if it has multiplicity 1.*

14 Higher Geometry in the Projective Plane

14.1 Projective Curves

We define curves in the *projective plane* $\mathbb{P}^2(\mathbb{R})$ analogously to curves in the affine plane $\mathbb{R}^2$. The difference is that we can not use ordinary polynomials in two variables, but have to work with *homogeneous polynomials in three variables* instead. We have seen this in Section 12.8, for conics.

Thus the polynomial

$$X_0 + 5X_0 X_1^2$$

is not homogeneous, since one monomial which occurs is X_0, and another is $5X_0 X_2^2$. They are of degrees 1 and 3, respectively. On the other hand, the polynomial

$$X_0^3 + 5X_0 X_1^2$$

is *homogeneous*, the two monomials which occur are both of degree 3.

Now assume that we have a homogeneous polynomial with real coefficients

$$F(X_0, X_1, X_2) = \Sigma_{I \in S} c_I X_0^{i_0} X_1^{i_1} X_2^{i_2}$$

where $I = (i_0, i_1, i_2)$, $d = i_0 + i_1 + i_2$, and the symbol $\Sigma_{I \in S}$ means that we have a sum where I runs through a finite subset S of triples of non-negative integers, when no confusion is possible we usually write just Σ_I. c_I is a real number, called the *coefficient* of the *monomial* $X_0^{i_0} X_1^{i_1} X_2^{i_2}$. Let $(a_0, a_1, a_2) \in \mathbb{R}^3$. Then we have

$$F(ta_0, ta_1, ta_2) = \Sigma_I c_I (ta_0)^{i_0} (ta_1)^{i_1} (ta_2)^{i_2} =$$

$$t^d (\Sigma_I c_I a_0^{i_0} a_1^{i_1} a_2^{i_2}) = t^d F(a_0, a_1, a_2)$$

since $d = i_0 + i_1 + i_2$. Thus we find that whenever $t \neq 0$, then

$$F(ta_0, ta_1, ta_2) = 0 \text{ if and only if } F(a_0, a_1, a_2) = 0.$$

It follows that the zero locus for a homogeneous polynomial in X_0, X_1 and X_2 is well defined in $\mathbb{P}^2(\mathbb{R})$. Moreover, we also note the

Theorem 21. *In the irreducible factorization of a homogeneous polynomial, given by Remark 1, all the irreducible polynomials occuring are also homogeneous.*

Proof. The proof is by induction on $d = \deg(F)$. For $d = 1$ the claim is immediate. Suppose that the claim is true for all homogeneous polynomials of degree $< d$, and let F be homogeneous of degree d. If F is irreducible, there is nothing to prove. Otherwise we may write

$$F = F_1 F_2$$

where F_1 and F_2 are polynomials of degrees $< d$. We may write

$$F_i = H_i + G_i, \text{ for } i = 1, 2$$

where H_i is the homogeneous piece of highest degree of F_i. Thus

$$F = H_1 H_2 + G_1 H_2 + G_2 H_1 + G_1 G_2 = H_1 H_2 + G$$

but since F and $H_1 H_2$ are homogeneous of the same degree, and G, if it were non zero would be of degree $< d$, it follows that

$$F = H_1 H_2$$

and the claim follows by induction.   □

Definition 18 (Projective Algebraic Curve). *A plane projective curve $C \subset \mathbb{P}^2(\mathbb{R})$ is the zero locus of a homogeneous polynomial in X_0, X_1 and X_2, with coefficients from $\mathbb{R}$. The irreducible components of C, as well as their multiplicities, are defined analogously to the affine case by means of Theorem 21.*

14.2 Projective Closure and Affine Restriction

Given an affine curve $K \subset \mathbb{R}^2$, with equation $f(x, y) = 0$. In the same way as we did for curves of degree 2, we may define the *projective closure $C \subset \mathbb{P}^2(\mathbb{R})$* of K. It is defined by the equation $F(X_0, X_1, X_2) = 0$ where $F(X_0, X_1, X_2)$ is constructed by putting $x = \frac{X_1}{X_0}$ and $y = \frac{X_2}{X_0}$ and substituting this in $f(x, y)$, and writing the result as

$$f\left(\frac{X_1}{X_0}, \frac{X_2}{X_0}\right) = \frac{F(X_0, X_1, X_2)}{X_0{}^m}.$$

where X_0 does not divide the numerator. Here $F(X_0, X_1, X_2)$ is a homogeneous polynomial with real coefficients, uniquely determined by $f(x, y)$ as follows: If

$$f(x, y) = \sum_{I=(i_1, i_2) \in \Phi} a_I x^{i_1} y^{i_2}$$

where Φ denotes a finite set of tuples of non-negative integers (i_1, i_2), then the degree of K is $d = \max\{i_1 + i_2 | (i_1, i_2) \in \Phi\}$, and the projective closure is given by the equation

$$F(X_0, X_1, X_2) = \sum_{I=(i_1, i_2) \in \Phi} a_I X_0^{d-i_1-i_2} X_1^{i_1} X_2^{i_2} = 0.$$

d is the degree of the original affine curve K as well as of its projective closure C.

Definition 19. *The homogeneous polynomial $F(X_0, X_1, X_2)$ as defined above is denoted by $f^h(X_0, X_1, X_2)$, and referred to as the homogenization of the (non-homogeneous) polynomial $f(x, y)$.*

The key to understanding the relation between an affine curve and its projective closure lies in the simple and beautiful relation

$$f(a, b) = f^h(1, a, b)$$

which holds for all a and b.

Thus if K is the affine curve defined by $f(x, y) = 0$, then the projective closure C of K is defined by the equation $f^h(X_0, X_1, X_2) = 0$. Conversely, if we are given a projective curve C by the equation $F(X_0, X_1, X_2) = 0$, then we may define its *affine restriction* to $D_+(X_0)$ as identified with $\mathbb{R}^2$ as the curve given by the equation $F(1, x, y) = 0$. But this affine restriction is not always defined: Namely, if $F(X_0, X_1, X_2) = X_0^d$, then C is the line $L_\infty = V_+(X_0)$, the line at infinity, with multiplicity d. Of course the affine restriction of this curve to $D_0(X_0)$ is given by the equation $1 = 0$, so we might say that the *affine restriction of this curve to $D_+(X_0)$ is empty*. On the other hand, if we chose to take the affine restriction to $D_+(X_1)$ instead, and put $x = \frac{X_0}{X_1}$ and $y = \frac{X_2}{X_1}$, then the affine restriction is the curve given by $x^d = 0$, in other words the y-axis counted with multiplicity d.

So the concepts of projective closure and affine restriction are not independent of the coordinate system. The change to another projective coordinate system in $\mathbb{P}^2(\mathbb{R})$ has been described in Section 11.3. The equations defining the new coordinate system may also be used to define a bijective mapping of $\mathbb{P}^2(\mathbb{R})$ onto itself, known as *a projective transformation*. This was also explained in Section 11.3, and will not be repeated here.

Even though the concepts of projective closure and affine restriction do depend on the coordinate system, they are very useful in the investigation of properties and concepts which *are* coordinate independent. Normally we perform the projective closure by letting $V_+(X_0)$ contain the added points at infinity, and identify the affine plane $\mathbb{R}^2$ with $D_+(X_0)$. When an alternative procedure is used, this will be explicitely stated. Also, if $V_+(aX_0+bX_1+cX_2)$ is a projective line in $\mathbb{P}^2(\mathbb{R})$, then we may identify $D_+(aX_0+bX_1+cX_2)$ with $\mathbb{R}^2$ and carry out affine restrictions to $\mathbb{R}^2$ by restricting to $D_+(aX_0+bX_1+cX_2)$. Again, if this non-standard procedure is used we shall explicitely state so. The most convenient method is to choose a new projective coordinate system by putting

$$\overline{X}_0 = aX_0 + bX_1 + cX_2$$

and choosing linear forms $a_1X_0+b_1X_1+c_1X_2$ and $a_2X_0+b_2X_1+c_2X_2$ such that the determinant of the coefficients of the three forms is non-zero, so letting

$$\overline{X}_1 = a_1X_0 + b_1X_1 + c_1X_2 \text{ and } \overline{X}_2 = a_2X_0 + b_2X_1 + c_2X_2$$

we get a new projective coordinate system. Then the affine restriction is carried out in the standard fashion with respect to it.

Using an affine restriction we are able to study local properties of a curve, like questions of tangency or singularity, with greater precision. Taking projective closure we obtain information on how the curve behaves very far away from the origin, *at infinity*, information crucial to a global understanding of the affine curve itself. An example of this which we shall return to later is the determination of all the *asymptotes* of a curve in $\mathbb{R}^2$. We conclude this section on projective closure and affine restriction with the

Proposition 15. *1. Let K be an affine curve in $\mathbb{R}^2$, and let C be its projective closure. Then the affine restriction of C is equal to K.*

2. Let C be a projective curve, and let K be its affine restriction. If C is just a multiple of $V_+(X_0)$ then K is empty.[1] Otherwise K is an affine curve, and its projective closure C' consist of all irreducible components of C, with the same multiplicity as before, except possibly for the component $V_+(X_0)$, which is removed when passing from C to C'.

Proof. To prove 1., let K be given by

$$f(x,y) = \sum_{I=(i_1,i_2)\in\Phi} a_I x^{i_1} y^{i_2}$$

Then the projective closure is given by

$$F(X_0, X_1, X_2) = \sum_{I=(i_1,i_2)\in\Phi} a_I X_0^{d-i_1-i_2} X_1^{i_1} X_2^{i_2} = 0.$$

Substituting $X_0 = 1$, $X_1 = x$ and $X_2 = y$ clearly gives us back $f(x,y)$, and 1. is proven.

---

[1] Or, as we shall say here, K does not exist.

As for 2., assume that C is given by the homogeneous polynomial

$$F(X_0, X_1, X_2) = X_0^r \left(\sum_{I=(i_1,i_2)\in\Phi} a_I X_0^{d-i_1-i_2} X_1^{i_1} X_2^{i_2} \right)$$

where the polynomial inside the parenthesis is not divisible by X_0. Denoting the latter by $G(X_0, X_1, X_2)$, we find that

$$F(1, x, y) = G(1, x, y)$$

and the affine restriction of C is defined by $G(1, x, y) = 0$. So the projective closure C' of the affine restriction is defined by $G(X_0, X_1, X_2)$. This completes the proof. $\square$

14.3 Smooth and Singular Points on Affine and Projective Curves

Let C be given by the equation

$$F(X_0, X_1, X_2) = 0.$$

Moreover, let $P = (a_0 : a_1 : a_2)$ be a point on C.

Definition 20. *We say that the point P is a smooth point on C if*

$$\left(\frac{\partial F}{\partial X_0}(a_0, a_1, a_2), \frac{\partial F}{\partial X_1}(a_0, a_1, a_2), \frac{\partial F}{\partial X_2}(a_0, a_1, a_2) \right) \neq (0, 0, 0)$$

Whenever this condition is not satisfied, the point is referred to as a singular point. Correspondingly, a smooth point is also referred to as a non-singular point.[2]

Earlier we defined the term *smooth point* for affine curves $K \subset \mathbb{R}^2$. Even if this previous definition is similar to the one we have given here, we need to show that they do not contradict one another. Namely, when we form the *projective closure* of the affine curve K, we obtain a *projective curve* $C \subset \mathbb{P}^2(\mathbb{R})$. A point $p \in K$ should then be smooth as a point of the affine curve K if and only if it is smooth as a point on the projective curve C.

This problem is disposed of by means of the following proposition:

Proposition 16. *With notations as in as in Section 14.2 we have*

$$\left(\frac{\partial f}{\partial x} \right)^h (x, y) = \frac{\partial f^h}{\partial X_1}(X_0, X_1, X_2)$$

and

$$\left(\frac{\partial f}{\partial y} \right)^h (x, y) = \frac{\partial f^h}{\partial X_2}(X_0, X_1, X_2)$$

---

[2] In more advanced texts on algebraic geometry, the terms "smooth" and "non-singular" have slightly different meanings.

Proof. We put

$$f(x,y) = \sum_{I=(i_1,i_2)\in\Phi} a_I x^{i_1} y^{i_2}$$

then $F = f^h$ is given by

$$F(X_0, X_1, X_2) = \sum_{I=(i_1,i_2)\in\Phi} a_I X_0^{d-i_1-i_2} X_1^{i_1} X_2^{i_2} = 0.$$

The verification of the claim is immediate from this. □

Corollary 4. *Let K be an affine curve, and let C be the projective closure of K, where $V_+(X_0)$ is the points at infinity. Then (a,b) is a smooth point on the affine curve K if and only if $(1:a:b)$ is a smooth point on the projective curve C.*

Proof. We apply the relation

$$g(a,b) = g^h(1,a,b)$$

to the partial derivatives. □

The second important observation concerning smooth or singular points is contained in the

Proposition 17. *The concept of smooth point on a projective curve is independent of the projective coordinate system.*

Proof. We may write the transition from one coordinate system to another as a matrix multiplication as follows:

$$\left\{\begin{array}{ccc} \alpha_{0,0} & \alpha_{0,1} & \alpha_{0,2} \\ \alpha_{1,0} & \alpha_{1,1} & \alpha_{1,2} \\ \alpha_{2,0} & \alpha_{2,1} & \alpha_{2,2} \end{array}\right\} \cdot \left\{\begin{array}{c} Y_0 \\ Y_1 \\ Y_2 \end{array}\right\} = \left\{\begin{array}{c} X_0 \\ X_1 \\ X_2 \end{array}\right\}$$

where the matrix has determinant $\neq 0$,

$$\left|\begin{array}{ccc} \alpha_{0,0} & \alpha_{0,1} & \alpha_{0,2} \\ \alpha_{1,0} & \alpha_{1,1} & \alpha_{1,2} \\ \alpha_{2,0} & \alpha_{2,1} & \alpha_{2,2} \end{array}\right| \neq 0.$$

Clearly

$$\frac{\partial X_i}{\partial Y_j} = \alpha_{i,j}.$$

Moreover, if the curve C is given in the original coordinate system as

$$F(X_0, X_1, X_2) = 0$$

then it will be given in the new coordinate system by

$$G(Y_0, Y_1, Y_2) = 0$$

where

$$G(Y_0, Y_1, Y_2) =$$
$$F(\alpha_{0,0}Y_0 + \alpha_{0,1}Y_1 + \alpha_{0,2}Y_2, \alpha_{1,0}Y_0 + \alpha_{1,1}Y_1 + \alpha_{1,2}Y_2,$$
$$\alpha_{2,0}Y_0 + \alpha_{2,1}Y_1 + \alpha_{2,2}Y_2)$$

Now let the point P be expressed as $(a_0 : a_1 : a_2)$ and $(b_0 : b_1 : b_2)$ in the two coordinate systems. Then we get by the chain rule

$$\frac{\partial G}{\partial Y_0}(b_0, b_1, b_2) = \alpha_{0,0}\frac{\partial F}{\partial X_0}(a_0, a_1, a_2) + \alpha_{0,1}\frac{\partial F}{\partial X_1}(a_0, a_1, a_2) +$$
$$\alpha_{0,2}\frac{\partial F}{\partial X_2}(a_0, a_1, a_2)$$

$$\frac{\partial G}{\partial Y_1}(b_0, b_1, b_2) = \alpha_{1,0}\frac{\partial F}{\partial X_0}(a_0, a_1, a_2) + \alpha_{1,1}\frac{\partial F}{\partial X_1}(a_0, a_1, a_2) +$$
$$\alpha_{1,2}\frac{\partial F}{\partial X_2}(a_0, a_1, a_2)$$

$$\frac{\partial G}{\partial Y_2}(b_0, b_1, b_2) = \alpha_{2,0}\frac{\partial F}{\partial X_0}(a_0, a_1, a_2) + \alpha_{2,1}\frac{\partial F}{\partial X_1}(a_0, a_1, a_2) +$$
$$\alpha_{2,2}\frac{\partial F}{\partial X_2}(a_0, a_1, a_2)$$

Since the determinant of the matrix of the α's is non-zero, it follows by *Cramer's Theorem* 4 in Section 5.6 that the vector of the evaluated partials to the left will not all vanish if and only if the vector of the evaluated partials to the right do not all vanish. Thus smoothness or singularity for a point on C is independent of the coordinate system in which the corresponding condition is expressed.   □

14.4 The Tangent of a Projective Curve

Before we deduce the equation for *the tangent line to a projective curve*, we need to make some comments on *lines and other curves on parametric form* in $\mathbb{P}^2(\mathbb{R})$. We first consider the case of lines. A line $L \subset \mathbb{P}^2(\mathbb{R})$ which passes through the points $(a_0 : a_1 : a_2)$ and $(b_0 : b_1 : b_2)$ may be expressed as follows, on parametric form:

$$L = \left\{ (X_0 : X_1 : X_2) \left| \begin{array}{l} X_0 = ua_0 + vb_0 \\ X_1 = ua_1 + vb_1 \\ X_2 = ua_2 + vb_2 \end{array} \right. \right\}$$

Here u and v are two real parameters which yield all the points on the line L, but as we see, it is only the ratio $(u : v)$ which distinguish between the points. In particular we have that $(u : v) = (1 : 0)$ yields the point $(a_0 : a_1 : a_2)$, while $(u : v) = (0 : 1)$ yields $(b_0 : b_1 : b_2)$.

More generally we may consider a curve in $\mathbb{P}^2(\mathbb{R})$ given on parametric form:

$$C = \left\{ (X_0 : X_1 : X_2) \left| \begin{array}{l} X_0 = \xi_0(u, v) \\ X_1 = \xi_1(u, v) \\ X_2 = \xi_2(u, v) \end{array} \right. \right\},$$

Here we assume that the polynomials $\xi_0(u, v), \xi_1(u, v)$ and $\xi_2(u, v)$ are *homogeneous of the same degree* in the variables u and v. The class of curves which may be so described do not contain all projective curves in $\mathbb{P}^2(\mathbb{R})$, there are curves which are *not* parameterizable by polynomials. But it does include all lines in $\mathbb{P}^2(\mathbb{R})$.

If we choose

$$\xi_0(u, v) = u^2, \xi_1(u, v) = uv \text{ and } \xi_2(u, v) = v^2$$

we get a curve of degree 2, in other words *a projective conic section*: It has the equation

$$X_1^2 - X_0 X_2 = 0.$$

If we choose $\xi_0(u, v), \xi_1(u, v)$ and $\xi_2(u, v)$ as general homogeneous polynomials of degree 2, then we get general projective curves of degree 2 in $\mathbb{P}^2(\mathbb{R})$: The class of projective curves of degree 2 consists of parameterizable ones. But even if we let $\xi_0(u, v), \xi_1(u, v)$ and $\xi_2(u, v)$ be general homogeneous polynomials of degree 3, we only obtain a special class of degree 3 projective curves in $\mathbb{P}^2(\mathbb{R})$, namely the *rational cubics* in $\mathbb{P}^3(\mathbb{R})$.

In general, the curves are parameterizable as described above by homogeneous polynomials of the same degree d are referred to as *the rational degree d-curves in $\mathbb{P}^2(\mathbb{R})$*. Here an explanation should be interjected: This number d, the common degree of the polynomials $\xi_0(u, v), \xi_1(u, v)$ and $\xi_2(u, v)$, turns out to be the degree of the equation

$$F(X_0, X_1, X_2) = 0.$$

which expresses the relation between the polynomials $\xi_0(u, v), \xi_1(u, v)$ and $\xi_2(u, v)$.

We now come to the concept of *tangent line of a general projective curve* in $\mathbb{P}^2(\mathbb{R})$. We consider a point $P = (a_0 : a_1 : a_2)$ on the curve C defined by the homogeneous polynomial $F(X_0, X_1, X_2)$,

$$F(X_0, X_1, X_2) = 0.$$

As we did in the affine case, we consider the collection of all lines passing through P, as we saw above these lines are all given on parametric form as

$$L = \left\{ (X_0 : X_1 : X_2) \left|
\begin{array}{l}
X_0 = ua_0 + vb_0 \\
X_1 = ua_1 + vb_1 \\
X_2 = ua_2 + vb_2
\end{array}
\right. \right\}$$

where $P = (a_0 : a_1 : a_2)$ is the fixed point on C, and $Q = (b_0 : b_1 : b_2)$ is another point $\neq P$ in $\mathbb{P}^2(\mathbb{R})$ and u and v are the parameters describing the line L passing through P and Q, the points on L corresponding to the ratio $u : v$. The point P corresponds to $u : v = 1 : 0$, while Q corresponds to $u : v = 0 : 1$. We wish to examine the points of intersection of the line L with C, as well as the multiplicities with which they occur. We then have to find all u and v which satisfy the equation

$$F(ua_0 + vb_0, ua_1 + vb_1, ua_2 + vb_2) = 0.$$

But our objective now is *not* to find all the other points of intersection between L and C. Instead, we are interested in examining *how the line intersects the curve in the point P*, in other words we wish to study the solution $(u, v) = (1, 0)$ of the equation, and since only the ratios count, this amounts to studying the solution $t = 0$ of the equation

$$\varphi(t) = F(a_0 + tb_0, a_1 + tb_1, a_2 + tb_2) = 0.$$

Since $P \in C$, $t = 0$ certainly is a solution. As in the affine case the multiplicity of the solution $t = 0$ is referred to as the *multiplicity with which the line L intersects C at P*.

Expanding $\varphi(t)$ in a Taylor series around $t = 0$ we actually get a polynomial of degree d, the degree of the curve C. We get

$$\varphi(t) = \varphi(0) + \varphi'(0)t + \frac{1}{2}\varphi''(0)t^2 + \cdots + \frac{1}{i!}\varphi^{(i)}(0)t^i + \cdots + \frac{1}{d!}\varphi^{(d)}(0)t^d$$

Here $\varphi(0) = 0$, and using the general Chain Rule we obtain

$$\varphi'(t) = b_0 \frac{\partial}{\partial X_0} F(a_0 + tb_0, a_1 + tb_1, a_2 + tb_2) +$$

$$b_1 \frac{\partial}{\partial X_1} F(a_0 + tb_0, a_1 + tb_1, a_2 + tb_2) +$$

$$b_2 \frac{\partial}{\partial X_2} F(a_0 + tb_0, a_1 + tb_1, a_2 + tb_2)$$

$$= ((b_0 \frac{\partial}{\partial X_1} + b_1 \frac{\partial}{\partial X_1} + b_2 \frac{\partial}{\partial X_2})F)(a_0 + tb_0, a_1 + tb_1, a_2 + tb_2)$$

and hence

$$\varphi'(0) = ((b_0 \frac{\partial}{\partial X_1} + b_1 \frac{\partial}{\partial X_1} + b_2 \frac{\partial}{\partial X_2})F)(a_0, a_1, a_2)$$

Taking the derivative of $\varphi'(t)$ and using the Chain Rule again, we similarly get the expression

$$\varphi''(0) = ((b_0 \frac{\partial}{\partial X_0} + b_1 \frac{\partial}{\partial X_1} + b_2 \frac{\partial}{\partial X_2})^2 F)(a_0, a_1, a_2)$$

The expression

$$(b_0 \frac{\partial}{\partial X_0} + b_1 \frac{\partial}{\partial X_1} + b_2 \frac{\partial}{\partial X_2})^2 F$$

is short term for the more elaborate

$$b_0^2 \frac{\partial^2 F}{\partial X_0^2} + b_1^2 \frac{\partial^2 F}{\partial X_1^2} + b_2^2 \frac{\partial^2 F}{\partial X_2^2}$$

$$+2b_0 b_1 \frac{\partial^2 F}{\partial X_0 \partial X_1} + 2b_0 b_2 \frac{\partial^2 F}{\partial X_0 \partial X_2} + 2b_1 b_2 \frac{\partial^2 F}{\partial X_1 \partial X_2}$$

The point is that we have the general formula

$$\varphi^{(m)}(0) = ((b_0 \frac{\partial}{\partial X_0} + b_1 \frac{\partial}{\partial X_1} + b_2 \frac{\partial}{\partial X_2})^i F)(a_0, a_1, a_2)$$

where the expression

$$(b_0 \frac{\partial}{\partial X_0} + b_1 \frac{\partial}{\partial X_1} + b_2 \frac{\partial}{\partial X_2})^m F$$

has a similar meaning as in the case $i = 2$: We multiply out the polynomial in D_0, D_1 and D_2,

$$\Delta_m(D_0, D_1, D_2) = (b_0 D_0 + b_1 D_1 + b_2 D_2)^m$$

and then replace the monomials

$$D_0^{j_0} D_1^{j_1} D_2^{j_2}$$

by

$$\frac{\partial^{j_0 + j_1 + j_2} F}{\partial X_0^{j_0} \partial X_1^{j_1} \partial X_2^{j_2}}$$

It is a reasonably straightforward exercise to prove this by induction on the exponent i. We thus have the following formula:

$$\varphi(t) = F(a_0, a_1, a_2) + (D_{(b_0, b_1, b_2)} F)(a_0, a_1, a_2)t +$$

$$\frac{1}{2}(D_{(b_0, b_1, b_2)}^2 F)(a_0, a_1, a_2)t^2 + \cdots + \frac{1}{i!}(D_{(b_0, b_1, b_2)}^i F)(a_0, a_1, a_2)(0)t^i$$

$$+ \cdots + \frac{1}{d!}(D_{(b_0, b_1, b_2)}^d F)(a_0, a_1, a_2)(0)t^d$$

where

$$D_{(b_0, b_1, b_2)} = b_0 \frac{\partial}{\partial X_0} + b_1 \frac{\partial}{\partial X_1} + b_2 \frac{\partial}{\partial X_2}$$

Actually we can give a precise formula for $(D_{(b_0, b_1, b_2)}^m F)(a_0, a_1, a_2)$. In fact, there is a generalization of the familiar *binomial formula*

$$(D_0 + D_1)^m = \sum \frac{m!}{i_0! i_1!} D_0^{i_0} D_1^{i_1}$$

where the sum runs over all non-negative i_0, i_1 such that $i_0 + i_1 = m$, to the case of any number of indeterminates $D_0, \ldots, D_r$. Indeed, we have the formula

$$(D_0 + D_1 + \cdots + D_r)^m = \sum \frac{m!}{i_0! i_1! \cdots i_r!} D_0^{i_0} D_1^{i_1} \cdots D_r^{i_r}$$

where the sum runs over all non-negative $i_0, i_1, \ldots, i_r$ such that $i_0 + \cdots + i_r = m$. We may prove this formula by induction by first noting that it holds for $m = 0$ or 1. Then assuming it for $m - 1$ we need only verify the multiplication

$$\left(\sum_{i_1 + \cdots + i_r = m-1} \frac{(m-1)!}{i_0! i_1! \cdots i_r!} D_0^{i_0} D_1^{i_1} \cdots D_r^{i_r} \right)(D_0 + D_1 + \ldots D_r) =$$

$$\sum_{i_1 + \cdots + i_r = m} \frac{m!}{i_0! i_1! \cdots i_r!} D_0^{i_0} D_1^{i_1} \cdots D_r^{i_r}$$

which we leave to the reader.

Using this *Multinomial Formula* we obtain the important identity, valid for any number of variables but stated here only for three:

$$
(D^m_{(b_0,b_1,b_2)}F)(a_0, a_1, a_2) =
$$
$$
\sum b_0^{i_0} b_1^{i_1} b_2^{i_2} \frac{m!}{i_0!i_1!i_2!} \left(\frac{\partial^m}{\partial X_0^{i_0} \partial X_1^{i_1} \partial X_2^{i_2}} F \right)(a_0, a_1, a_2)
$$

where the sum runs over all non-negative i_0, i_1, i_2 such that $i_0 + i_1 + i_2 = m$.

We have $F(a_0, a_1, a_2) = 0$, so the constant term of $\varphi(y)$ is zero for all choices of (b_0, b_1, b_2). It may happen that the coefficient of t vanishes as well, for all choices of (b_0, b_1, b_2), and so on, up to a certain t^m. We make the following definition:

Definition 21 (Multiplicity of Points on Projective Curves). *The point $P = (a_0 : a_1 : a_2)$ on the projective curve C given by $F(X_0, X_1, X_2) = 0$ is said to be of multiplicity m if for all $n < m$ and all i_0, i_1, i_2*

$$
\left(\frac{\partial^n F}{\partial X_0^{i_0} \partial X_1^{i_1} \partial X_2^{i_2}} \right)(a_0, a_1, a_2) = 0,
$$

while for at least one choice of i_0, i_1, i_2

$$
\left(\frac{\partial^m F}{\partial X_0^{i_0} \partial X_1^{i_1} \partial X_2^{i_2}} \right)(a_0, a_1, a_2) \neq 0.
$$

This definition is independent of the projective coordinate system, the proof is straightforward but a little complicated. We omit it here. Clearly we have the following result:

Proposition 18. *The point $P = (a_0 : a_1 : a_2)$ on the projective curve C is of multiplicity 1 if and only if it is smooth.*

We also note the following:

Proposition 19. *The point $P = (a_0 : a_1 : a_2)$ on the projective curve C given by $F(X_0, X_1, X_2) = 0$ is of multiplicity m if and only if for all $n < m$ and all $(b_0, b_1, b_2) \notin C$*

$$
\sum_{i_0+i_1+i_2=n} b_0^{i_0} b_1^{i_1} b_2^{i_2} \frac{n!}{i_0!i_1!i_2!} \left(\frac{\partial^n F}{\partial X_0^{i_0} \partial X_1^{i_1} \partial X_2^{i_2}} \right)(a_0, a_1, a_2) = 0.
$$

while for at least one tuple $(b_0, b_1, b_2) \notin C$

$$
\sum_{i_0+i_1+i_2=m} b_0^{i_0} b_1^{i_1} b_2^{i_2} \frac{m!}{i_0!i_1!i_2!} \left(\frac{\partial^m F}{\partial X_0^{i_0} \partial X_1^{i_1} \partial X_2^{i_2}} \right)(a_0, a_1, a_2) \neq 0.
$$

This last proposition shows that any line L through the point P will intersect C at P with multiplicity at least equal to m, the multiplicity of the point P on C. And there exists at least one line through P which intersects C at P with multiplicity m.

We need the following result:

Proposition 20. *Let $F(X_0, X_1, X_2)$ be a homogeneous polynomial of degree d. Then the following identity holds:*

$$X_0\frac{\partial F}{\partial X_0} + X_1\frac{\partial F}{\partial X_1} + X_2\frac{\partial F}{\partial X_2}$$

$$= dF(X_0, X_1, X_2)$$

Proof. We consider the following identity which holds for the variables $X_0, \ldots, X_n$ and t

$$F(tX_0, tX_1, \ldots, tX_n) = t^d F(X_0, X_1, \ldots, X_n)$$

This identity is verified by substituting tX_i for X_i in the polynomial, and observing that by definition all the monomials which occur are of the same degree d.

We now compute the derivative with respect to t. The right hand side yields

$$dt^{d-1}F(X_0, X_1, \ldots, X_n)$$

while the chain rule applied to the left hand side yields

$$\frac{\partial F}{\partial X_0}(tX_0, tX_1, tX_2)X_0 + \frac{\partial F}{\partial X_1}(tX_0, tX_1, tX_2)X_1 + \frac{\partial F}{\partial X_2}(tX_0, tX_1, tX_2)X_2$$

These two polynomials are equal, and putting $t = 1$ we get the claimed formula. $\square$

Now assume that $P = (a_0 : a_1 : a_2) \in C$ has multiplicity m on C. We define the curve $T_{C,P}$ by the equation

$$H_{C,P}(X_0, X_1, X_2) =$$

$$\sum_{i_0+i_1+i_2=m} X_0^{i_0} X_1^{i_1} X_2^{i_2} \frac{m!}{i_0!i_1!i_2!}\left(\frac{\partial^m F}{\partial X_0^{i_0}\partial X_1^{i_1}\partial X_2^{i_2}}\right)(a_0, a_1, a_2) = 0.$$

This equation actually does define a curve of degree m in $\mathbb{P}^2(\mathbb{R})$, as it is non-zero and homogeneous of degree m. We have the following important result:

Theorem 22. $T_{C,P}$ *is the union of a finite number of lines through P. The irreducible components of $T_{C,P}$ are exactly the lines through P which intersect C with multiplicity $> m$.*

Proof. We start out by noticing that $P \in TC_P(C)$. Indeed, we have the identity

$$d(d-1)\ldots(d-m+1)F(X_0, X_1, X_2) =$$
$$\sum_{i_0+i_1+i_2=m} X_0^{i_0} X_1^{i_1} X_2^{i_2} \frac{m!}{i_0! i_1! i_2!} \left(\frac{\partial^m F}{\partial X_0^{i_0} \partial X_1^{i_1} \partial X_2^{i_2}} \right) = 0$$

as is easily seen by repeated application of Proposition 20, first to F, then to the first order partial derivatives, so to the second order ones and so on, up to the partials of order m.

Since by definition of the multiplicity m there exists a point $(b_0 : b_1 : b_2)$ such that $H_{C,P}(b_0, b_1, b_2) \neq 0$, $T_{C,P}$ is a curve. Moreover, if $H_{C,P}(b_0, b_1, b_2) = 0$ then the line through P and $Q = (b_0, b_1, b_2)$ intersects C at P with multiplicity $> m$, since the corresponding $\varphi(t)$ has $t = 0$ as a root occuring with multiplicity $> m$. Thus any point $Q' = (b_0', b_1', b_2')$ on that line will also satisfy $H_{C,P}(b_0', b_1', b_2') = 0$. Thus the curve $T_{C,P}$ consists of lines passing through P. $\square$

Definition 22. *The curve $T_{C,P}$ is referred to as the projective tangent cone to C at P.*

We have earlier defined the concept of multiplicity and tangent cone in the affine case. These concepts are completely compatible under affine restriction and under projective closure:

Proposition 21. *Let K be an affine curve, and $p = (a, b) \in K$. Let C be the projective closure, and put $P = (1 : a : b) \in C$. Then the point p, as a point on an affine curve, is of the same multiplicity as the point P on the projective curve C. Moreover the affine restriction of $T_{C,P}$ is equal to the affine tangent cone of K at p, as given in Definition 16.*

Proof. We may change the projective coordinate system to one in which $P = (1 : 0 : 0)$, this corresponds to a change of affine coordinate system to one in which $(a, b) = (0, 0)$. In this case the claim is easily checked. $\square$

The irreducible components of the curve $T_{C,P}$ are referred to as the *lines of tangency* to C at P. If the point $P \in C$ is smooth, then $m = 1$ and there is only one line of tangency, which we refer to as the *tangent line to C at P*, and denote as before by $T_{C,P}$. The equation is

$$X_0 \frac{\partial F}{\partial X_0}(a_0, a_1, a_2) + X_1 \frac{\partial F}{\partial X_1}(a_0, a_1, a_2) + X_2 \frac{\partial F}{\partial X_2}(a_0, a_1, a_2) = 0.$$

We finally compute an example. Consider the projective curve given by

$$F(X_0, X_1, X_2) = X_0 X_2^2 - X_1^3 - X_0 X_1^2 = 0.$$

which is the projective closure of the affine curve defined by

$$y^2 - x^3 - x^2 = 0,$$

in other words, the nodal cubic curve. To find the projective tangent cone at the point $(1 : 0 : 0)$, it is most convenient to pass to the affine restriction. Then we immediately see that the affine tangent cone is defined by

$$y^2 - x^2 = 0,$$

which has the projective closure given by

$$X_2^2 - X_1^2 = 0.$$

This is the fastest way to proceed. But we could also use the definition directly. Then we compute

$$\frac{\partial F}{\partial X_0} = X_2^2 - X_1^2, \frac{\partial F}{\partial X_1} = 3X_1^2 - 2X_0 X_1, \frac{\partial F}{\partial X_2} = 2X_0 X_2$$

They all evaluate to zero at $(1 : 0 : 0)$, so this point is singular (as we know from the affine restriction). We differentiate again, and obtain

$$\frac{\partial^2 F}{\partial X_0^2} = 0, \frac{\partial^2 F}{\partial X_1^2} = 6X_1 - 2X_0, \frac{\partial^2 F}{\partial X_2^2} = 2X_0$$
$$\frac{\partial^2 F}{\partial X_0 \partial X_1} = -2X_1, \frac{\partial^2 F}{\partial X_0 \partial X_2} = 2X_2, \frac{\partial^2 F}{\partial X_1 \partial X_2} = 0$$

Evaluating at P, we get

$$\frac{\partial^2 F}{\partial X_0^2} = 0, \frac{\partial^2 F}{\partial X_1^2} = -2, \frac{\partial^2 F}{\partial X_2^2} = 2$$
$$\frac{\partial^2 F}{\partial X_0 \partial X_1} = 0, \frac{\partial^2 F}{\partial X_0 \partial X_2} = 0, \frac{\partial^2 F}{\partial X_1 \partial X_2} = 0$$

and thus according to our formula the equation for $T_{C,P}$ is

$$\frac{2!}{0!2!0!}(-2)X_1^2 + \frac{2!}{0!0!2!}2X_2^2 = 0,$$

confirming what we found above.

14.5 Projective Equivalence

We have now arrived at a point where it is appropriate to attempt using the techniques introduced to create some system, or order, in the vast *menagerie* of different algebraic curves, affine or projective, which exist in $\mathbb{R}^2$ and in $\mathbb{P}^2(\mathbb{R})$.

For this we introduce the notion of *projectively equivalent plane curves.* We make the following

Definition 23. *Two irreducible projective curves C and C' are called projectively equivalent if C is mapped to C' by a projective transformation. Two irreducible affine curves K and K' are said to be projectively equivalent if their projective closures are projectively equivalent, and an affine curve K is projectively equivalent to its projective closure C.*

Remark Frequently we replace the sentence C *is mapped to* C' *by a projective transformation* by C *becomes* C' *by a projective change of coordinate system.* This way of expressing the equivalence is perfectly legitimate when we think of the curves as given by explicite equations.

Thus for example the affine version of the nodal cubic

$$y^2 - x^3 - x^2 = 0.$$

is projectively equivalent to the projective version defined by

$$X_0 X_2^2 - X_1^3 - X_0 X_1^2 = 0.$$

However, note that *we do not assert* that its affine tangent cone at the origin

$$y^2 - x^2 = 0.$$

is projectively equivalent to the projective tangent cone at $(1 : 0 : 0)$,

$$X_2^2 - X_1^2 = 0.$$

as we use this terminology for irreducible curves only.

We first study non-degenerate conic sections in light of the above definition. In Theorem 18 we proved the following:

The equation

$$Ax^2 + 2Bxy + Cy^2 + 2Dx + 2Ey + F = 0.$$

yields a non-degenerate conic section if and only if

$$\begin{vmatrix} A & B & D \\ B & C & E \\ D & E & F \end{vmatrix} \neq 0$$

A non-degenerate conic section is irreducible, as it would otherwise be two lines or one line with multiplicity 2, both cases being degenerate conics. The projective closure is given by the equation

$$AX_1^2 + 2BX_1X_2 + CX_2^2 + 2DX_0X_1 + 2EX_0X_2 + FX_0^2 = 0.$$

This equation, with the condition that the above determinant be non-zero, yields a general non-degenerate conic section: Any non-degenerate conic section may be expressed in this way. We have carried out an investigation of conic sections in $\mathbb{P}^2(\mathbb{R})$ in Section 12.8. Now we are going to make a suitable choice of projective coordinate system which gives this general equation a very simple form. For this we use Corollary 2. We proceed as follows:

First of all, since the conic section is assumed non-degenerate, any four points Q_1, Q_2 Q_3 and Q_4 on it form *an arc of four*. No three points among them are collinear. By the corollary referred to above there is a projective transformation G of $\mathbb{P}^2(\mathbb{R})$ onto itself mapping Q_1 to $(1 : 0 : 0)$, Q_2 to $(1 : 1 : 0)$, Q_3 to $(1 : 0 : 1)$ and Q_4 to $(1 : 1 : 1)$ Thus after a projective transformation we may assume that these four points lie on the projective conic section C. Taking the affine restriction of the situation, we get the points $(0,0)$, $(1,0)$, $(0,1)$ and $(1,1)$ in $\mathbb{R}^2$, they lie on the affine conic section given by

$$Ax^2 + 2Bxy + Cy^2 + 2Dx + 2Ey + F = 0.$$

Thus we obtain

$$F = 0, A + 2D = 0, C + 2E = 0 \text{ and } A + 2B + C + 2D + 2E = 0.$$

Thus $B = 0$ and the equation becomes

$$Ax^2 - Ax + Cy^2 - Cy = 0.$$

So completing the square we obtain the equation as

$$\lambda_1 x^2 + \lambda_2 y^2 + \lambda_3 = 0.$$

and going back to $\mathbb{P}^2(\mathbb{R})$ by the projective closure we get

$$\lambda_3 X_0^2 + \lambda_1 X_1^2 + \lambda_2 X_2^2 = 0.$$

Since the conic section is non-degenerate, λ_1, λ_2 and λ_3 are all non-zero. Since we are dealing with real theory, we also may exclude the case that λ_1, λ_2 and λ_3 have the same sign. Thus, again after a projective transformation if needed, we may assume that $\lambda_3 < 0$, and $\lambda_1, \lambda_2 > 0$. Dividing by $|\lambda_3|$ the equation becomes

$$\mu_1 X_1^2 + \mu_2 X_2^2 - X_0^2 = 0.$$

which after the projective transformation

$$(a_0 : a_1 : a_2) \mapsto (a_0 : \sqrt{\mu_1} a_1 : \sqrt{\mu_2} a_2)$$

becomes

$$X_1^2 + X_2^2 - X_0^2 = 0.$$

We have completed the proof of the following

Theorem 23. *Up to projective equivalence there is only one non-degenerate conic section in* $\mathbb{P}^2(\mathbb{R})$.

There is a very close connection between *projective transformations* on one hand and actual *projections* on the other.

In fact, it is fair to say that these two concepts are practically equivalent. We shall now explain this.

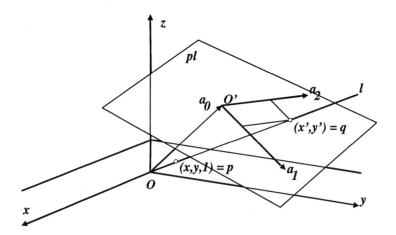

Fig. 14.1. A projective transformation from the plane $z = 1$ to the plane defined by the three vectors α_0, α_1 and α_2 as explained in the text.

We identify the affine plane $\mathbb{R}^2$ with the plane $z = 1$, as shown in Figure 14.1. Then, as was explained in Section 8.3, the line through the origin O and the point $(x, y, 1)$ identifies the point (x, y) with the point $(1 : x : y)$ which corresponds to it under the identification of $D_+(X_0)$ with $\mathbb{R}^2$, and the lines contained in the xy-plane then correspond to the points at infinity, in other words to the points in $V_+(X_0)$.

The projection mapping with center at the origin O from the plane $z = 1$ to a general plane pl, subject only to the condition that $O \notin pl$, may be described as follows: A point $(x, y) \in \mathbb{R}^2$ is identified with $p = (x, y, 1)$, and the line ℓ through O and p is produced to intersect the plane pl in the point q. This point q is then the projection of p from the center O to the plane pl. To describe this mapping in terms of coordinates, we need to find a suitable description of the plane pl. Any plane in $\mathbb{R}^3$, the xyz-space, which does not pass through the origin O is uniquely determined by three linearly independent vectors,

$$
\begin{aligned}
a_0 &= (a_0, b_0, c_0) \\
a_1 &= (a_1, b_1, c_1) \\
a_2 &= (a_2, b_2, c_2)
\end{aligned}
$$

The first vector a_0 determines a point O' in pl, which we will use as the origin in a coordinate system to be introduced in pl, see Figure 14.1. Further, to specify the plane we need the two vectors a_1 and a_2, which together with a_0 form a basis for the vector space $\mathbb{R}^3$. On (double) parametric form the plane pl is then given as

$$
pl = \left\{ (X, Y, Z) \in \mathbb{R}^3 \;\middle|\; \begin{aligned} X &= a_0 + ua_1 + va_2 \\ Y &= b_0 + ub_1 + vb_2 \\ Z &= c_0 + uc_1 + vc_2 \end{aligned} \quad \text{where } (u, v) \in \mathbb{R}^2 \right\}
$$

With this description of pl we let the origin O' of a coordinate system in pl be the end point of a_0, in other words the point (a_0, b_0, c_0). The vector a_1 is affixed to O', and determines the x'-axis, similarly a_2 determines the y'-axis. Their lengths are taken to be unit lengths along the respective axes, thus the point q is represented by the coordinates (x', y') which are the values of u and v defining q in the parametric form of pl.

On the other hand the parametric form of ℓ is

$$
\ell = \left\{ (X, Y, Z) \in \mathbb{R}^3 \;\middle|\; \begin{aligned} X &= xt \\ Y &= yt \\ Z &= t \end{aligned} \quad \text{where } t \in \mathbb{R} \right\}
$$

Thus we have

$$
\begin{aligned}
tx &= a_0 + x'a_1 + y'a_2 \\
ty &= b_0 + x'b_1 + y'b_2 \\
t &= c_0 + x'c_1 + y'c_2
\end{aligned}
$$

Thus if we identify (x, y) with $(1 : x : y) = (x_0 : x_1 : x_2)$ where, as usual, $z = x_0$, and similarly identify (x', y') with $(1 : x' : y') = (\overline{x}_0 : \overline{x}_1 : \overline{x}_2)$, then we get

$$
\begin{aligned}
x_0 &= c_0\overline{x}_0 + c_1\overline{x}_1 + c_2\overline{x}_2 \\
x_1 &= a_0\overline{x}_0 + a_1\overline{x}_1 + a_2\overline{x}_2 \\
x_2 &= b_0\overline{x}_0 + b_1\overline{x}_1 + b_2\overline{x}_0
\end{aligned}
$$

from which we conclude that the projection is really a projective transformation.

We now see that Theorem 23 immediately implies the following result, which shows that Conic Sections as defined by Apollonius are precisely the curves in $\mathbb{R}^2$ of degree 2:

Corollary 5. *The curves of degree two in $\mathbb{R}^2$ are precisely those which can be obtained as the intersection between a fixed circular cone and a varying plane.*

We conclude this section with an examination of two classes of curves studied in Sections 13.1 and 13.1. A semi cubical parabola may, by a change of affine coordinate system, be brought on the form

$$x^3 - y^2 = 0.$$

The usual projective closure of this curve is given by

$$X_0 X_2^2 - X_1^3 = 0.$$

in $\mathbb{P}^2(\mathbb{R})$. Taking the affine restriction to $D_+(x_2)$ and letting $\bar{y} = \frac{X_0}{X_2}, \bar{x} = \frac{X_1}{X_2}$, we get the equation

$$\bar{y} - \bar{x}^3 = 0.$$

which is a cubical parabola. Thus cubical and semi cubical parabolas are projectively equivalent.

14.6 Asymptotes

We may now give a simple treatment of a subject which often appears rather mysterious. An *asymptote* to a given curve is defined as a line such that the distance from a point on the curve to the line tends to zero as the point on the curve moves further and further away from the origin.

This definition renders it quite mysterious how to actually compute all asymptotes to a given curve. Another drawback is that it defines the concept in terms of *distance*, thus the concept defined in this way is not an algebraic one.

The following definition is equivalent to the one given above for algebraic affine curves in $\mathbb{R}^2$:

Definition 24. *Let K be the affine curve defined by*

$$f(x, y) = 0.$$

Let C be the projective closure in $\mathbb{P}^2(\mathbb{R})$ obtained by letting $x = \frac{X_1}{X_0}, y = \frac{X_2}{X_0}$ as usual. Let $P_1, \ldots, P_m$ be the points at infinity of C, and let $L_1, \ldots, L_r$ be all lines in $\mathbb{P}^2(\mathbb{R})$ different from $V_+(X_0)$ and appearing as a line of tangency to C at one of the points $P_1, \ldots, P_m$. Let $\ell_1, \ldots, \ell_r$ be the affine restrictions of $L_1, \ldots, L_r$. Then $\ell_1, \ldots, \ell_r$ are all the asymptotes of K in $\mathbb{R}^2$.

The Trisectrix of Maclaurin has equation $x^3 + xy^2 + y^2 - 3x^2 = 0$. It is treated in Section 13.1, and its appearance makes one wonder if it might have a vertical asymptote, crossing the x-axis somewhere to the left of the origin. We shall now check this.

The projective closure of the trisectrix is given by the equation $X_1^3 + X_1X_2^2 + X_0X_2^2 - 3X_0X_1^2 = 0$. We find the points at infinity by substituting $X_0 = 0$ into this equation, we get $X_1^3 + X_1X_2^2 = 0$. This yields one real point, given by $X_1 = 0$, and two complex points determined by $X_1^2 + X_2^2 = 0$, which do not concern us as we are dealing with the real points only. Thus the one (real) point at infinity is $(0 : 0 : 1)$. We now take the affine restriction to $D_+(X_2)$ by putting $x' = \frac{X_0}{X_2}$ and $y' = \frac{X_1}{X_2}$. This affine restriction is given by $y'^3 + y' + x' - 3x'y'^2 = 0$. Hence the origin is a smooth point, the tangent there is given by $x' + y' = 0$. Going back to the projective plane, this line has the equation $X_0 + X_1 = 0$, and taking the affine restriction to the original affine xy-plane, we get the equation $x = -1$: This, then, is the asymptote of the curve, affirming our suspicion that such a line might exist.

An even simpler example, but an important one, is to verify the asymptotes of a general hyperbola. Assume it is given on standard form, as

$$\left(\frac{x}{a}\right)^2 - \left(\frac{y}{b}\right)^2 = 1$$

To show is that the asymptotes are given by

$$\left(\frac{x}{a}\right)^2 - \left(\frac{y}{b}\right)^2 = 0.$$

We leave this verification as an exercise.

14.7 General Conchoids

In Section 4.6 we explained the construction of the Conchoid of Nicomedes. Now we treat his great invention in more detail. First we deduce the equation of the conchoid. We choose the coordinate system with origin at the fixed point P, y-axis parallel to ℓ and x-axis normal to ℓ as shown in Figure 14.2. A line through P has the equation $y = tx$, and a point (x, y) on it, at distance b from its intersection with ℓ must satisfy

$$y = tx$$
$$(x - a)^2 + (y - ta)^2 = b^2$$

which when $t = \frac{y}{x}$ from the former is substituted in the latter, yields

$$(x - a)^2 x^2 + y^2 (x - a)^2 = b^2 x^2$$

or

$$(x - a)^2(x^2 + y^2) = b^2x^2$$

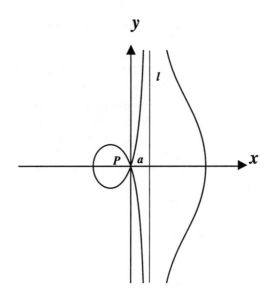

Fig. 14.2. The Conchoid of Nicomedes with $b = 3 > a = 1$.

The line ℓ is an asymptote for the conchoid. This discovery is attributed to Nicomedes himself. We check the result with our method for finding all asymptotes. The projective closure of the conchoid is given by

$$(X_1 - aX_0)^2(X_1^2 + X_2^2) - b^2X_0^2X_1^2 = 0.$$

The points at infinity are determined by

$$X_1^2(X_1^2 + X_2^2) = 0.$$

thus as we only consider real points, the only point at infinity is $(0 : 0 : 1)$. We now take the affine restriction to $D_+(X_2)$ by letting $x' = \frac{X_0}{X_2}$ and $y' = \frac{X_1}{X_2}$. The equation in the $x'y'$-plane becomes

$$(y' - ax')^2(y'^2 + 1) - b^2x'^2y'^2 = 0.$$

The homogeneous part of lowest degree of this polynomial is $H(x', y') = (y' - ax')^2$, so the tangent cone at the point $(0,0)$ is the line $y' = ax'$, with multiplicity 2. Taking the projective closure again we get the projective line $X_1 - aX_0 = 0$, and its affine restriction to our original affine plane $\mathbb{R}^2 = D_+(X_0)$ is $x = a$. Thus we have proved Nicomedes' theorem on the asymptote of the conchoid.

The Conchoid of Nicomedes is a special case of a general class of curves. We make the following

Definition 25. *Given an affine curve K and a fixed point P. Consider the collection of all lines through P. The conchoid of K for the pole P and constant b is then the locus of all points Q such that Q lies on one of these lines at a distance b from its intersection with K.*

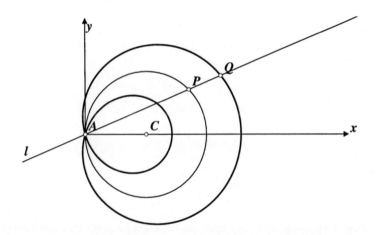

Fig. 14.3. The Conchoid of a circle, called a Limaçon. The circle about C with radius AC is fixed. The line ℓ rotates about A, and the point Q is on ℓ at the fixed distance from the circle $b = PQ$. Of course $d = 2AC$, here $d > b$.

A conchoid of *a circle for a fixed point on it* is called a *Limaçon of Pascal,* the first part of the name picked by Étienne Pascal, the name meaning *snail* in French. When b is equal to the diameter d of the circle, the curve is called the *cardioid,* in other words the *heart curve,* and if the constant b is equal to the radius of the circle, we get a curve which may be used to trisect an angle in equal parts, often referred to as a *trisectrix* (but not to be confused with the Trisectrix of Maclaurin, treated earlier). We shall now analyze the different cases. Depending on the relation between b and d we get three versions of the Limaçon, one being shown in Figure 14.3, the two others in Figure 14.4.

These curves are simple to describe in polar coordinates, as

$$r = b + d\cos(\varphi)$$

where d is the diameter of the circle and b is the constant, see Figure 14.3.

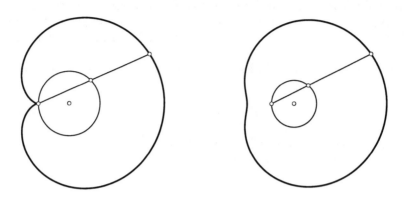

Fig. 14.4. The two other versions of the limaçon. To the left $d = b$, to the right $d < b$.

A simple computation yields

$$x^2 + y^2 - dx = db \cos(\varphi) + b^2$$

and hence

$$(x^2 + y^2 - dx)^2 = d^2b^2 \cos^2(\varphi) + 2db^3 \cos(\varphi) + b^4 = b^2(x^2 + y^2)$$

Thus we obtain the somewhat less transparent form in usual xy-coordinates

$$(x^2 + y^2 - dx)^2 - b^2(x^2 + y^2) = 0.$$

In Figure 14.4 we see some limaçons.

14.8 The Dual Curve

In projective algebraic geometry the *principle of duality* acquires a very precise meaning. For $\mathbb{P}^2(\mathbb{R})$ we have the following:

Every line in $\mathbb{P}^2(\mathbb{R})$ is given by an equation

$$a_0X_0 + a_1X_1 + a_2X_2 = 0.$$

If we multiply each coefficient by a common non-zero constant, then we get the same equation. Thus the line may be associated with a uniquely determined point of another copy of the projective plane,

$$L^\vee = (a_0 : a_1 : a_2) \in \mathbb{P}^2(\mathbb{R})$$

Conversely, to any point $P \in \mathbb{P}^2(\mathbb{R})$ we may associate a *line* $P^\vee \subset \mathbb{P}^2(\mathbb{R})$. The correspondence $(\ )^\vee$ preserves *incidence*, as already explained in Section 12.2

We now extend this to *projective, algebraic curves*. We get the following concept of duality: For any projective curve $C \subset \mathbb{P}^2(\mathbb{R})$, consider the subset

$$C^\vee = \{(L)^\vee \,|\, L \text{ is a line of tangency to } C\}$$

We denote this set by $C^\vee$, and refer to it as the *dual curve of* C. Indeed, it turns out that this subset of $\mathbb{P}^2(\mathbb{R})$ is actually a projective curve, in $\mathbb{P}^2(\mathbb{R})$, except for the case when C is a projective line, in which case $C^\vee$ consists of just one point.

Assume that C has the equation $F(X_0, X_1, X_2) = 0$. The equation for the dual curve is then expressed in terms of the indeterminates Y_0, Y_1 and Y_2 when we eliminate X_0, X_1, X_2 in the system

$$\frac{\partial F}{\partial X_0}(X_0, X_1, X_2) = Y_0$$
$$\frac{\partial F}{\partial X_1}(X_0, X_1, X_2) = Y_1$$
$$\frac{\partial F}{\partial X_2}(X_0, X_1, X_2) = Y_2$$
$$F(X_0, X_1, X_2) = 0.$$

Here we have six variables X_0, X_1, X_2 and Y_0, Y_1, Y_2, and four relations among them. In general we may then *eliminate* any three of them, and obtain *one relation* between the remaining variables. We now eliminate X_0, X_1 and X_2. This will give as result a single equation

$$G(Y_0, Y_1, Y_2) = 0.$$

which defines the dual curve $C^\vee$.

We may think of what we are doing here in the following way: We write

$$\frac{\partial F}{\partial X_0}(X_0, X_1, X_2) = Y_0(X_0, X_1, X_2)$$
$$\frac{\partial F}{\partial X_1}(X_0, X_1, X_2) = Y_1(X_0, X_1, X_2)$$
$$\frac{\partial F}{\partial X_2}(X_0, X_1, X_2) = Y_2(X_0, X_1, X_2)$$

We then *solve this system of equations* for X_0, X_1 and X_2:

$$X_0 = X_0(Y_0, Y_1, Y_2)$$
$$X_1 = X_1(Y_0, Y_1, Y_2)$$
$$X_2 = X_2(Y_0, Y_1, Y_2)$$

and then get

$$G(Y_0, Y_1, Y_2) = F(X_0(Y_0, Y_1, Y_2), X_1(Y_0, Y_1, Y_2), X_2(Y_0, Y_1, Y_2))$$

Of course, usually we are not able to find X_0, X_1 and X_2 as homogeneous polynomials in the Y's, not even as single valued functions. To put these considerations on a mathematically sound basis, it was necessary to develop the machinery of *elimination theory*. But we shall bypass this, and work for a while with such ficticious entities as the X_0, X_1 and X_2 as functions of Y_0, Y_1 and Y_2. In the end they are gone, and only the $G(Y_0, Y_1, Y_2)$, which does exist thanks to elimination theory, remains. But in some happy cases the X_0, X_1 and X_2 *do exist* as homogeneous polynomials in Y_0, Y_1 and Y_2, and then they simplify the situation considerably.

Since questions of tangency are independent of projective coordinate system, the same is true for questions of duality.

We shall use this important observation in proving the following

Theorem 24. *The dual curve of a non-degenerate conic section in $\mathbb{P}^2(\mathbb{R})$ is again a non-degenerate conic section.*

Proof. We showed in Theorem 23 that any non-degenerate conic section is projectively equivalent to the one given by

$$X_1^2 + X_2^2 - X_0^2 = 0.$$

thus we may assume that $F(X_0, X_1, X_2) = X_1^2 + X_2^2 - X_0^2$. Then

$$\frac{\partial F}{\partial X_0} = -2X_0$$
$$\frac{\partial F}{\partial X_1} = 2X_1$$
$$\frac{\partial F}{\partial X_2} = 2X_2$$

So we have to eliminate X_0, X_1, X_2 in

$$-2X_0 = Y_0$$
$$2X_1 = Y_1$$
$$2X_2 = Y_2$$
$$X_1^2 + X_2^2 - X_0^2 = 0$$

In this case we may solve for X_0, X_1 and X_2, which yields

$$Y_1^2 + Y_2^2 - Y_0^2 = 0.$$

which is a non-degenerate conic section. $\square$

As we see, the equation is the same as the one we started with, only the indeterminates have different names. Just looking at this one example, one might be tempted to draw the conclusion that $C = C^\vee$. But this is far from true, even for general conic sections. The point is that the property of *having a non-degenerate conic section as dual* is independent of the coordinate system, while the property of *having a dual which is defined by a fixed homogeneous polynomial* certainly very much depends on the coordinate system. However, we have the following important theorem, which is true in much greater generality than the version we give here:

Theorem 25. *Let C be a curve given by an irreducible homogeneous polynomial in $\mathbb{P}^2(\mathbb{R})$. Then if we dualize the dual curve of C, we get C back:*
$$C^{\vee\vee} = C$$

Proof. We use the simplified form given above, avoiding elimination theory. Then G is really nothing but the original F, but expressed in terms of the variables Y_0, Y_1, Y_2 instead of X_0, X_1, X_2. We therefore only have to prove that if we put

$$
\begin{aligned}
\overline{X}_0 &= \tfrac{\partial F}{\partial Y_0} \\
\overline{X}_1 &= \tfrac{\partial F}{\partial Y_1} \\
\overline{X}_2 &= \tfrac{\partial F}{\partial Y_2}
\end{aligned}
$$

then

$$\overline{X}_0 = \alpha X_0, \overline{X}_1 = \alpha X_1 \text{ and } \overline{X}_2 = \alpha X_2$$

for some real number $\alpha \neq 0$. This is done as follows: By Proposition 20, we have that

$$X_0 Y_0 + X_1 Y_1 + X_2 Y_2 = dF(X_0, X_1, X_2)$$

where d is the degree of F in X_0, X_1, X_2. Hence

$$\overline{X}_i = \frac{1}{d} \frac{\partial}{\partial Y_i}(X_0 Y_0 + X_1 Y_1 + X_2 Y_2) = X_i$$

for $i = 0, 1$ and 2. Thus the claim follows. $\square$

14.9 The Dual of Pappus' Theorem

The degenerate case of Theorem 19 is known as *Pappus' theorem*. It is shown to the right in Figure 14.5.

We shall now show how we may obtain a new, apparently completely different, theorem simply by dualizing Pappus' theorem.

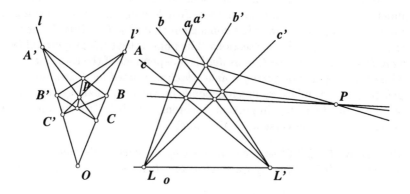

Fig. 14.5. Pappus' theorem, to the left, and its dual to the right.

To the left we have a point O, with two lines ℓ and ℓ' passing through it. To the right there is a *line o*, with two points L and L' on it. To the left we have three points, all different and different from O, on each of the lines. They are labeled A, B and C for the points on ℓ, and A', B' and C' for the points on ℓ'. Dually, to the right there are lines a, b and c through L, and a', b' and c' through L', all of them different from o and from one another.

Now we draw the lines AB' and BA', and mark their point of intersection. We do the same for the pairs AC' and $A'C$, BC' and $B'C$. Then, as Pappus' Theorem tells us, these three points are *collinear*, they lie on one line, labeled p.

Dually, to the right, we take the point of intersection between a and b', and the point of intersection between a' and b. We draw the line through these two points. Similarly we find the point of intersection of a and c', as well as the point of intersection of a' and c. We draw the line between these two points as well. Finally, we find the point of intersection of b and c', as well as the point of intersection of b' and c. And, for the third time, draw the line between these two points. We now have drawn altogether three lines. The theorem is that *these three lines pass through a common point*, labeled P in the right part of the figure.

14.10   Pascal's Mysterium Hexagrammicum

In Figure 14.6 the ellipse to the left illustrates Pascal's Theorem. Six points A, B, C and A', B', C' are given on the non-degenerate conic section. Then we draw lines connecting each point to two of the other points, we have the lines AB', AC', $A'B$, $A'C$ and BC', $B'C$. We form three pairs of these lines by pairing the ones labeled by the same letters, primed or unprimed, the three pairs determine three points of intersection labeled D, E and F. Then by the theorem these three points lie on one line, labeled p.

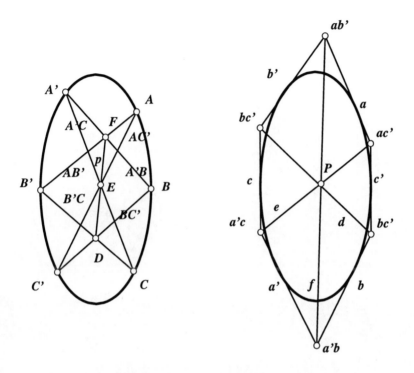

Fig. 14.6. Pascal's *Mysterium Hexagrammicum.*

To the right we dualize the situation. The dual of a non-degenerate conic section is again a non-degenerate conic section. Choosing points on the conic section corresponds to selecting tangents on the dual conic section, they are labeled a, b, c and a', b', c. To draw lines connecting each point to two of the other points, correspond dually to taking the points of intersection between one tangent with two of the other ones. We then have the points ab', ac', $a'b$, $a'c$ and bc', $b'c$. Taking the points of intersection between similarly labeled lines dually corresponds to drawing lines between similarly labeled points. In Pascal's Theorem the points of intersection lie on the same line p, which dually corresponds to the lines in the dual situation passing through the same point P.

We have proved the following:

Theorem 26. *Let there be given six tangent lines to a non-degenerate conic section. In the resulting circumscribed hexagon connect diametrically opposite corners. The resulting lines then pass through a common point.*

15 Sharpening the Sword of Algebra

15.1 On Rational Polynomials

A *rational polynomial in the variable* x is an expression

$$f(x) = a_n x^n + a_{n-1} x^{n-1} + \ldots + a_0, a_n \neq 0,$$

where $a_n, \ldots, a_0$ are rational numbers. Similarly we define integral, real or complex polynomials. The degree of $f(x)$ is the integer n. For the *zero polynomial*, the one where all coefficients are zero, it is convenient to define the degree as $-\infty$.

We add polynomials by adding the coefficients of the same powers of x, and multiplication is carried out by the formula

$$(a_m x^m + a_{m-1} x^{m-1} + \ldots + a_0) \cdot (b_n x^n + b_{n-1} x^{n-1} + \ldots + b_0)$$

$$= c_N x^N + c_{N-1} x^{N-1} + \ldots + c_0.$$

where $N = m + n$, and $c_N = a_m b_n$, $c_{N-1} = a_m b_{n-1} + a_{m-1} b_m$, and so on, down to $c_0 = a_0 b_0$.

The concepts of addition, subtraction and multiplication of polynomials are probably reasonably well known to the readers already, but less familiar, perhaps, is the concept of *polynomial division*.

Suppose that we wish to divide the polynomial $x^5 + 2x^4 + x^2 + 4x + 2$ by $x^2 + 1$. This will give a quotient $q(x)$ as well as a *remainder* $r(x)$. This is very similar to the situation for *division of integers*. We proceed as follows:

$$
\begin{array}{l}
x^5 + 2x^4 + x^2 + 4x + 2 : x^2 + 1 = x^3 + 2x^2 - x - 1 \\
\underline{-(x^5 + x^3)} \\
2x^4 - x^3 + x^2 + 4x + 2 \\
\underline{-(2x^4 + 2x^2)} \\
-x^3 - x^2 + 4x + 2 \\
\underline{-(-x^3 - x)} \\
-x^2 + 5x + 2 \\
\underline{-(-x^2 - 1)} \\
5x + 3
\end{array}
$$

So here the quotient is

$$q(x) = x^3 + 2x^2 - x - 1,$$

and the remainder is

$$r(x) = 5x + 3$$

We may also express this by writing

$$\frac{x^5 + 2x^4 + x^2 + 4x + 2}{x^2 + 1} = x^3 + 2x^2 - x - 1 + \frac{5x + 3}{x^2 + 1}$$

In this form most students of calculus will need polynomial division as a tool for computing integrals of rational functions.

In the general case we proceed exactly as in the example above: Suppose that we have polynomials $a(x)$ and $b(x)$, both with rational coefficients. Then we may find polynomials $q(x)$ and $r(x)$, where $r(x)$ is either the zero polynomial or is of degree $<$ the degree of $b(x)$, such that

$$\frac{a(x)}{b(x)} = q(x) + \frac{r(x)}{b(x)}$$

If $a(x)$ is of degree *less than* the degree of $b(x)$, we take $q(x) = 0$, and $a(x) = r(x)$. If, on the other hand, $a(x)$ is of degree $\geq$ the degree of $b(x)$, we find $q(x)$ as in the example: Assume that

$$a(x) = a_n x^n + a_{n-1} x^{n-1} + \ldots + a_0, a_n \neq 0,$$

and

$$b(x) = b_m x^m + b_{m-1} x^{m-1} + \ldots + b_0, b_m \neq 0.$$

where we assume $n \geq m$. Then the first term (*the leading term*) of $q(x)$ is $\frac{a_n}{b_m} x^{n-m}$. We multiply this term by $b(x)$, the result is subtracted from $a(x)$ and we obtain a new polynomial $a_1(x)$ which is of degree $< n$. Then the process is repeated with $a_1(x)$ instead of $a(x)$, this yields the next term of $q(x)$, and so on, until we get a polynomial $a_h(x)$ which is of degree $< m$. Then it is no longer possible to repeat the procedure, and we have found $q(x)$, while $r(x)$ is the polynomial $a_h(x)$. We have shown the following:

Proposition 22. *Let there be given two polynomials $a(x)$ and $b(x)$, both with rational coefficients. Then there exists two other rational polynomials $q(x)$ and $r(x)$, where $r(x)$ is either the zero polynomial or is of degree $<$ the degree of $b(x)$[1], such that*

---

[1] Strictly speaking the first part of the sentence is redundant, as we have defined the zero polynomial to be of degree $-\infty$

$$\frac{a(x)}{b(x)} = q(x) + \frac{r(x)}{b(x)}$$

We shall need a result about *rational roots of polynomial equations*. This is also useful in other situations, for instance in computing integrals involving rational functions.

Proposition 23. *Let*

$$f(x) = a_n x^n + a_{n-1} x^{n-1} + \ldots + a_0 = 0, \ \text{where } a_n, a_0 \neq 0,$$

be an equation with integral coefficients,

$$a_n, \ldots, a_0 \in \mathbb{Z}$$

Let $x = \frac{r}{s}$ be a rational root, where r and s are mutually prime integers, i.e., the expression for x cannot be simplified. Then there exist integers a and b such that

$$a_0 = ra, a_n = sb$$

In particular if $a_n = 1$, which we express by saying that the polynomial is monic, then $s = \pm 1$, and every rational root of the equation must be an integer dividing the constant term a_0.

Proof. It suffices to show that there exist an integer a such that $a_0 = ra$: Indeed, $y = \frac{s}{r}$ is a root of the equation

$$a_0 y^n + \ldots + a_n = 0,$$

and the first part of the claim applied to this situation yields the second part of the claim.

We get

$$a_n \left(\frac{r}{s}\right)^n + a_{n-1}\left(\frac{r}{s}\right)^{n-1} + \ldots + a_0 = 0,$$

thus

$$a_n r^n + a_{n-1} r^{n-1} s + \ldots + a_0 s^n = 0.$$

This gives

$$a_0 s^n = r(a_n r^{n-1} + \ldots + a_1 s^{n-1}) = rA,$$

where A is an integer. We therefore obtain

$$a_0 = \frac{rA}{s^n}$$

So the rational expression $\frac{rA}{s^n}$ is, therefore, *an integer*, which means that any prime number dividing s must divide rA. But no prime factor of s divides r, so that it has to divide A. Repeating the argument, if necessary, we find that all prime factors of s^n divide A, and raised to the power to which they occur in s^n, so

$$ a = \frac{A}{s^n} $$

has to be an integer, and the claim follows.   □

Remark In this proof we have taken certain fundamental properties of the integers for granted. These are properties which we normally use without thinking twice about it, but nevertheless the terms involved need to be defined, and the facts need proofs. They are as follows:

1. A prime number is a positive integer with no divisors except 1 and itself. The first prime numbers are $2, 3, 5, 7, 11, 13, \ldots$

2. If a prime number p divides a product of two integers ab, then p either divides a or it divides b.

3. Every positive integer n may be written uniquely as

$$ n = p_1^{i_1} p_2^{i_2} \cdots p_r^{i_r}, $$

where $p_1 < \ldots < p_r$ are prime numbers.

We shall not pursue the issue here, but refer instead to any introductory text on algebra or number theory. The material outlined here was important to Euclid, and forms part of Books VII, VIII and IX of his Elements.

15.2 The Minimal Polynomial

If a real number α satisfy an equation $p(\alpha) = 0$, where $p(x)$ is a polynomial with rational coefficients, then we frequently need to find *all* rational polynomials which have α as a root. It turns out that the answer is very simple and elegant.

Let $p_0(x)$ denote a polynomial with coefficients from $\mathbb{Q}$. We assume that $p_0(\alpha) = 0$, and that $p_0(x)$ is not the zero polynomial. Assume also that it is of minimal degree among the polynomials which have α as a root. We may assume that *the coefficient of the highest power of x which occurs in $p_0(x)$ is equal to 1*, this may be accomplished by dividing the polynomial by this coefficient. The following important result implies that $p_0(x)$ is uniquely determined, and we call $p_0(x)$ the *minimal polynomial of α over $\mathbb{Q}$*.

Theorem 27. *If $p(x)$ is a rational polynomial such that $p(\alpha) = 0$, and $p_0(x)$ denotes the minimal polynomial of α over $\mathbb{Q}$, then $p(x)$ is equal to $p_0(x)h(x)$ for some rational polynomial $h(x)$.*

Proof. By Theorem 22 we find

$$p(x) = q(x)p_0(x) + r(x),$$

where the polynomial $r(x)$ is of degree $<$ the degree $p_0(x)$. If $r(x)$ is not the zero polynomial, then $r(\alpha) = 0$ will contradict that $p_0(x)$ is of minimal degree among the non-zero polynomials which have α as root. Thus $r(x)$ must be the zero polynomial, and the claim follows. $\square$

In particular this theorem implies that $p_0(x)$ is an irreducible polynomial, that is to say that it cannot be written as a product of other non-constant polynomials with rational coefficients: In fact,

$$p_0(x) = a(x)b(x)$$

implies that

$$a(x) \text{ or } b(x) \text{ are constant polynomials.}$$

Indeed, if $a(\alpha)b(\alpha) = 0$, then at least one of the factors vanish, say $a(\alpha) = 0$. But then

$$a(x) = p_0(x)q(x),$$

so that

$$p_0(x) = a(x)b(x) = p_0(x)q(x)b(x)$$

Abbreviating this expression we get

$$q(x)b(x) = 1,$$

thus in particular we obtain that both $b(x)$ and $q(x)$ are constant polynomials.

Remark Above we have used the fact that in a polynomial expression we may always perform abbreviations. This follows from the observation that the product of two non-zero polynomials with coefficients from $\mathbb{Z}$, $\mathbb{Q}$, $\mathbb{R}$ or $\mathbb{C}$ can not be the zero polynomial. In fact, if

$$a(x) = a_m x^n + a_{m-1}x^{m-1} + \cdots + a_0.$$

and

$$b(x) = b_n x^n + b_{n-1}x^{n-1} + \cdots + b_0.$$

where a_m and b_n are $\neq 0$, then $a_m b_n \neq 0$, thus the product $a(x)b(x)$ is not the zero polynomial.

15.3 The Euclidian Algorithm

Euclid's Algorithm is a method for finding *the greatest common divisor* of two integers or of two polynomials. We first illustrate the method for two integers.

So let there be given two integers a and b, which we may assume to be > 0. Division of a by b yields a quotient $q \geq 0$ and a remainder r such that $0 \leq r < b$:

$$a = bq + r, 0 \leq r < b$$

If $r = 0$, then $a = bq$, so that the greatest common divisor of a and b is b itself. If on the other hand $r > 0$, then we repeat the division as follows:

$$b = rq_1 + r_1, 0 \leq r_1 < r$$

If now $r_1 = 0$, then the greatest common divisor of a and b is r: Clearly r divides both a and b. For we have $a = bq_1 + r, b = rq_1$, so that $a = r(qq_1 + 1)$. On the other hand if d is a common divisor of a and b, then d will have to divide r, since $r = a - bq$.

If $r_1 > 0$ we repeat the process. Sooner or later we get a remainder $r_i = 0$: Indeed, we have that

$$r > r_1 > r_2 > \ldots \geq 0,$$

and clearly such a sequence of strictly decreasing non-negative integers cannot be infinite, so at some point $r_i = 0$. We get the following, where we have put $r = r_0$ and $q = q_0$:

$$a = bq_0 + r_0, 0 < r_0 < b$$

$$b = r_0 q_1 + r_1, 0 < r_1 < r_0.$$

$$r_0 = r_1 q_2 + r_2, 0 < r_2 < r_1$$

$$\vdots$$

$$r_{i-3} = r_{i-2} q_{i-1} + r_{i-1}, 0 < r_{i-1} < r_{i-2}$$

$$r_{i-2} = r_{i-1} q_i$$

and we have $r_i = 0$ in the last step, after $i + 1$ steps. This sequence of divisions, which ends when the remainder becomes 0, is referred to as *the Euclidian Algorithm*. The point is the following result:

Theorem 28. *The greatest common divisor for the positive integers a and b is the number r_{i-1}, the last non-vanishing remainder in the Euclidian Algorithm for the numbers a and b.*

Proof. By the last line in the algorithm r_{i-1} divides r_{i-2}, which by the next to the last line implies that r_{i-1} divides r_{i-3}, and so on. In the end we find that r_{i-1} divides b, which by the first line finally yields that r_{i-1} divides a. So r_{i-1} is a common divisor of a and b.

Conversely assume that d is a divisor of a and b. By the first line d divides r_0, so by the second line it also divides r_1, and so on. The next to the last line shows that d divides r_{i-1}, and thus r_{i-1} is the greatest common divisor of a and b. □

Remark When we speak of the greatest common divisor of the positive integers a and b, the naive interpretation is to mean the largest integer c which divides both a and b. But this turns out to be the same as to require that the positive integer c shall divide both a and b, and that any integer d which divides a and b, should also divide c. It is easily seen that these two definitions of *greatest common divisor* for positive integers are equivalent, and we have used the latter form in the above proof.

It is also this definition which is used for greatest common divisor of two polynomials $a(x)$ and $b(x)$:

Definition 26. *Let a and b be either positive integers, or polynomials with coefficients from $\mathbb{Q}$, $\mathbb{R}$ or $\mathbb{C}$ in the variable x, different from the zero polynomial. Then d is called a divisor in a if there exists q, a positive integer or a polynomial as the case may be, such that $a = dq$. c is called the greatest common divisor of a and b if c is a divisor in a and b, and if every divisor in a and b also is a divisor in c. We put*

$$c = (a, b)$$

Note that c is positive when a and b are positive integers. We shall require that c(x) is a monic polynomial if $a = a(x)$ and $b = b(x)$ are polynomials in x: This means that the coefficient of the highest power of x which occurs in c(x) is 1.

The lines constituting the Euclidian Algorithm contain much useful information. For example we may do the following: Putting $A_0 = 1$ and $B_0 = -q_0$, we get from the first line

$$r_0 = aA_0 + bB_0,$$

which by line 2 yields

$$r_1 = aA_1 + bB_1,$$

where $A_1 = -A_0 q_1$, $B_1 = 1 - B_0 q_1$. In the same way we get from line 3 that

$$r_2 = aA_2 + bB_2,$$

where $A_2 = A_0 - A_1 q_2$, $B_2 = B_0 - B_1 q_2$. From the line numbered i we find that

$$r_{i-1} = r_{i-3} - r_{i-2}q_{i-1} = aA_{i-3} + bB_{i-3} - (aA_{i-2} + bB_{i-2})q_{i-1},$$

which gives

$$r_{i-1} = aA_{i-1} + bB_{i-1}$$

where

$$A_{i-1} = A_{i-3} - A_{i-2}q_{i-1}, B_{i-1} = B_{i-3} - B_{i-2}q_{i-1}$$

These computations are valid both if a and b are positive integers, and when they are polynomials different from the zero polynomial. In particular we have proved the following:

Theorem 29. *Let a and b be either positive integers or polynomials in x different from the zero polynomial, with coefficients from $\mathbb{Q}$, $\mathbb{R}$ or $\mathbb{C}$. Then there exist A and B, integers (not necessarily positive) or polynomials of the type in question, such that*

$$(a, b) = aA + bB$$

The identity in Theorem 29 is referred to as *Bézout's Identity*. Etienne Bézout, 1730 – 1783, was a French mathematician who among other things is known for some excellent textbooks. He gave the first satisfactory proof of the assertion that two projective curves in $\mathbb{P}^2(\mathbb{R})$ or in $\mathbb{P}^2(\mathbb{C})$ of degrees m and n, without any common components, intersect in exactly mn points, real or complex, and counted with multiplicity. This fact was asserted by Maclaurin, but is credited to Bézout as *Bézout's Theorem*.

15.4 Number Fields and Field Extensions

We have encountered numbers with various different properties: The set of the *natural* numbers, *the integers*, or *the rational numbers*. The *complex numbers* represent the most extensive algebraic system of objects which we may refer to as *numbers* without stretching the concept. But we may go on, if we are willing to abandon some of the fundamental properties: The *Hamiltonian Quaternions* resemble the complex numbers in many ways, but multiplication is no longer commutative: In general $ab \neq ba$. The *Cayley Numbers* or the *octonians* are even more weird, not only is multiplication non-commutative, it is *non-associative* as well: In general $a(bc) \neq (ab)c$. As $\mathbb{C}$ is based on a multiplication introduced in $\mathbb{R}^2$, so the quaternions is based on $\mathbb{R}^4$ and the octonians on $\mathbb{R}^8$. There are also "number systems" in which there are only *finitely many integers*, in that we may have

$$1 + 1 + \cdots + 1 + 1 = 0.$$

if we add 1 to itself sufficiently many times. The smallest positive N such that adding N 1's together yields 0 must be a prime number (an ordinary prime number, of course), it is called the *characteristic* of the "number system". Actually computers internally work with a system where $N = 2$, and if N is very, very large, we might perhaps consider living with a number system like that.

Here we shall stay within the framework of the *the rational numbers*. However, as we shall see the rationals encompass a rich and interesting structure. Many questions, appearing simple, concerning *integers* remain unanswered despite intensive research. Other problems, which have occupied mathematicians over *millenias*, have been answered only in modern times. The answers have become possible only by invoking the finer structure provided by *the system of real numbers*. This applies to the so called *classical problems*, namely *the trisection of an angle, the doubling of a cube* and *the squaring of a circle*. These geometric problems may be answered once and for all using the material we are about to explain. It is not easy, but that has to be expected. After all, the sharpest brains humanity has produced tried in vain to find the solution for more than 2000 years!

We start out with the following fundamental definition.

Definition 27. *A (real) number field K is a set of real numbers which contains 0 and 1 and which is such that if $r, s \in K$ then $r \pm s \in K, rs \in K$, and $\frac{r}{s} \in K$ if $s \neq 0$.*

Clearly, if K is a number field, then $K \supset \mathbb{Q}$. It is also clear that $\mathbb{R}$ itself is a number field. It is the study of the number fields between these two extremes which yields the insights into the *classical problems*.

Definition 28. *If $K \subset L$ are two number fields, then we refer to L as an extension of K.*

In the first two paragraphs of this chapter we have treated polynomials with coefficients from $\mathbb{Q}$ and $\mathbb{R}$. But everything we did there applies equally well to polynomials over K, where K is any number field. In particular we have the following:

Proposition 24. *Let there be given two polynomials $a(x)$ and $b(x)$ over the number field K. We assume that $b(x)$ is not the zero polynomial. Then there exists two other polynomials over K $q(x)$ and $r(x)$, where $r(x)$ either is the zero polynomial or is of degree $<$ the degree of $b(x)$, such that*

$$\frac{a(x)}{b(x)} = q(x) + \frac{r(x)}{b(x)}$$

Assume that the real number α satisfies an equation $p(\alpha) = 0$, where $p(x)$ is a polynomial over K. As we did for $K = \mathbb{Q}$ we consider the non-zero polynomial $p_0(x)$ with coefficients from K which have α as a root, and which is of minimal degree among the non-zero polynomials which have this property. As before we also specify that the highest power of x which occurs in $p_0(x)$ has coefficient 1, i.e., that the polynomial be *monic*. We refer to $p_0(x)$ as *the minimal polynomial of α over K*. As before we have the

Theorem 30. *If $p(x)$ is a polynomial over K such that $p(\alpha) = 0$, and $p_0(x)$ denotes the minimal polynomial of α over K, then $p(x) = h(x)p_0(x)$ for some polynomial $h(x)$ over K.*

Let K be a number field, and let α be a real number. If there exists a polynomial $p(x)$ over K, different from the zero polynomial, such that $p(\alpha) = 0$, then we say that α is *algebraic over K*. Otherwise we say that α is *transcendental over K*. If $K = \mathbb{Q}$ then we say only that α is an algebraic or a transcendental number. Clearly all numbers in K are algebraic over K, as we may take simply $p_0(x) = x - \alpha$. Moreover, $\sqrt{2}$ is algebraic, being root in the equation $x^2 - 2 = 0$. On the other hand it is known that the number e – base for the natural logarithms – and the number π both are transcendental. The proofs are absolutely non-trivial. We return to this subject in Section 16.6.

Now let α be a number and let K be a number field. We let $K[\alpha]$ denote all polynomial expressions in the number α with coefficients from K:

$$K[\alpha] = \{\beta \mid \beta = f(\alpha) = a_n\alpha^n + \ldots + a_1\alpha + a_0, \text{ where } a_n, \ldots, a_0 \in K\}$$

Assume first that α is algebraic over K. If the minimal polynomial $p(x) = b_m x^m + \ldots + b_0$ of α over K is of degree m, then we may assume that $n < m$ in the definition of $K[\alpha]$ above: Indeed, we have that

$$f(x) = q(x)p(x) + r(x), \text{ where } \deg(r(x)) < m,$$

so that if necessary we may replace $f(x)$ by $r(x)$.
We shall prove that

Proposition 25. *If α is algebraic, then the set $K[\alpha]$ actually is a number field.*

Proof. The proof is in no way obvious: We must show that if we have a polynomial $f(x)$ over K such that $f(\alpha) \neq 0$, then there exists another polynomial $g(x)$, also over K, such that $f(\alpha)g(\alpha) = 1$. We show this as follows: We claim that $(p_0(x), f(x)) = 1$. Indeed, if $p_0(x)$ and $f(x)$ should have a common factor of degree > 0, then this factor would have to be a constant multiple of $p_0(x)$ itself, since $p_0(x)$ an irreducible polynomial by Theorem 30. But then $f(\alpha) = 0$, against the assumption. Thus $(p_0(x), f(x)) = 1$, and by Theorem 29 there exist polynomials $A(x)$ and $B(x)$ such that

$$1 = p_0(x)A(x) + f(x)B(x),$$

which gives

$$1 = p_0(\alpha)A(\alpha) + f(\alpha)B(\alpha) = f(\alpha)B(\alpha)$$

Thus we may take $g(x) = B(x)$, and the claim is proven.   □

Now let α be any number, algebraic or transcendental over K. We consider the set $K(\alpha)$ of all rational expressions in α:

$$K(\alpha) = \left\{ \beta \,\middle|\, \beta = \frac{a_m\alpha^m + \ldots + a_0}{b_n\alpha^n + \ldots + b_0}, b_n\alpha^n + \ldots + b_0 \neq 0 \right\}$$

Clearly this is a number field, and it is the smallest number field which contains K and α. We have the following:

Proposition 26. $K(\alpha) = K[\alpha]$ *if and only if* α *is algebraic over* K.

Proof. If α is algebraic over K then $K[\alpha]$ is already a number field, as we have shown in Proposition 25. Thus $K[\alpha] = K(\alpha)$. If conversely $K[\alpha] = K(\alpha)$, then in particular $\frac{1}{\alpha} \in K[\alpha]$, thus there exists $a_0, \ldots, a_m \in K$ such that $\frac{1}{\alpha} = a_m\alpha^m + \cdots + a_1\alpha + a_0$. But this implies that α is a root in a polynomial equation,

$$a_m x^{m+1} + \cdots a_1 x^2 + a_0 x - 1 = 0,$$

so that α must be algebraic over K. This completes the proof.   □

We finally consider some simple examples. First let $\alpha = \sqrt{2}$. Then the minimal polynomial is $p_0(X) = X^2 - 2$. We express the situation as follows:

$$\left.\begin{array}{c} \mathbb{Q}(\sqrt{2}) \\ | \\ \mathbb{Q} \end{array}\right] X^2 - 2$$

Next consider the extension $\mathbb{Q}(\sqrt[4]{2})$ of $\mathbb{Q}$. Since $\sqrt{2} = (\sqrt[4]{2})^2 \in \mathbb{Q}(\sqrt[4]{2})$, we have $\mathbb{Q}(\sqrt[4]{2}) \supset \mathbb{Q}(\sqrt{2})$. We thus have a small *"tower"* of extensions which looks like this:

$$\begin{array}{l} \mathbb{Q}(\sqrt[4]{2}) \\ \quad | \quad] \; X^2 - \sqrt{2} \\ \mathbb{Q}(\sqrt{2}) \\ \quad | \quad] \; X^2 - 2 \\ \quad \mathbb{Q} \end{array} \Bigg] \; X^4 - 2.$$

More generally we have

$$\begin{array}{l} \mathbb{Q}(\sqrt[mn]{2}) \\ \quad | \quad] \; X^n - \sqrt[m]{2} \\ \mathbb{Q}(\sqrt[m]{2}) \\ \quad | \quad] \; X^m - 2 \\ \quad \mathbb{Q} \end{array} \Bigg] \; X^{mn} - 2.$$

These examples show that the minimal polynomial changes when the base number field changes: The number $\sqrt[4]{2}$ has different minimal polynomials over the number fields $\mathbb{Q}(\sqrt{2})$ and $\mathbb{Q}$.

15.5 More on Field Extensions

This section presupposes a basic knowledge of linear algebra. Nevertheless we have chosen to give a self contained treatment, in that the material which is needed and used will be explained. But the treatment is brief, and a reader may well omit the section at the first reading.

Definition 29. *A number field L is referred to as an extension of another number field K if $L \supset K$. If all the numbers in L are algebraic over K, then we say that L is an algebraic extension of K.*

In particular such a field extension L is a *vector space* over K. This concept will not be given a full explanation here, since we need only one important aspect of it, namely *the order* or *the dimension* of L over K.

Let $\beta_1, \dots, \beta_m \in L$. We say at these elements are *linearly dependent* over K if there exist elements $a_1, \dots, a_m \in K$ which are not all zero, such that

$$a_1 \beta_1 + \cdots + a_m \beta_m = 0.$$

Otherwise we say that $\beta_1, \dots, \beta_m$ are linearly *independent* over K.

We have seen an important example of this concept in the previous section: Let $L = K(\alpha)$, where α is algebraic over K, with minimal polynomial $p_0(X)$. Let $d = \deg(p_0(X))$. Then

$$1, \alpha, \alpha^2, \dots, \alpha^{d-1}$$

are linearly independent, while

$$1, \alpha, \alpha^2, \ldots, \alpha^{d-1}, \alpha^d$$

are linearly dependent over K. For the latter of the two claims, $p_0(\alpha) = 0$ yields the relation

$$a_0 1 + a_1 \alpha + a_2 \alpha^2 + \cdots + a_{d-1} \alpha^{d-1} + \alpha^d = 0.$$

thus $1, \alpha, \alpha^2, \ldots, \alpha^{d-1}$ and α^d are linearly dependent. The former claim, that $1, \alpha, \alpha^2, \ldots, \alpha^{d-1}$ be linearly independent, follows since d is the degree of the minimal polynomial of α over K: No polynomial of lower degree than d with coefficients from K may have α as a root.

In this example we have a further property: Namely that all elements β from L may be written as

$$\beta = b_0 1 + b_1 \alpha + b_2 \alpha^2 + \cdots + b_{d-1} \alpha^{d-1},$$

where $b_i \in K$. We say that β is a linear combination in the linearly independent elements $1, \alpha, \alpha^2, \ldots, \alpha^{d-1}$ with coefficients from K. Such a set of elements in L is called a *basis* for L over K:

Definition 30. *Let the number field L be an extension of the number field K. The elements $\alpha_1, \ldots, \alpha_m \in L$ is called a basis for L over K if the following two conditions are satisfied:*

1. *$\alpha_1, \ldots, \alpha_m$ are linearly independent.*

2. *Every element in L may be written as a linear combination in $\alpha_1, \ldots, \alpha_m$ with coefficients from K.*

In this definition we may combine the two conditions into one *single condition*, so that we get the following definition of a basis for L over K:

3. Every element in L may be written *uniquely* as a linear combination $\alpha_1, \ldots, \alpha_m$ with coefficients from K.

We need the following important result:

Theorem 31. *Let $\alpha_1, \ldots, \alpha_n$ and $\beta_1, \ldots, \beta_m$ be two bases for L over K. Then $m = n$.*

Proof. We may assume that $m \geq n$, if necessary by interchanging the two bases. To show is that $m = n$. We have:

$$\beta_1 = a_{1,1} \alpha_1 + \cdots + a_{1,n} \alpha_n$$

$$\beta_2 = a_{2,1}\alpha_1 + \cdots + a_{2,n}\alpha_n$$

$$\cdots$$

$$\beta_m = a_{m,1}\alpha_1 + \cdots + a_{m,n}\alpha_n$$

Our method of proof is the following: We replace the base $\beta_1, \ldots, \beta_m$ by a new one, $\beta_1', \ldots, \beta_m'$, where we know that the number of elements still is m, but where it is easier to compare with the base $\alpha_1, \ldots, \alpha_n$. This is repeated until the comparison becomes so simple that we may read off the conclusion immediately, namely that $n = m$. The method is known by the name *Gaussian elimination*.

Step 1. By renumbering $\alpha_1, \ldots, \alpha_n$, we may assume that $a_{1,1} \neq 0$. In fact, if this were not possible, then we would have $\beta_1 = 0$, and the β's would not be linearly independent. By putting $\beta_1' = \frac{1}{a_{1,1}}\beta_1$ we still have a base $\beta_1', \beta_2, \ldots, \beta_m$, and we may replace β_1 by β_1'. Thus we get $a_{1,1} = 1$. We now replace β_2 by $\beta_2' = \beta_2 - a_{2,1}\beta_1$. $\beta_1, \beta_2', \beta_3, \ldots, \beta_m$ still is a base, and α_1 does not occur in β_2'. Thus we have accomplished that $a_{2,1} = 0$. Repeating the procedure we may assume that $a_{3,1} = \cdots = a_{m,1} = 0$. We now proceed to the next step.

Step 2. By renumbering $\alpha_2, \ldots, \alpha_n$, we may assume that $a_{2,2}$ is $\neq 0$. In fact, if this were not possible, then we would have $\beta_2 = 0$, and the β's would not be linearly independent. Putting $\beta_2' = \frac{1}{a_{2,2}}\beta_2$ we get $\beta_1, \beta_2', \ldots, \beta_m$ which still is a base, and we may replace β_2 by β_2'. Then $a_{2,2} = 1$. As above we may simplify further, and assume that $a_{3,2} = \cdots = a_{m,2} = 0$.

We continue in this manner. After one further step we get the following situation:

$$
\begin{array}{llll}
\beta_1 = & \alpha_1 + & a_{1,2}\alpha_2 + & a_{1,3}\alpha_3 + & a_{1,4}\alpha_4 + \cdots + a_{1,n}\alpha_n \\
\beta_2 = & & \alpha_2 + & a_{2,3}\alpha_3 + & a_{2,4}\alpha_4 + \cdots + a_{2,n}\alpha_n \\
\beta_3 = & & & \alpha_3 + & a_{3,4}\alpha_4 + \cdots + a_{3,n}\alpha_n \\
\beta_4 = & & & & a_{4,4}\alpha_4 + \cdots + a_{3,n}\alpha_n \\
& & \cdots & & \\
\beta_m = & & & & a_{m,4}\alpha_4 \cdots + a_{m,n}\alpha_n
\end{array}
$$

By now it is clear where this process is heading. After $r \leq n$ steps we will be left with the situation

$$\beta_1 = \quad \alpha_1 + \quad a_{1,2}\alpha_2 + \quad a_{1,3}\alpha_3 + \quad \cdots + a_{1,r}\alpha_r + \cdots + a_{1,n}\alpha_n$$
$$\beta_2 = \quad\quad\quad \alpha_2 + \quad a_{2,3}\alpha_3 + \quad \cdots + a_{2,r}\alpha_r + \cdots + a_{2,n}\alpha_n$$
$$\beta_3 = \quad\quad\quad\quad\quad\quad \alpha_3 + \quad \cdots + a_{3,r}\alpha_r + \cdots + a_{3,n}\alpha_n$$
$$\cdots$$
$$\beta_r = \quad\quad\quad\quad\quad\quad\quad\quad\quad\quad\quad\quad\quad\quad \alpha_r + \cdots + a_{r,n}\alpha_n$$
$$\cdots$$
$$\beta_m = \quad\quad\quad\quad\quad\quad\quad\quad\quad\quad\quad\quad\quad\quad a_{m,r}\alpha_r + \cdots + a_{m,n}\alpha_n$$

and when finally $r = n$ we have

$$\beta_1 = \quad \alpha_1 + \quad a_{1,2}\alpha_2 + \quad a_{1,3}\alpha_3 + \quad \cdots + a_{1,r}\alpha_r + \cdots + a_{1,n}\alpha_n$$
$$\beta_2 = \quad\quad\quad \alpha_2 + \quad a_{2,3}\alpha_3 + \quad \cdots + a_{2,r}\alpha_r + \cdots + a_{2,n}\alpha_n$$
$$\beta_3 = \quad\quad\quad\quad\quad\quad \alpha_3 + \quad \cdots + a_{3,r}\alpha_r + \cdots + a_{3,n}\alpha_n$$
$$\cdots$$
$$\beta_n = \quad\quad\quad\quad\quad\quad\quad\quad\quad\quad\quad\quad\quad\quad a_{n,n}\alpha_n$$
$$\cdots$$
$$\beta_m = \quad\quad\quad\quad\quad\quad\quad\quad\quad\quad\quad\quad\quad\quad a_{m,n}\alpha_n$$

This is a base for L over K. But that is absurd unless $n = m$, and the claim is proved. □

Definition 31. *Let L be a number field which contains the number field K. If L has a finite base $\alpha_1, \ldots, \alpha_n$ over K then we put $[L : K] = n$. Otherwise we write $[L : K] = \infty$.*

With this notation we have shown the following important result above:

Theorem 32. *Let α be a number which is algebraic over the number field K. Then $[K(\alpha) : K]$ is equal to the degree of the minimal polynomial for α over K. If α is transcendental over K, then $[K(\alpha) : K] = \infty$.*

We finally prove a very important theorem, which is the key to understanding which constructions we may legally perform using compass and straightedge:

Theorem 33. *Let $M \supseteq L \supseteq K$ be three number fields. Then the following equality holds:*

$$[M : K] = [M : L][L : K]$$

Proof. Let $\alpha_1, \ldots, \alpha_m$ be a base for L over K, and $\beta_1, \ldots, \beta_n$ be a base for M over L. We claim that then

$$\alpha_1\beta_1, \dots, \alpha_1\beta_n,$$
$$\alpha_2\beta_1, \dots, \alpha_2\beta_n,$$
$$\dots$$
$$\alpha_m\beta_1, \dots, \alpha_m\beta_n$$

is a base for M over K. This will of course suffice to prove the theorem.

We first show that all the elements in M may be expressed as a linear combination in these elements with coefficients from K: Let $\gamma \in M$. Since $\beta_1, \dots, \beta_n$ is a base for M over L, we have

$$\gamma = \delta_1\beta_1 + \cdots + \delta_n\beta_n,$$

where $\delta_1, \dots, \delta_n \in L$. Since $\alpha_1, \dots, \alpha_m$ is a base for L over K, we have $a_{i,j} \in K$ such that

$$\delta_1 = a_{1,1}\alpha_1 + \cdots + a_{1,m}\alpha_m,$$
$$\delta_2 = a_{2,1}\alpha_1 + \cdots + a_{2,m}\alpha_m,$$
$$\dots$$
$$\delta_n = a_{n,1}\alpha_1 + \cdots + a_{n,m}\alpha_m$$

This gives

$$\begin{aligned}
\gamma = \quad & (a_{1,1}\alpha_1 + \cdots + a_{1,m}\alpha_m)\beta_1 \\
+ \ & (a_{2,1}\alpha_1 + \cdots + a_{2,m}\alpha_m)\beta_2 \\
+ \ & (a_{3,1}\alpha_1 + \cdots + a_{3,m}\alpha_m)\beta_3 \\
+ \ & \quad \cdots \\
+ \ & (a_{n,1}\alpha_1 + \cdots + a_{n,m}\alpha_m)\beta_n \\
= \ & \Sigma a_{i,j}\alpha_i\beta_j
\end{aligned}$$

We next show that the elements $\alpha_i\beta_j, 1 \le i \le m, 1 \le j \le n$, are linearly independent. Assume that we have

$$\Sigma a_{i,j}\alpha_i\beta_j = 0,$$

we shall prove that then all $a_{i,j} = 0$. We obtain that

$$\begin{aligned}
0 = \quad & (a_{1,1}\alpha_1 + \cdots + a_{1,m}\alpha_m)\beta_1 \\
+ \ & (a_{2,1}\alpha_1 + \cdots + a_{2,m}\alpha_m)\beta_2 \\
+ \ & (a_{3,1}\alpha_1 + \cdots + a_{3,m}\alpha_m)\beta_3 \\
+ \ & \quad \cdots \\
+ \ & (a_{n,1}\alpha_1 + \cdots + a_{n,m}\alpha_m)\beta_n.
\end{aligned}$$

Since $\beta_1, \dots, \beta_n$ are linearly independent over L, this gives that

$$a_{1,1}\alpha_1 + \cdots + a_{1,m}\alpha_m = 0,$$
$$a_{2,1}\alpha_1 + \cdots + a_{2,m}\alpha_m = 0,$$
$$\dots$$
$$a_{n,1}\alpha_1 + \cdots + a_{n,m}\alpha_m = 0.$$

But since $\alpha_1, \ldots, \alpha_m$ are linearly independent over K, it follows that all the coefficients $a_{i,j} = 0$.

This completes the proof. $\square$

Example. We shall look at an example, which will be used later. Let $\alpha = \sqrt[3]{2}$, and let $L = \mathbb{Q}(\alpha)$. Then

$$[L : \mathbb{Q}] = 3.$$

Indeed, we have that $\alpha^3 = 2$, so that α is a root of the equation

$$x^3 - 2 = 0.$$

We prove that $p(x) = x^3 - 2$ is *the minimal polynomial* of $\sqrt[3]{2}$ over $\mathbb{Q}$. Assume that this is not the case. Then the minimal polynomial would have to be a proper factor of $p(x)$, and thus be of degree either 1 or 2. In both cases there would exist rational numbers a, b and c such that

$$x^3 - 2 = (x^2 + ax + b)(x + c)$$

Thus the equation would have a rational root, namely $-c$.

We now use Proposition 23: In fact, according to this proposition any rational root of $x^3 - 2 = 0$ would have to be an *integer* dividing the constant term, and thus the only possibilities would be the numbers $\pm 1, \pm 2$. Since none of these are solutions to the equation, the claim is proved.

This example may be viewed as a special case of a general fact, which we formulate in the proposition below:

Proposition 27. *A polynomial $p(x)$ of degree 3 with coefficients from $\mathbb{Q}$ is irreducible if and only if it has no rational root, in other words there is no number $\alpha \in \mathbb{Q}$ such that $p(\alpha) = 0$.*

Proof. If $p(x)$ factors as a product of two polynomials with coefficients from $\mathbb{Q}$, one of the factors is of degree 1. Thus $p(x)$ has a rational root. For the converse, we use the following:

If an arbitrary polynomial $p(x)$ with coefficients from some number fields K has a root α from K, then $x - \alpha$ divides $p(x)$:

$$p(x) = q(x)(x - a)$$

Indeed, we perform the division of $p(x)$ by $x - \alpha$ and get a quotient and a remainder,

$$p(x) = q(x)(x - a) + r,$$

where the remainder r is a number in K. But since α is a root in $p(x)$, we get $p(\alpha) = r = 0$. This completes the proof. $\square$

16 Constructions with Straightedge and Compass

16.1 Review of Legal Constructions

The usual meaning of the term *construction* throughout the history of Geometry, is to *draw a figure*, usually in the plane, such that the figure possesses certain properties specified *a priori*. In doing so one is required to start out from a given set of points, in some cases also certain fixed curves. This is referred to as the *start data* for the construction. One is required to use only certain tools, which have been specified as allowable, and to use them in certain prescribed ways only. Normally one is allowed only a finite number of steps in the construction. If an infinite number of steps is needed, then we speak of an *asymptotic* version of the construction. To carry out such an asymptotic construction is, of course, not humanly possible. But the method may be used to create constructions which approximate the required one arbitrarily well.

Recall from Section 3.6 that when we speak of *constructions by straightedge and compass*, or constructions by the *Euclidian tools*, we mean the following: Starting from a finite number of points, a construction of a figure with the required properties should be achieved using straightedge and compass only, in such a way that

1. The straightedge may be used to draw a line through two different points which are given or already have been constructed, and this line may be produced arbitrarily in both directions.
2. The compass may be used to draw a circle with center in a point which is given or already has been constructed, passing through another point which is given or already has been constructed.

In Chapter 3 we have seen how Greek geometers used more powerful tools than the Euclidian ones to solve the classical problems. This included devising mechanical instruments which could perform needed constructions like finding *a double mean proportionality*, required in doubling a cube as explained in Section 3.8. These gadgets were scorned by the purists, as representing a despicable *mechanization* of geometry. Such purists were more approving when various curves, including conic sections, were used as start data in constructions which solved these problems. And, of course, Archytas' space-geometric construction was praised as being one of the high points of Greek geometry. The gadgets referred to *are* interesting, and they do indeed form part of our geometric heritage. We have treated some of them in the appropriate sections.

Another direction taken by people interested in constructions has been *not to strengthen* the tools, but *to weaken them*, at least prescribe tools which are *apparently weaker* than the Euclidian ones.

Thus for instance, around 980 the Arabian geometer *Abû'l-Wefâ* had the idea of performing constructions by means of a straightedge and a *rusty compass*, that is to say a compass with which one is allowed to draw circles with a radius fixed once and for all. It may be surprising that using only such a deficient compass in addition to the straightedge, one may still perform *all* constructions which are possible by the Euclidian tools. But this is not the end: In fact, in 1822 the French mathematician *Jean Victor Poncelet*, (1788 – 1867) found the remarkable result that with *one circle* of any fixed radius added to the start data, a straightedge suffices to perform all constructions possible by the Euclidian tools! The rusty compass needs to be used only once, and may then be discarded.

Such questions still beckon mathematicians and students. This is manifest by all the papers continuing to appear on similar subjects. Thus for instance the interesting and readable paper [40] by *Peter Y. Woo* of the University of Hong Kong, shows by elementary geometric means that with start data including a fixed parabola, with its focus and directrix, all Euclidian constructions may be performed with a straightedge.

Finally, it has been shown that all points obtainable by Euclidian tools can also be obtained by compass alone. This result, which was surprising at the time, was found by the Italian geometer *Lorenzo Mascheroni*, (1750 – 1800).

We shall return to some of these findings when we have developed the powerful algebraic tools which can settle these questions with one stroke.

There is an entirely different notion of *"construction"*, namely that of *Construction by Folding*. Then we draw no line between two points A and B. Instead, we simply *fold the paper*, thus producing the line as the resulting straight indentation in the paper. We return to this in Section 16.8.

16.2 Constructible Points

A point is *constructible* if it is one of the points of the start data, or a point of intersection between two constructible lines or circles.

In Section 3.6 we showed that we are allowed to use the compass in the following way:

3. If A, B and C are any three points which are given or have already been constructed, then we may draw a circle through A with radius BC.

This is the assertion that in the presence of the straightedge, the Euclidian compass is equivalent to the modern compass.

As we saw in Section 3.9, we may solve the *Verging problem* to which the Trisection Problem may be reduced by the use of a *marked straightedge*. This means that we allow ourselves to move a distance, not only by means of the compass, but also by means of the straightedge, and to *insert* the distance between any lines or circles in the construction, while at the same time having the straightedge pass through, or *verge towards*, a suitable point:

4. Two points, the distance between which are equal to the distance between two given or already constructed points may be marked on the straightedge and lines may be drawn through a point which is given or already constructed in such a way that the two marked points on the straightedge fall on constructed lines or circles.

In the presence of the Euclidian or the modern compass this is equivalent to the *Rule of a (fixed) Marked Straightedge:* There is a *fixed distance* marked off on the straightedge, with which the operation 4. above is allowed. We return to this in Section 16.8.

16.3 What is Possible?

We are now going to determine *which points one may construct* by the procedures 1. and 2. above, or equivalently, by 1. and 3., starting from a given set of points $P_0, P_1, \ldots, P_n$.

We introduce a coordinate system in the plane by taking the origin in P_0, and letting the x-axis pass through P_1. Moreover, we chose the scale so that such that $P_1 = (1,0)$. Put $P_i = (a_i, b_i)$, for $i = 1, \ldots, n$. Let $Q = (a, b)$ be a point which is constructed in *one operation* from the points $P_0, \ldots, P_n$. If it is operation 1. which has been employed, then Q is the intersection between two lines, $P_i P_j$ produced and $P_k P_\ell$ produced. This means that $x = a$ and $y = b$ is a solution of the following system of equations

$$(a_i - a_j)(y - b_j) = (b_i - b_j)(x - a_j)$$
$$(a_k - a_\ell)(y - b_\ell) = (b_k - b_\ell)(x - a_\ell)$$

Clearly this yields a and b which are *rational expressions* in the given coordinates, thus that a and b may be computed from $a_i, a_j, a_k, a_\ell, b_i, b_j, b_k$ and b_ℓ by repeated use of addition, subtraction, multiplication and division.

Next assume that it is operation 3. which is employed. Then Q will either be a point of intersection between a line $P_i P_j$ produced and a circle about P_k with radius r equal to the distance between two points P_ℓ and P_m, or between two such circles.

In the former case this means that $x = a$ and $y = b$ are solutions of the system of equations

$$(a_i - a_j)(y - b_j) = (b_i - b_j)(x - a_j)$$
$$(x - a_k)^2 + (y - b_k)^2 = r^2$$

where $r^2 = (a_m - a_\ell)^2 + (b_m - b_\ell)^2$. Here we see that a and b may be expressed by the a_i and b_i, $i = 1, \ldots, n$ by repeated use of the operations of addition, subtraction, multiplication, division *and square root*.

In the latter case Q is a point of intersection between two circles, centered at two of the points and with radii equal to distances between two pairs of points among the $P_0, \ldots, P_n$. Then $x = a$ and $y = b$ are solutions of

$$(x - a_j)^2 + (y - b_j)^2 = r_1^2$$
$$(x - a_k)^2 + (y - b_k)^2 = r_2^2$$

where r_1 and r_2 are distances between two pairs of points as asserted above. This system may be simplified to one in which there is only one equation of degree two: Multiplying out, we obtain

$$x^2 - 2a_j x + y^2 - 2b_j a y = r_1^2 - a_j^2 - b_j^2$$
$$x^2 - 2a_k x + y^2 - 2b_k y = r_2^2 - a_k^2 - b_k^2$$

which is equivalent to the system

$$x^2 - 2a_j x + y^2 - 2b_j y = r_1^2 - a_j^2 - b_j^2$$
$$(a_k - a_j)x + (b_k - b_j)y = \tfrac{1}{2}(r_1^2 - r_2^2 + a_k^2 + b_k^2 - a_j^2 - b_j^2)$$

Again we get that a and b may be computed from $a_i, a_j, a_k, a_\ell, b_i, b_j, b_k$ and b_ℓ by repeated use of addition, subtraction, multiplication, division *and square root*. We have shown the following:

Proposition 28. *If $Q = (a, b)$ may be constructed from $P_0, \ldots, P_n$ in one step by the operations 1. or 3., then a and b may be computed from the coordinates of these points by repeated use of addition, subtraction, multiplication, division and square root.*

We next proceed one step further, and ask which points we may construct in a finite number of steps, at each stage including the previously constructed points among the allowable ones. We shall say that the real number a *is constructible* if the point $(a, 0)$ is constructible. Clearly the point (a, b) is constructible if and only if a and b are constructible real numbers.

Moreover, it is clear that if a is constructible in *one step* from $P_0, \ldots, P_n$ by the operations 1. and 3., then we may express a by $a_1, \ldots, a_n, b_1, \ldots, b_n$, using operations $+, -, \cdot, :,$ and $\sqrt{}$.

Actually, if a is constructible in one step, then we get at most *simple* $\sqrt{}$'s: there are no expressions of the type

$$\sqrt{\alpha + \sqrt{\beta}}$$

We say that a may be expressed rationally by $a_1, \ldots, a_n, b_1, \ldots, b_n$ and simple $\sqrt{\ }$'s.

But if a were constructible in *two steps*, then we would get such double root-expressions as well: Indeed, then there would be a constructible point $P_{n+1} = (a_{n+1}, b_{n+1})$ such that a would be expressed in $a_1, \ldots, a_n, a_{n+1}$, $b_1, \ldots, b_n, b_{n+1}$ using operations $+, -, \cdot, :$, and simple $\sqrt{\ }$'s, while a_{n+1} and b_{n+1} would be expressed by $a_1, \ldots, a_n, b_1, \ldots, b_n$ using operations $+, -, \cdot$, and simple $\sqrt{\ }$'s. When the latter expressions are substituted into the former, we obtain a, expressed by $a_1, \ldots, a_n, b_1, \ldots, b_n$ using operations $+, -, \cdot, :$, as well as single *and* double $\sqrt{\ }$'s.

Continuing like this, we obtain that if the number a is constructible from the given start data above, then it may be expressed in $a_1, \ldots, a_n, b_1, \ldots, b_n$ using operations $+, -, \cdot$, and *multiple* $\sqrt{\ }$'s.

We have the following key result. Note that the numbers $a_1, \ldots, a_n$, $b_1, \ldots, b_n$ given by the start data of course are constructible:

Theorem 34. *a is constructible from the start data* $P_0, \ldots, P_n$ *given above, with* $a_1, \ldots, a_n, b_1, \ldots, b_n$ *as above, if and only a may be expressed by* $a_1, \ldots, a_n, b_1, \ldots, b_n$ *using operations* $+, -, \cdot, :$, *and* multiple $\sqrt{\ }$*'s.*

Proof. It only remains to prove that the criterion implies constructibility. For this, it suffices to show the following

Proposition 29. *Assume that* α *and* β *are constructible. Then so are* $\alpha + \beta$, $\alpha - \beta$, $\frac{\alpha}{\beta}$ *(if* $\beta \neq 0$), $\alpha\beta$ *and* $\sqrt{\alpha}$ *(if* $\alpha \geq 0$).

We shall first see that this proposition suffices to prove the theorem. The method of proof is best illuminated by an example. Assume that $n = 3$ and that a looks like this:

$$a = \frac{a_2 + b_1\sqrt{b_2 - \sqrt{a_3 + b_3}}}{b_2 - b_3\sqrt{b_1 + \sqrt{a_1 - b_1}}}$$

By the proposition the number $a_4 = a_3 + b_3$ is constructible, and so is $b_4 = a_1 - b_1$. Then we also have that $a_5 = \sqrt{a_4}$ and $b_5 = \sqrt{b_4}$ are constructible. We now have

$$a = \frac{a_2 + b_1\sqrt{b_2 - a_5}}{b_2 - b_3\sqrt{b_1 + b_5}}$$

where all the numbers which occur on the right hand side are constructible. Furthermore, we find that $a_6 = \sqrt{b_2 - a_5}$ and $b_6 = \sqrt{b_1 + b_5}$ are constructible, so that

$$a = \frac{a_2 + b_1 a_6}{b_2 - b_3 b_6}$$

where again the numbers occuring on the right hand side are constructible. The proposition again implies that $a_7 = b_1 a_6$ and $b_7 = b_3 b_6$ are constructible, and we have

$$a = \frac{a_2 + a_7}{b_2 - b_7} = \frac{a_8}{b_8}$$

where the numbers a_i and b_i involved are all constructible. Thus, with a final use of the proposition, a is constructible. This completes the example.

To give a formal proof, we proceed by induction. In order to make the induction work properly, we need to phrase the statement $P(N)$, which is to be proven by induction, a little carefully. We formulate the following:

P(N): Assume that the real number a may be expressed in terms of constructible numbers by the operations $+, -, \cdot, :$, and multiple $\sqrt{\ }$'s. Assume that the total number of operations $+, -, \cdot, :$, and $\sqrt{\ }$'s which are needed is less than or equal to the natural number N. Then a is constructible for all N.

To prove this assertion we proceed by induction on the number N.

If $N = 0$, then a itself is constructible and there is nothing to prove. Assume the claim for all numbers b which may be expressed by $\leq N - 1$ operations in some constructible numbers.

As in the example we now go to one of the innermost expressions in the formula giving a in terms of some constructible numbers, say $a_1, \ldots, a_m$. This innermost expression must be of the simple kind covered by the proposition, otherwise it would not be innermost, hence it is constructible by the proposition. Thus including this number in the set of the $a_1, \ldots, a_m$, labeling it a_{m+1}, we now have a expressed in terms of $a_1, \ldots, a_m, a_{m+1}$ with $\leq N - 1$ operations. This reduction in the number of operations has been achieved at the expense of increasing m, the size of the set of constructible numbers involved in the expression for a, but that is OK: It is covered by our induction assumption, the statement $P(N)$ says nothing about the size of the set of constructible numbers involved. So now we may take a as b above, and the claim follows by the induction assumption.

Proof of the proposition. With the normalizing assumptions we have made, $P_0 = (0,0)$ in our coordinate system and $P_1 = (0,1)$. Thus the x-axis as well as the $y - axis$ are constructible lines. In particular 0 and 1 are constructible numbers.

We now refer to Figure 16.1. We first prove that $\alpha + \beta$ and $\alpha - \beta$ are constructible. This is shown in the upper left construction: We use Rule 3., and draw the circle about O with radius α. It intersects the positive x-axis in the point A. About A we then draw the circle with radius β, finding the points B and C. Then $OC = \alpha + \beta$ and $OB = \alpha - \beta$. It follows in particular that all integers are constructible.

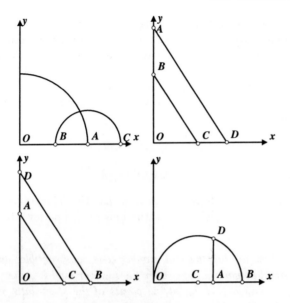

Fig. 16.1. The constructions used in the proof of Proposition 29.

Next, the upper right construction shows that $h = \frac{\alpha}{\beta}$ is constructible: Using Rule 3., we set off A and B on the y-axis, and C on the x-axis, so that $OA = \alpha$, $OB = \beta$ and $OC = 1$. Constructing the line through A parallel to the line BC, we find the point of intersection with the x-axis, D. From the similar triangles $\triangle BOC \sim \triangle AOD$ we then find that

$$\frac{OD}{1} = \frac{\alpha}{\beta}$$

hence $OD = h$, so that h is constructible.

Proving that $h = \alpha\beta$ is constructible follows the same lines, the construction is shown in the lower left corner of Figure 16.1. We now mark A on the y-axis such that $OA = \alpha$, and B and C on the x-axis such that $OB = \beta$ and $OC = 1$. Through B we then draw the parallel to the line AC, it intersects the y-axis in the point D. From the similar triangles we find

$$\frac{OD}{\alpha} = \frac{\beta}{1}$$

and hence $h = OD$ is constructible.

We finally prove that $h = \sqrt{\alpha}$ is constructible. We have

$$h^2 = \alpha \text{ or } \frac{h}{\alpha} = \frac{1}{h}$$

The construction is given in the lower right corner of Figure 16.1. We set off α along the x-axis from O, finding A, and 1 from A further along the x-axis finding B. Bisecting OB we find C, about C we draw the circle through O, which of course also passes through B. Now the normal to the x-axis erected at A intersects the circle in D, and we then have the similar triangles $\triangle ODA \sim \triangle DBA$. From this it follows that AD is the mean proportional between OA and AB, whence

$$AD : 1 = OA : AD$$

thus $AD^2 = \alpha$, so $AD = \sqrt{\alpha}$ is constructible.   □

The theorem which we have just proved, has the following consequence which is the key to deciding which constructions we can perform by Euclidian tools:

Theorem 35. *Assume that the real number α is constructible, and let $f(z)$ denote the minimal polynomial of α over the number field K generated by the coordinates $a_1, \ldots, a_n, b_1, \ldots, b_n$ of the points of the start data, namely the points $P_1, \ldots, P_n$. Then $f(z)$ is of degree 2^m for a suitable integer $m \geq 0$.*

Proof. By Theorem 34 we may express α by $a_1, \ldots, a_n$, $b_1, \ldots, b_n$ and operations $+$, $-$, $\cdot$ and $:$ as well as $\sqrt{\ }$. Thus α is contained in a field extension L of K, see Section 15.5, which we may obtain by a finite number of field extensions

$$K = K_0 \subset K_1 \subset K_2 \subset \ldots \subset K_{M-1} \subset K_M = L$$

where K_i comes from K_{i-1} by adjoining to K_{i-1} *one square root* of an element from K_{i-1}. By Theorem 33 we therefore obtain that

$$[L : K] = 2^M,$$

since $[K_i : K_{i-1}] = 2$, by Theorem 32. But this means that

$$2^M = [L : K] = [L : K(\alpha)][K(\alpha) : K],$$

and since $[K(\alpha) : K]$ is the degree of the minimal polynomial of α, again by Theorem 32, this degree as well must be a power of 2, and the proof is complete.   □

16.4 Trisecting any Angle

The assignment is to divide an arbitrary angle in three equal parts, using compass and straightedge according to the rules 1. and 2. from Section 16.1, or equivalently using 1. and 3. from Section 16.2. It may be explained as follows:

Let u be an arbitrary angle, and let $v = \frac{u}{3}$. The angle u is given, in the form of *three points* P_0, P_1 and P_2. These three points represent our start data, and we choose a coordinate system in the plane $\mathbb{R}$ in such a way that

$$P_0 = (0,0), P_1 = (1,0), P_2 = (\alpha,0),$$

where $\alpha = \cos(u) \in [-1,1]$. Now we have

$$\alpha = \cos(3v) = 4\cos^3(v) - 3\cos(v),$$

$\cos(v)$ is thus root in the equation

$$4x^3 - 3x - \alpha = 0.$$

This equation has three real roots. In fact this is clear from the way we found it: Namely, letting

$$v_0 = v, v_1 = v_0 + \frac{2\pi}{3}, v_2 = v_0 + \frac{4\pi}{3},$$

we have

$$3v_0 = u, 3v_1 = u + 2\pi, 3v_2 = u + 4\pi,$$

thus the three real roots are

$$\cos(v_0), \cos(v_1), \cos(v_2)$$

It is clear that if we are able to construct one of these roots, then we may construct the remaining two as well. Therefore we have reduced the problem of trisecting u in three equal parts using Euclidian tools to the following:

Problem. Given start data P_0, P_1 and P_2 as above, so $P_0 = (0,0), P_1 = (1,0)$ and $P_2 = (\alpha,0)$, where $\alpha = \cos(u) \in [-1,1]$. Then construct, with Euclidian tools, one root of

$$4x^3 - 3x - \alpha = 0.$$

For certain values of u or α this si fully possible: For example if $u = \frac{\pi}{2}$, such that $\alpha = 0$, then the equation becomes

$$4x^3 - 3x = 0$$

which gives

$$x = 0, x = \pm\frac{1}{2}\sqrt{3},$$

and these roots are all constructible from our start data, which in this case are just

$$P_0 = (0,0), P_1 = (1,0)$$

But for most choices of α the roots will not be constructible from the given start data $P_0 = (0,0), P_1 = (1,0), P_2 = (\alpha,0)$. Thus for instance, take $\alpha = \frac{1}{4}$. Then the equation becomes

$$4x^3 - 3x - \frac{1}{4}.$$

Putting $z = 4x$, we get the equation

$$p(z) = z^3 - 12z - 4 = 0.$$

It suffices to show that no root of this equation is constructible.

Let z_0 denote a root in the equation, and let $m(z)$ denote the minimal polynomial of z_0 over $\mathbb{Q}$. Let g denote the degree of $m(z)$. Then clearly $g \leq 3$, we shall prove that $g = 3$.

Suppose that $m(z)$ is of degree 2. Then Theorem 27 implies that

$$p(z) = q(z)m(z),$$

where $q(z)$ must be of degree 1. If $g = 1$, then we find in the same manner that

$$p(z) = q(z)m(z),$$

where now $m(z)$ is of degree 1 while $q(z)$ is of degree 2. In either case we see that $p(z)$ must have a *rational root*, namely the root given by the factor of degree 1. But by Proposition 23 we have that such a root must be one of the integers $\pm 4, \pm 2, \pm 1$. But none of these integers are roots in the equation $p(z) = 0$, and thus the claim that $g = 3$ is proved. By the powerful Theorem 35 this of course suffices to settle the question: z_0 *is not constructible*, and thus the first of the famous classical problems is settled in the negative: The trisection of any angle in equal parts by Euclidian tools is impossible.

16.5 Doubling the Cube

Having developed the tools needed to settle the trisection problem, it is simple to decide the problem of *Doubling the Cube* by Euclidian tools as well. The problem now consists in deciding whether or not the number $z_0 = \sqrt[3]{2}$ is constructible. This number is a root of the equation

$$p(z) = z^3 - 2 = 0.$$

This polynomial is also the *minimal polynomial* for z_0 over $\mathbb{Q}$: If not, then it would be possible to factorize $p(z)$ as a product of a rational polynomial of degree 1 and another of degree 2. But as we saw in the example at the end of Section 15.5 is this not possible.

Thus again, the second one of the classical problems is settled in the negative: A cube may not be doubled by a construction using Euclidian tools.

16.6 Squaring the Circle

We may decide the question of *Squaring the Circle* as well. A circle of radius r have area $A = \pi r^2$. A square with the same area must, therefore, have side equal to $a = r\sqrt{\pi}$. The problem then is to decide if the number $\alpha = \sqrt{\pi}$ is constructible from the start data $P_0 = (0,0)$ and $P_1 = (1,0)$. But it is known that the number π is *transcendental*, in other words it does not satisfy any polynomial equation with coefficients from $\mathbb{Q}$. This result is not so easy to prove, but nevertheless quite within reach of a reasonably complete course in calculus at the college level. We have already quoted this fact, in Section 15.4. Then the number $\alpha = \sqrt{\pi}$ can also not be root in an algebraic equation, since that would imply

$$[\mathbb{Q}(\sqrt{\pi}) : \mathbb{Q}] = m < \infty$$

by Theorem 32. On the other hand we have

$$[\mathbb{Q}(\pi) : \mathbb{Q}(\sqrt{\pi})] = 2,$$

so that Theorem 33 implies that

$$[\mathbb{Q}(\pi) : \mathbb{Q}] = 2m < \infty$$

This of course contradicts that π is a transcendental number, and we have shown that the Squaring of the Circle is impossible by Euclidian tools.

This proof hinges on the fact that the number π is transcendental. This was shown by the German mathematician *Carl Louis Ferdinand von Lindemann*, (1852 – 1939) in 1882, the year before he joined *Hurwitz*[1] and *Hilbert* as professor at the University of *Königsberg*, now the Russian enclave named *Kaliningrad*. Lindemann gave a remarkable proof of the fact that π is transcendental, tying it to the much simpler fact that the number e, base of the natural logarithms, is transcendental, and heavily using complex numbers.

In 1873 the French mathematician *Charles Hermite* (1822-1901) proved that the number e is transcendental. His proof was long and difficult, but

---

[1] The German mathematician Adolph Hurwitz (1859 – 1919) had been a student of Felix Klein, and taught 8 years at Königsberg.

soon simpler proofs were found. In fact, in 1893 the journal *Mathematische Annalen* contained three different and simpler proofs of the transcendency of e. One of them was due to Hilbert, another was by Hurwitz. Hilbert also gave a proof of the transcendency of π in the same journal. Hilbert's very elegant proof uses basic calculus, and the computations seem to work by pure magic. We refer to [24] for his proof.

Now, in 1882 von Lindemann had shown that the assertion that e is transcendental may be generalized to the following result:

Theorem 36 (Lindemann's Theorem). *Assume that $b_0, \ldots, b_n$ are distinct real or complex algebraic numbers, and let $a_0, \ldots, a_n$ be real or complex algebraic numbers, not all zero. Then*

$$a_0 e^{b_0} + \cdots + a_n e^{b_n} \neq 0.$$

Again we are not going to give the proof of this theorem, but shall use it to prove the transcendency π. Assume that π *were not* transcendental, in other words that it were algebraic. Then we get a contradiction by taking $a_0 = 1, a_1 = -1$, and $b_0 = i\pi, b_1 = 0$: Indeed, we have the famous formula, due to *Euler*, which ties together the five most important constants in mathematics, namely $0, 1, e, i = \sqrt{-1}$ and π:

$$e^{i\pi} + 1 = 0.$$

16.7 Regular Polygons

The last construction problem which we shall treat, is that of subdividing the circumference of a circle in n equal parts. It amounts to the same as to construct the *regular n-gon*. In Figure 16.2 we see the subdivision in n equal parts for $n = 3, 4, 5, 6, 8, 10$.

The construction, by Euclidian tools, of regular n-gons were known for several values of n by Greek geometers. We shall treat these constructions as that of *subdividing the circumference of a given circle* in n equal parts. Thus our start data is two points, A and B, the aim is to subdivide the circle about A through B in n equal parts.

Some of these constructions are quite simple, of course. Thus the regular 6-gon is constructed by drawing the circle about A through B, then drawing a circle about B through A, finding C and D, and so on, as shown in Figure 16.3. The construction leads to a beautiful ornament.

The regular 3-gon, or triangle, may be obtained by choosing every second point of the regular 6-gon, the regular 4-gon, or the *square* by erecting the

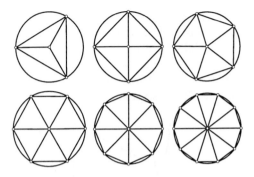

Fig. 16.2. The circle divided into 3,4,5,6,8 and 10 equal parts.

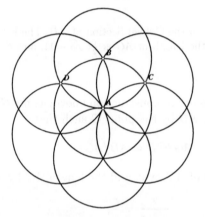

Fig. 16.3. The beautiful ornament of the construction of the regular 6-gon.

normal to AB at A, AB produced and this normal produced then intersect the circle in the points dividing its circumference in four equal parts.

In general it is clear that once we have subdivided the circumference of the circle in n_0 equal parts, then we obtain the subdivision into $2n_0$ equal parts simply by bisecting all the angles coming from the n_0-subdivision. The procedure may be continued, and we find that once the n_0-subdivision has been achieved, then the $n = 2^r n_0$-subdivision follows for all positive integers r.

For the 3-subdivision of the circle we found, however, that the simplest construction is to perform the 6-subdivision first, and then take every second point to get the 3-subdivision.

This is also the case for the construction of subdividing the circumference into 5 equal parts: The simplest and fastest construction is to find the 10-subdivision first.

The construction of the regular pentagon is one of the high points of Euclid's Elements. The construction is given in Book IV, Proposition 11. We shall not examine Euclid's treatment, as before we present the themes in a modern setting. For an explanation of Euclid's arguments, we refer to Hartshorne's beautiful book [14]. However, the construction we give here contains the same ideas as Euclid's version.

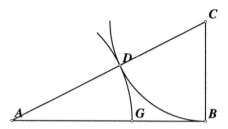

Fig. 16.4. The construction of the Golden Section of AB. The length of AB is r, and that of BC is $\frac{r}{2}$. Thus the length of AG is $\frac{r}{2}(\sqrt{5} - 1)$, and G is the point of the Golden Section of AB.

The construction hinges on the construction known as *the Golden Section*: To divide the line segment AB by the point G such that

$$AG/GB = AB/AG$$

If we denote the length of AB by r, and the length AG by x, then we get

$$\frac{x}{r - x} = \frac{r}{x},$$

or

$$x^2 + rx - r^2 = 0$$

and solving this equation we obtain

$$x = \frac{r}{2}(-1 + \sqrt{5}).$$

Actually the term *Golden Ratio* or *Golden Mean* usually refers to the ratio $\frac{x}{r-x}$ above, which is

$$\frac{x}{r - x} = \frac{-1 + \sqrt{5}}{3 - \sqrt{5}} = \frac{1}{2}(1 + \sqrt{5})$$

Now a Golden Section for the segment AB may be constructed by Euclidian tools by a very simple construction. We refer to Figure 16.4, where the construction of the point G such that $AG/GB = AB/AG$ is explained.

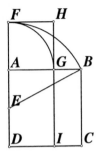

Fig. 16.5. The construction of the Golden Section of AB according to Euclid.

In Figure 16.5 we show how the construction of the Golden Section is carried out in Euclid's Elements. The square $ABCD$ has side $AB = r$. E is the mid point of DA, about E a circle through B is drawn, it intersects DA produced in F, the square $AGHF$ is constructed. Then it is shown, in the Elements, that the latter square has area equal to the rectangle $GBCI$. Thus, in our present day notation, $AG^2 = AB(AB - AG)$, from which it follows that G is the Golden Section of AB. Of course we easily verify the claim about the areas today, by computing as we did using Figure 16.4.

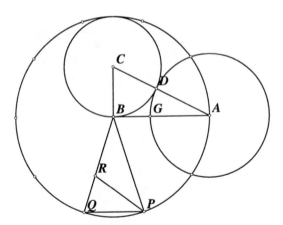

Fig. 16.6. The construction of the regular 10-gon, inscribed in the circle of center B through A: The Golden Section of the radius BA, which is of length r, is carried out by erecting BC, normal to BA at B of length $\frac{1}{2}r$. The circle about C through B intersects CA at D, the circle about A through D intersects AB at G, the Golden Section. We now find the subdivision of the circumference of the circle about B through A by setting off AG as chords around the circle. The triangle BPQ is used in the text to show that the construction is correct.

To subdivide the circumference of the circle about B through A in 10 equal parts, we observe that if the side of the inscribed 10-gon is set off from A along AB, then we find the Golden Section G of the radius AB. Using this, all we have to do is to find the Golden Section of the radius. See Figure 16.6.

To see that the construction is correct, we look at the triangle BPQ. The angle at B is $36°$, hence the angles at P and Q are both $72°$. Bisecting the angle at P, we find the point R, and now have $\triangle BPQ \sim \triangle PQR$. Putting $PQ = x$, we then have $RB = x$ and thus from the similar triangles that

$$\frac{r}{x} = \frac{x}{r-x}$$

and the claim follows. We note that this assertion is XIII.9, Proposition 9 in Euclid's Elements, Book XIII. It is equivalent to the following formula for s_{10}, the side of the regular 10-gon inscribed in a circle of radius r:

$$s_{10} = \frac{r}{2}(-1 + \sqrt{5}).$$

Moreover, the assertion Euclid XIII.10, which is equivalent to

$$s_5^2 = r^2 + s_{10}^2,$$

may be shown as follows, in modern language: Letting q denote the height from B to $s_{10} = QP$ we find

$$q^2 = r^2 - \frac{r^2}{16}(-1 + \sqrt{5})^2 = \frac{r^2}{16}(10 + 2\sqrt{5})$$

thus

$$q = \frac{r}{4}\sqrt{10 + 2\sqrt{5}}$$

A simple consideration of similar triangles then yields

$$\frac{\frac{s_5}{2}}{q} = \frac{s_{10}}{r}$$

i.e.,

$$s_5 = \frac{2}{r}qs_{10}$$

which after a short computation yields a formula for the side of the regular pentagon inscribed in a circle of radius r,

$$s_5 = \frac{r}{2}\sqrt{10 - 2\sqrt{5}}.$$

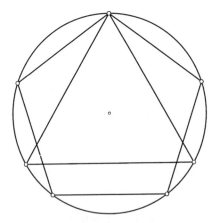

Fig. 16.7. The subdivision of the circumference of a circle in 15 equal parts, thereby constructing the regular 15-gon.

We finally make the following observation: If we may subdivide the circumference of a circle in n_1 and in n_2 equal parts, by separate constructions, and if there is no integer greater than 1 which divides both n_1 and n_2 , then we may subdivide the circumference in n_1n_2 equal parts. In fact, we carry out the two subdivisions of the same circle, starting at the same point A. Since there is no common factor other than 1 in n_1 and n_2, we get altogether $n_1 + n_2 - 1$ points, spaced at different distances around the circle. But now find two adjacent points where this distance is minimal. Draw the cord connecting them. Setting this minimal chord off around the circumference, we get the subdivision in n_1n_2 equal parts. In Figure 16.7 we carry this out to find the regular $15 = 5 \cdot 3$-gon.

We may sum up what we have learned so far by the following list of numbers n for which the regular n-gon may be constructed:

$$3, 4, 5, 6, 8, 10, 12, 15, 16, 20, \ldots$$

The first positive integer not on the list is 7. In Section 4.4 we have given Archimedes' construction of the regular 7-gon, the regular *heptagon*, by a *verging construction*. We asserted there that the construction is not possible by legal use of compass and straightedge, in other words by the Euclidian tools. We shall now prove this.

Our start data are the points $P_0 = (0,0)$ and $P_1 = (1,0)$. The assignment is to construct the angle $\frac{2\pi}{7}$, which evidently is equivalent to constructing the angle $\frac{\pi}{14}$. Letting $v = \frac{\pi}{14}$, we find $4v + 3v = \frac{\pi}{2}$, thus

$$\cos(4v) = \sin(3v)$$

The formula

$$\sin(3v) = 3\sin(v) - 4\sin^3(v)$$

yields, together with the formula

$$\cos(4v) = 1 - 8\sin^2(v) + 8\sin^4(v)$$

that $a = \sin(v)$ is a root in the equation

$$8x^4 + 4x^3 - 8x^2 - 3x + 1 = 0.$$

This polynomial is divisible by by $x + 1$, and we get that a is a root in the equation

$$8x^3 - 4x^2 - 4x + 1 = 0.$$

Letting $y = 2x$ we get

$$y^3 - y^2 - 2y + 1 = 0.$$

But this polynomial is *irreducible* by Proposition 23: Indeed, suppose that it could be factored, then one of the factors would have to be linear, thus the polynomial would have a rational root. But by the proposition, the only possibilities are ± 1, and none of them fits the equation. Therfore the minimal polynomial of $b = 2a$ over $\mathbb{Q}$ is

$$p(y) = y^3 - y^2 - 2y + 1$$

and so b is not constructible by Theorem 35, hence a is not constructible and the claim is proven.

In this way one could go on, and resolve the question for one integer n at the time. But *Gauss* realized how all the equations one obtains in this way may be given a unified treatment, and proved the following remarkable result:

Theorem 37. *The regular n-gon may be constructed by Euclidian tools if and only if*

$$n = 2^r p_1 p_2 \cdots p_s,$$

where $p_1 < p_2 < \ldots < p_s$ are prime numbers which are all of the form

$$p = 2^m + 1,$$

and where $r \geq 0$.

We see that $m = 1$ gives $p = 3$, $m = 2$ gives $p = 5$, $m = 3$ gives $p = 9$ which is not a prime number. The regular 9-gon can of course not be constructed with Euclidian tools by the theorem. This also follows by the results we obtained for the *Trisection Problem*: In fact, in the situation of Section 16.4, let $u = 60°$. The problem is equivalent to trisecting u. Then $\alpha = \frac{1}{2}$, and the equation from Section 16.4 becomes

$$4x^3 - 3x - \frac{1}{2}$$

Letting $y = 2x$, we get the equation

$$y^3 - 3y - 1,$$

Exactly as in the case of the equation coming from the regular heptagon, we see that this equation has no rational roots, hence is irreducible, and thus its root 2α can not be constructed by the Euclidian tools, so neither can α.

But $m = 4$ gives $p = 17$, which is the next constructible case. The construction is carried out in Hartshorne's book [14].

So the problem is reduced to the study of prime numbers of the form $2^m + 1$, the so called *Fermat - primes*. Unfortunately we only know of *five* such primes, and we do not even know if the total set of such prime numbers is finite. We do know this, however:

First of all we must have $m = 2^\nu$ for $2^m + 1$ to be a prime. In fact if m has an odd factor, then $2^m + 1$ cannot be prime. For assume that $m = m_1 m_2$, where m_2 is odd. The formula

$$1 + k + k^2 + \cdots + k^{m_2 - 1} = \frac{1 - k^{m_2}}{1 - k},$$

yields for $k = -2^{m_1}$ that

$$1 - 2^{m_1} + 2^{2m_1} - \cdots + 2^{(m_2 - 1)m_1} = \frac{1 + 2^{m_1 m_2}}{1 + 2^{m_1}}$$

since m_2 is odd. This yields

$$2^{m_1 m_2} + 1 = (2^{m_1} + 1)(1 - 2^{m_1} + \cdots + 2^{m_1(m_2 - 1)})$$

Now put $F_\nu = 2^{2^\nu} + 1$. Then

$$F_0 = 3, F_1 = 5, F_2 = 17, F_3 = 257, F_4 = 65537$$

are all primes. But then

$$F_5 = 4294967297$$

is not prime, as it has the factor 641. After F_4 there are no further Fermat primes known. So the problem of which regular n-gons we can construct by the Euclidian tools is still open. But if a new Fermat prime should be found, the actual construction would be way outsider what is humanly possible, as indeed is the case already for the last known case F_4. But the F_3-gon construction is supposed to have been carried out by enthusiasts.

16.8 Constructions by Folding

There is an amusing way of performing "constructions" by *folding the paper*. The lines obtained are thus given as the *fold* left when the paper is flattened after having been folded according to certain rules. This activity is called *Origami*, and may be carried out in various ways. Here we follow the article [30], but the reader may also find more information in [14].

As with compass and straightedge constructions, we start out from a certain set of points, our *start data*. Then we construct new points by either of the following procedures:

A new point is:
1. A point P of intersection between two previously constructed different lines ℓ_1 and ℓ_2,

or

2. A point Q obtained by folding the paper along a previously constructed line ℓ, from a previously constructed or given point P. In other words, a new point Q is constructed by reflecting P in ℓ.

Furthermore,

A new line is:
3. The line obtained by folding the paper along two given or previously constructed points P and Q. In other words, the line PQ produced in both directions.

or

4. The line ℓ obtained by folding the paper as follows: Two given or previously constructed points P_1 and P_2 are selected, as well as two previously constructed lines ℓ_1 and ℓ_2. Then the paper is folded in such a way that P_1 falls on ℓ_1 and P_2 falls on ℓ_2.

The four basic rules are illustrated in Figure 16.8.

As for Rule 4., there may be several such lines, and all of them are constructible. There may also, in special cases, be no such line. If P_1 is on ℓ_2 and P_2 is on ℓ_1, then one allowable fold is along the mid normal of P_1P_2. Thus we have

5. The mid normal: For two given or constructed points P and Q, the fold sending P to Q is allowed. In other words, the mid normal of PQ is constructible.

Further, we have

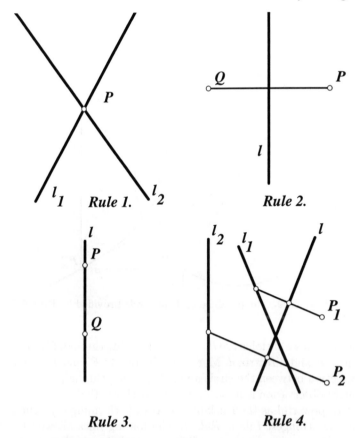

Fig. 16.8. The four basic rules of Paper Folding.

6. Dropping the normal: For a given or constructed point P outside a constructed line ℓ, the fold which leaves both P and ℓ fixed. In other words, the normal to ℓ may be dropped from P.

The following rule is some times taken as one of the basic rules:

7. The middle line of two lines: For two different constructed lines ℓ_1 and ℓ_2, it is permitted to fold the paper in such a way that the two lines coincide.

Finally we have the following:

8. "Parabolas and circles": For two points C and F which are given or already have been constructed, and a line d which does not pass through F, the paper may be folded such that C is fixed and F is sent to d.

There are infinitely many such folds, all of them are permitted. The phenomenon is illustrated in Figure 16.9.

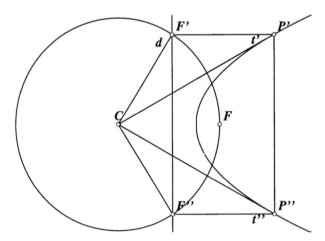

Fig. 16.9. The parabola and the circle provided by Rule 8.

The two allowable folds are *the tangents* passing through C to the parabola with focus F and directrix d. Moreover, the points F' and F'' are the points of intersection between the circle about C passing through F, and the line d. This latter observation is important, since it shows that the folding-rules are at least as powerful as the Euclidian tools in performing constructions.

But the strongest rule is *Rule 4*. The folding lines allowed by this rule are the *common tangents to two different* parabolas. In fact, if we are given a line d and a point F only, then the folds sending F to d are precisely all the tangents to the parabola with focus F and directrix d.

It turns out that Rule 4. is equivalent to working with a *cubic curve*, but we shall not go into the details here. Instead, we refer to the very interesting article [30], already referred to above.[2]

The process of construction by Rules 1., 2., 3. and 4. is exactly as powerful as the process of construction with compass and a *marked straightedge*. Thus any angle may be trisected in equal parts, the regular 7-gon (*heptagon*) may be constructed and the cube may be doubled.

Since we may scale any figure, constructible by Euclidian Tools, up and down as we wish with these tools, we may perform any construction using the straightedge "illegally" to move any distance, by means of a straightedge *on which there is marked off a single, fixed, distance*. Now Nicomedes has achieved this by his Conchoid.

---

[2] Of course, the article is in Norwegian, but the computation leading to the cubic curve, on page 175 in the volume of the journal, stands out as clearly legible in any language!

Moreover, adding a single conchoid somewhere as part of the start data, we achieve the same as marking a fixed distance on the straightedge we are using.

Now the folding constructions leads to the appearance of a curve of degree 3. And in fact, it is a conchoid. Hence constructions by folding can achieve all constructions we can carry out by using a marked straightedge. But conversely, the constructions by folding may also be carried out using a marked straightedge. We shall not pursue this issue further, however.

17 Fractal Geometry

17.1 Fractals and Their Dimensions

Loosely speaking we may say that a *fractal* is an object which is so far from being smooth that its dimension is no longer an integer.

We shall not pursue the issue of defining the concept of *dimension*, however. Instead, we take a relatively naive point of view, and simply examine how the magnification process of selected pieces of a figure works in different dimensions. The concept of dimension which emerges, is a simple form of the concept developed by Hausdorff in [13]. So our point is this: We cut out a piece of our figure, and enlarge it by a scaling factor of s. This makes it multiply by a factor of m: For instance, assume that we do this with a line segment, with a piece of a plane or with a domain in 3-space. The results are summarized in Figure 17.1.

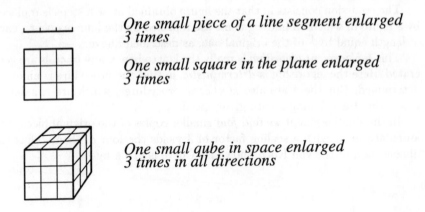

One small piece of a line segment enlarged 3 times

One small square in the plane enlarged 3 times

One small qube in space enlarged 3 times in all directions

Fig. 17.1. Subdivision and enlargement for three different figures. The simplest case.

As we see, for the line segment a scaling up by a factor of 3 of a small piece, yields a total of 3 copies of the original piece. But for the piece of the plane, a scaling factor of 3 yields altogether 9 new copies. And, for a small piece of space, the scaling factor of 3 yields altogether 27 new copies. The simple fact is, that letting d denote the dimension in our usual sense, we have

$$m = s^d.$$

We now take a very bold step, and proclaim this as the definition of *Fractal Dimension*:

> **Fractal Dimension.** Let F be a self-similar figure, which is to say, a figure such that whenever we cut out a certain small piece of it denoted by F', and enlarge it by a scaling factor of s, then we get a total of m identical copies of F'. Then the Fractal Dimension d of F is defined by the relation
>
> $$m = s^d.$$

17.2 The von Koch Snowflake Curve

We now examine the *von Koch Snowflake Curve*, which was introduced in Section 6.3. The curve of von Koch may be defined recursively, by a replacement algorithm. We start with a line segment, which is subdivided into three equal parts. The middle part is then replaced by *two* pieces, which together with the middle one, which has been removed, would have formed an equilateral triangle. See Figure 17.2.

The recursion consists in that the figure obtained after n steps is replaced by a figure in which all line segments are replaced by the four segments, each of length equal to $\frac{1}{3}$ of the original one, as described above.

This *replacement algorithm* is really the key to how many fractals are generated. Here the algorithm is *deterministic*, in that the procedure is uniquely determined. But there are also *stochastic* procedures, which are capable of generating fractal images with great speed.

In the small segment we find *four* smaller copies of the original piece, and an enlargement with a scaling factor of 3 yields the four of them. Thus the dimension d of the von Koch Snowflake Curve is given by

$$4 = 3^d$$

which gives

$$d = \frac{\log(4)}{\log(3)} \approx 1.262$$

From the construction of the Snowflake Curve we may also get an intuitive understanding of why this curve is continuous, but has no tangents. In fact, the curve may be approximated arbitrarily well by a *polygonal curve*, that is

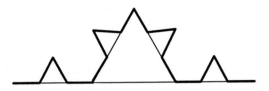

Fig. 17.2. First steps in the construction of the von Koch Snowflake Curve.

to say, a continuous curve consisting of line segments. It is not hard to come up with an ϵ-δ – type proof that the limit of such curves is a continuous curve. As for tangents, we choose two points P and Q on the Snowflake Curve, and select P_n, Q_n on the n-th polygonal curve C_n in the construction of the Snowflake Curve, such that P_n tends to P, Q_n tends to Q. The line P_nQ_n will then tend to the secant PQ of the Snowflake Curve. As Q approaches P, the secant PQ will tend to the tangent at P, if it exists. Assume that the tangent at P exists, denote it by T. We conclude that whenever P_n and Q_n are different points on C_n, such that P_n and Q_n tend to P as n increases, then the line P_nQ_n will tend towards the line T. However, a look at Figure 17.2 yields ample intuitive evidence that no common limiting position for P_nQ_n for all choices of P_n and Q_n tending to P can exist. Thus the Snowflake Curve has no tangents.

17.3 Fractal Shapes in Nature

Suppose we wish to apply these ideas to *fractal shapes in nature*. We might want to compute, or estimate, the dimension of a shoreline, of a cloud, and so on.

We are then faced with the difficulty that these objects are only *approximately* self similar.

There are several theoretical remarks one could make concerning this definition, at this point. However, we shall not go into the matter here. Except to note the following: It is obvious that a *different selection G'* from F may have the same property as F': Scaling it up by a factor of t would make it multiply by a factor of n. This certainly is possible as long as we do not demand an exact fit with the original, only an approximate equality of appearance. So we could get a different number e such that

$$n = t^e$$

Nevertheless, people do compute – or "compute" – the *"fractal dimensions"* of approximately self-similar objects occuring in nature.

According to this the typical shoreline would have a dimension of about 1.2. Similarly *clouds* would be objects with an estimated dimension of more than 2 but less than 3, most being deemed to be of dimension around 2.3.

17.4 The Sierpinski Triangles

Another simple *self similar* figure is the *Sierpinskis triangles*. This figure is not generated by a replacement algorithm, but rather by an *insertion algorithm*: We start with an equilateral triangle, where we draw three other equilateral triangles with side equal to half the side of the original one. Inside each one of these we repeat the process, and so on. The Sierpinski Triangles is shown in Figure 17.3.

Fig. 17.3. The Sierpinski Triangles.

We find the dimension of the Sierpinski Triangles as follows: In Figure 17.4 we indicate the enlargement with a scaling factor of 2.

As we see, a scaling factor of 2 yields four copies of the original. Thus

$$4 = 2^d$$

which gives

$$d = \frac{\log(4)}{\log(2)} \approx 1.585$$

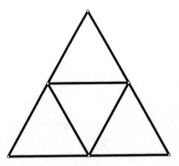

Fig. 17.4. First steps in the construction of the Sierpinski Triangles.

The Polish mathematician *Waclav Sierpinski* laid the foundation for an important school of topology in Poland.

17.5 A Cantor Set

We finally treat a figure obtained by an *excision algorithm*. We now start with an interval on a line, say of length 1. First remove the middle third of it. From the remaining two we also remove the middle third, and so on. In the end we obtain a set known as a *Cantor set*. The process is shown in Figure 17.5.

Fig. 17.5. The first four steps in the construction of the Cantor Set.

Enlarging with a scaling factor of 3 we get 2 copies, thus the dimension of the Cantor Set is $d = \frac{\log(2)}{\log(3)} \approx 0.631$

The *Cantor sets*, one of which is indicated in Figure 17.5, form a class of sets with this fractal nature. We have seen in earlier chapters how Georg Cantor has made significant contributions to our understanding of mathematics in general and set theory in particular.

18 Catastrophe Theory

18.1 The Cusp Catastrophe: Geometry of a Cubic Surface

We are going to treat, mathematically, the case of a Cusp Catastrophe, already encountered in Section 6.1. To start out, we consider a simple cubic surface in affine 3-space $\mathbb{R}^3$. Recall that an algebraic equation of degree 3 of the form

$$X^3 + aX + b = 0,$$

has its roots expressed by a beautiful formula involving cubic roots, usually referred to as *Cardano's Formula*. We shall not give the formula here, but we note that it follows from it that a number called the *discriminant of the equation* plays an important role: The discriminant is defined as

$$\Delta(a, b) = \left(\frac{b}{2}\right)^2 + \left(\frac{a}{3}\right)^3$$

Let x_1, x_2 and x_3 denote the roots of the equation, real or complex. Then we find that

$$\Delta(a, b) = -\frac{1}{108}(x_1 - x_2)^2(x_1 - x_3)^2(x_2 - x_3)^2,$$

using the relations

$$x_1 x_2 x_3 = -b$$
$$x_1 x_2 + x_1 x_3 + x_2 x_2 = a$$
$$x_1 + x_2 + x_3 = 0$$

This is a simple exercise with a symbolic calculator or with MAPLE, say.

We have the following result:

Proposition 30. *The equation $X^3 + aX + b = 0$, where a and b are real, has a complex root if and only if $\Delta(a, b) > 0$. Otherwise all its roots are real.*

Proof. If there are no complex roots[1] then the computation of $\Delta(a, b)$ above shows that it is non-positive.

Conversely, if there is one complex root, say x_1, then the complex conjugate $x_2 = \overline{x_1}$ is also a root, and x_3 must be real. Then $(x_1 - x_2)^2 < 0$ while $(x_1 - x_3)^2(x_2 - x_3)^2 > 0$, thus $\Delta(a, b) > 0$. □

The surface obtained from the cubic equation

$$x^3 - ux + v = 0,$$

is shown in Figure 18.1. We have plotted the curve given by

$$\Delta(-u, v) = 0, \text{ i.e., by } u^3 = \frac{27}{4}v^2$$

in the (u, v)-plane. The negative sign of the linear term in x is just to get a diagram similar to the ones discussed in Section 6.1.

Now we shall address a question which may have puzzled some readers: *Why is it that over the wedge shaped bifurcation area, only two of the points are possible? In other words, why is the middle piece of the fold "prohibited area"?* For this we need some rudiments from *Control Theory*.

18.2 Rudiments of Control Theory

We are given a set of free variables $u_1, u_2, \ldots, u_m$ called *control variables* representing a point in a certain domain in $\mathbb{R}^m$, and a set of variables depending on these, referred to as *state variables*, $x_1, x_2, \ldots, x_n$. The way in which the state variables depend on the control variables may be subtle, but in the cases we shall consider here there will be a finite number of possibilities for each of the state variables for any given choice of control variables.

Control theory now deals with the problem of how the state variables change when the control variables are altered along a continuous curve from a certain point in the control space $\mathbb{R}^m$ to another. The corresponding point in the state space $\mathbb{R}^n$ will then move along a track in the state space, a curve but possibly a curve with discontinuities.

Frequently the process by which the control variables determine the possible values of the state variables is that of an *optimization process*. This may be in the simple form of *minimizing* a certain function in all the variables, control and state:

$$V = V(x_1, x_2, \ldots x_m; u_1, u_2, \ldots, u_m).$$

---

[1] By a complex number we here understand a proper complex number, that is to say a number which is of the form $x = \alpha + \beta\sqrt{-1}$, where $\beta \neq 0$. Of course the set of complex numbers include the real ones as a subset, strictly speaking.

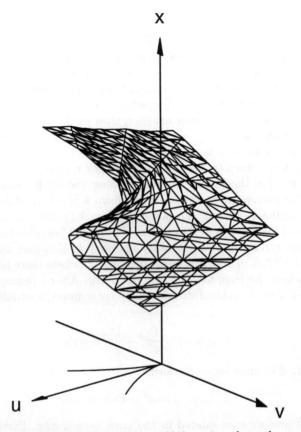

Fig. 18.1. The folded surface and the area of bifurcation above the area inside the semi cubical parabola. The equation of the surface is $x^3 - ux + v = 0$.

The semicolon instead of the comma in the functional notation serves no other purpose than to distinguish the two types of variables from each other.

The function V may appear in a variety of situations. In economic theory it would typically be a *cost function*, it could be an estimate of *exposure to risk*, in physics it might be the energy stored in some mechanical system, which tends to find an equilibrium where this energy is minimized. In some application to psychology it could be an estimate of *discomfort* suffered by an individual or by a group of people, like the inmates in a *prison*. For the dog analyzed in Zeeman's famous example, one might speculate that the function would be measuring the amount of certain *hormones* in the dog's bloodstream.

We now consider a simple situation, in which there are *two* control variables u and v, and just *one* state variable x. The function to be minimized is then denoted by

$$V = V(x; u, v).$$

For given values of u and v a possible value of x must then give a local minimum, assuming differentiability we therefore have that x must satisfy the equation $V'(x; u, v) = 0$, the derivative with respect to x must vanish. But not all such values of x are possible: If the *second* derivative is *positive*, x will give a local *maximum*, this corresponds to a *non-stable* equilibrium. If, however, the second derivative is *negative*, then we do have a local minimum, which corresponds to a *stable* equilibrium. Should the second derivative also vanish, a further analysis is required.

We may now return to the *Cusp Catastrophe* treated in Section 6.2. The point is that even though the surface defining the Cusp Catastrophe is of degree 3, the phenomenon actually arises from a problem related to a polynomial of degree 4: Indeed, the potential $V = V(x; u, v)$ entering into the situation is of degree 4 in x. After some simplifying considerations one finds that without serious loss of generality one may in this case assume an expression for $V(x; u, v)$ which is linear in u and v, where there is no constant term and where the term with x^3 does not occur. After a convenient scaling, the simplest way the control variables may enter into the situation is when

$$V(x, u, v) = \frac{1}{4}x^4 - \frac{1}{2}x^2 u + xv.$$

Thus (x, u, v) must lie on the surface with equation

$$V'(x, u, v) = x^3 - xu + v = 0,$$

exactly the one we encountered in the cusp catastrophe. Furthermore, we must have

$$v''(x, u, v) = 3x^2 - u \leq 0,$$

and we find the boundary of this area by eliminating x from the equations

$$x^3 - xu + v = 0,$$

$$3x^2 - u = 0,$$

which yields the curve given by the *discriminant* of $x^3 - xu + v$, which is $u^3 = \frac{27}{4}v^2$.

We now understand why the *"middle fold"* of the surface had to be cut out: Here the second derivative is *positive*, so the points there are non-stable equilibria, and therefore x can not remain at these values.

With this example of reasonably advanced algebra and geometry throwing light on phenomena in everyday life, we conclude this treatment of the ancient field of Geometry, so much part of our cultural heritage.

References

1. Archimedes. *The Method*. At Walters Art Gallery. http://www.thewalters.org/
2. V. Brun. *Alt er Tall* (Everything is Numbers). Universitetsforlaget, Oslo.
3. B.A. Cipra. Computer Search Solves an Old Math Problem. *Science*, Vol. 242, Nr. 4885,pp.1507-1508. Des. 1988.
4. M. Dzielska. Hypatia of Alexandria. Harvard University Press, 1996.
5. H. Eves. An Introduction to the History of Mathematics. Saunders College Publishing, 1992.
6. W. Fulton. *Algebraic Curves*. W. A. Benjamin, Inc., 1969.
7. T.Getty. *Tom Getty's Polyhedra Hyperpages*. http://www.home.teleport.com/~tpgettys/poly.shtml
8. E. Gibbon. *The Decline and Fall of the Roman Empire*. Edited and with a preface by D. A. Saunders. The Vicing Press, New York 1952.
9. M. Grant. *Cicero-On the Good Life*. Penguin Books, New York, 1971.
10. M. Greenberg. Euclidian and Non-Euclidian Geometries. Development and history. W. H. Freeman and Company. San Fransisco 1972.
11. A. Grothendieck with the collaboration of J. Dieudonné. Éléments de géométire algébrique *I* − *IV*. *Publications Mathématiques de l'IHES*, 4,8,11,17,20,24,28 and 32, 1961 − 1967.
12. T.W. Gray and J. Glynn *Exploring Mathematics with Mathematica*. Addison-Wesley Publishing Company, 1991.
13. F. Hausdorff. *Dimension und äußeres Maß*. Mathematische Annalen 79 (1919).
14. R. Hartshorne. *Geometry: Euclid and Beyond*. Springer Graduate Texts in Mathematics. Springer-Verlag, New York, Berlin, Heidelberg 2000.
15. T.L. Heath. *A History of Greek Mathematics*. New York: Dover 1981, 2 vols., paperback version of 1921.
16. T.L. Heath. *Euclid: The thirteen books of the Elements*. Translated from the text of Heiberg. Translated with introduction and commentary by Sir Thomas L. Heath. 3 vols. Dover Publications, New York 1956.
17. Herodotus. *The Histories*. Translated by A. de Sélincourt, revised by J. Marincola. 1954 and 1972, Penguin Books, London. 1996.
18. D. Hilbert. *Grundlagen der Geometrie*. 1899. Tenth Printing, Teubner Verlag 1968.
19. A. Holme. *Innføring i Geometri. Fra Euklid of Mandelbrot*. (Introduction to Geometry. From Euclid to Mandelbrot.) Alma Mater Forlag, Bergen 1996. In Norwegian.
20. A. Holme. *Innføring i lineær algebra med anvendelser*. (Introduction to linear algebra with applications.) Fagbokforlaget, Bergen 1999. In Norwegian.

21. A. Holme. *Matematikkens Historie. Fra Babylon of mordet på Hypatia.* (History of Mathematics. From Babylon to the Murder of Hypatia.) Fagbokforlaget, Bergen 2001. In Norwegian.

22. Iamblichus. *Life of Pythagoras.* Translated by T. Taylor. London, 1818. Our reference is a summary written anonymously by a student in 1905, reprinted on Kessinger Publishing, LLC.

23. E.W. McLemore, Past Present (we) Present Future (you). *Women in Mathematics Newsletter* 9 (6) 1979 pp. 11 – 15.

24. W.J. LeVeque. *Fundamentals of Number Theory.* Addison Wesley Publishing Company, 1977.

25. O. Neugebauer. *The Exact Sciences of Antiquity.* Second edition. Dover Publications, Inc. New York 1969. Reprinted from the original, published by Brown University, 1957.

26. O. Neugebauer and A. Sachs. *Mathematical Cuneiform Texts.* Yale University Press, New Haven, Conn. USA, 1945.

27. Plutarchus. *Lives of Illustrious Men.* 3 volumes. Translated from the Greek by J. Dryden and others. Hovendon Company, New York, 1880.

28. J. Richter-Gebert and U.H. Kortenkamp. *The Interactive Geometry Software Cinderella.* Springer-Verlag Berlin, Heidelberg, New York, 1999.

29. G. Robins and C. Shute. *The Rhind Mathematical Papyrus.* Published by the Trustees of British Museum, by British Museum Press. London 1997, reprinted 1990, 1998.

30. B. Scimemi. Algebra og geometri ved hjelp av papirbretting. (Algebra and geometry by folding paper.) Translated from the Italian with comments by Christoph Kirfel. *Nordisk Matematisk Tidsskrift,* 4, 1988.

31. P.T. Saunders. *An introduction to Catastrophe Theory.* Cambridge University Press 1980.

32. Robert R. Stoll. *Set Theory and Logic.* W. H. Freeman and Company, San Fransisco and London, 1961.

33. R.D.J. Struik. *A Concise History of Mathematics.* Fourth Revised Edition Dover Publications Inc, New York 1987.

34. W.W. Struve. Mathematisher Papyrus des Staatlichen Museums der Schönen Künste. *Part A, Quellen* (1930).

35. B.-L. van der Waerden. *Science Awakening.* English translation by Arnold Dresden, with additions by the author. P. Nordhoof Ltd - Groningen, Holland 1954.

36. Agnes Scott University *Women in Mathematics webcite* http://www.agnesscott.edu/lriddle/women

37. St. Andrews, University of: *History of Mathematics.* http://www-history.mcs.st-andrews.ac.uk/history/Mathematicians

38. *The Catholic Encyclopedia.* http://www.newadvent.org./cathen

39. University of Arizona, Women's study department. *Greek Women Philosophers.* http://info-center.ccit.arizona.edu/˜ws/

40. P. Y. Woo. Straightedge Constructions, Given a Parabola. *The College Mathematics Journal.* Vol. 31, No. 5, November 2000.

41. E.C. Zeeman. Catastrophe Theory. *Scientific American,* April 1976, pp. 65-83.

Index

Printing: Saladruck, Berlin
Binding: H. Stürtz AG, Würzburg